# ADVANCED COURSES OF MATHEMATICAL ANALYSIS IV
IN MEMORY OF PROFESSOR ANTONIO AIZPURU TOMÁS

EDITORS

# F Javier Pérez-Fernández
# Fernando Rambla-Barreno
*University of Cádiz, Spain*

PROCEEDINGS OF THE FOURTH INTERNATIONAL SCHOOL

# Advanced Courses of Mathematical Analysis IV
IN MEMORY OF PROFESSOR ANTONIO AIZPURU TOMÁS

Jerez de la Frontera, Spain    8 – 12 September 2009

NEW JERSEY • LONDON • SINGAPORE • BEIJING • SHANGHAI • HONG KONG • TAIPEI • CHENNAI

*Published by*

World Scientific Publishing Co. Pte. Ltd.
5 Toh Tuck Link, Singapore 596224
*USA office:* 27 Warren Street, Suite 401-402, Hackensack, NJ 07601
*UK office:* 57 Shelton Street, Covent Garden, London WC2H 9HE

**British Library Cataloguing-in-Publication Data**
A catalogue record for this book is available from the British Library.

**ADVANCED COURSES OF MATHEMATICAL ANALYSIS IV**
**Proceedings of the Fourth International School**
**In Memory of Professor Antonio Aizpuru Tomás**

Copyright © 2012 by World Scientific Publishing Co. Pte. Ltd.

*All rights reserved. This book, or parts thereof, may not be reproduced in any form or by any means, electronic or mechanical, including photocopying, recording or any information storage and retrieval system now known or to be invented, without written permission from the Publisher.*

For photocopying of material in this volume, please pay a copying fee through the Copyright Clearance Center, Inc., 222 Rosewood Drive, Danvers, MA 01923, USA. In this case permission to photocopy is not required from the publisher.

ISBN-13 978-981-4335-80-5
ISBN-10 981-4335-80-0

Printed in Singapore.

## PREFACE

The IV International Course of Mathematical Analysis in Andalusia, held in Jerez (Cádiz), 8-12 September 2009, follows the tradition of previous courses (Cádiz (2002), Granada (2004) and La Rábida (Huelva) (2007)). Seven years ago, representatives of several Andalusian universities made a concerted effort organizing a course to provide an extensive overview of the research in different areas of Mathematical Analysis. The friendly cooperation of many Andalusian research groups in those areas and the initiative of the research group of Functional Analysis of the University of Cádiz, led by Antonio Aizpuru, made possible the organization of the first course in Cádiz. New and wide collaborations by Andalusian research groups in real analysis, complex analysis and functional analysis and mainly the encouragement and hard work of María Victoria Velasco in 2004 and Juan M. Delgado and Tomás Domínguez in 2007 made the dream of the second and third editions in Granada and La Rábida (Huelva) come true. The main purposes of these encounters are:

- That the different research groups in Andalusia working in mathematical analysis have a common frame for meeting and collaboration.
- Contributing to the education of young researchers from these groups, thus creating a setting for communication and access to the most advanced research lines, by means of the joint efforts of all the Andalusian research groups in Mathematical Analysis.

The fourth edition was going to be held in Almería in 2009 under the responsibility of the local organizing committee composed of the research groups in Mathematical Analysis from that university and, particularly, guided by Amin El Kaidi and Juan Carlos Navarro. Unfortunately, on March 1st, 2008 Antonio Aizpuru, one of the pioneers of these events, passed away. The members of his Functional Analysis research group asked that the fourth edition to be moved to Cádiz and held it in memory of Antonio. We are really grateful to the remaining members of the Scientific and Organizing Committees and, particularly, to the main researchers in Almería who graciously and unanimously accepted such a change of place.

Financed by the Spanish National Government, the regional government, research project CONSOLIDER Ingenio MATHEMATICA (i-MATH) and the University of Cadiz, the IV Cidama is thus celebrated and leading researchers are invited to deliver three seminars (called "minicourses" in this volume) and several plenary talks. There were more than one hundred participants from different countries and 22 parallel comunications and posters were presented.

The purpose of the lectures (plenary talks and seminars) is to provide the opportunity to explore several subjects, posing open problems and presenting techniques and tools which could be effective in its resolution, guided by prominent researchers in their expertise fields. Every conference provided an overview of the subject in one hour, whereas the seminars, taking three hours along three days, gave room for a more detailed discussion on the presented topics.

As in previous editions, the high scientific quality of the conferences and seminars taught in this course inspires us to elaborate them in this collection. Thanks to the kindness of the lecturers we could agree with World Scientific Publishing Co. about the publication of this proceedings, which without doubt will be of great interest for graduate students and researchers in the several areas of Mathematical Analysis.

The present proceedings include the contributions corresponding to the seminars given by Pietro Aiena, Joe Diestel and Thomas Schlumprecht and the plenary conferences given by Jesús Araujo, Richard M. Aron, Francisco J. García-Pacheco, Mostafa Mbekhta, Eve Oja, Fernando Rambla-Barreno, Juan B. Seoane-Sepúlveda and Javier Soria.

We would like to seize the oportunity of this preface to give a special tribute to the memory of Antonio Aizpuru, to whom this edition of the courses is dedicated to; they were simultaneously a congress in his honor, and a biographical sketch of him was presented in the opening ceremony of the courses; which follows after this preface.

The talk by J. Araujo is an up-to-date account of the main results on isometric shifts. The author is one of the main specialists in this topic, which has several open problems nowadays.

R. Aron gives a survey (coauthored by P. Galindo) of results dealing with the Banach algebra structure of the set of uniformly continuous and symmetric holomorphic functions on the ball of some classical Banach spaces.

F. J. García-Pacheco reviews the results of Antonio Aizpuru in the

local theory of normed spaces, paying particular attention to Aizpuru's
$E$-property and Aron-Lohman's $\lambda$-property.

The survey by M. Mbekhta is articulated around two major axis. The first one concerns the Kaplansky problem; the history of the problem and several results are presented. The second one concerns some new preserver problems. Both situations are interesting only in the infinite-dimensional case. Several open questions are presented.

E. Oja surveys some connections of the bounded approximation property with Banach operator ideals, both historical and recent ones. The exposition is self-contained and includes an introduction to Banach operator ideals. Complete references are given for the work of A. Grothendieck and P. Enflo.

F. Rambla-Barreno studies several generalizations of the classical Banach-Stone theorem. Particularly, they deal with the four possible combinations of considering continuous or Lipschitz functions in the linear or in the bilinear setting. The main attribute of the mappings is not being an isometry but being separating, which is a related property. This is a joint, unfinished work with Antonio Aizpuru, who sadly passed away during the realization of it.

J. B. Seoane-Sepúlveda presents an overview of some of Prof. Aizpuru's results on the applications of lineability to Series and Summability Methods. He goes through his last papers and some of his results relating these two concepts.

J. Soria surveys some recent works concerning the study of the norm of the Hardy operator minus the Identity. In particular, optimal estimates are found for the cone of radially decreasing functions on the minimal Lorentz spaces (restricted type estimates), and a new Banach function space, closely related to them, is also defined. He also mentions a recent solution to a conjecture of Kruglyak and Setterqvist.

P. Aiena gives a seminar describing all Weyl type theorems, together with some of their generalized versions and shows the equivalences of these theorems for classes of operators which satisfy certain so-called "polaroid" conditions on the isolated points of the spectrum, or on the isolated points of the approximate point spectrum.

J. Diestel discusses in his seminar (coauthored by A. Spalsbury) most of the known results concerning finitely additive measures. He deals with the classical Nikodym, Grothendieck and Vitali-Hahn-Saks theorems. He also considers classical measures, orthogonal groups, Banach limits and several other results, all presented in detail.

T. Schlumprecht's seminar is devoted to study the sampling and recovery of bandlimited functions, and some of its applications. Although belonging to an applied field of Analysis, the abstraction of the results presented makes them interesting both for the pure and the applied mathematicians.

The editors want to express their gratitude to the previously cited authors for their careful preparation of their contributions to this volume, as well as to the corresponding referees.

During this fourth edition of the courses there was also leisure time, which permitted scientific exchanges between the participants and cultural and touristic enjoyment after the long mathematical sessions. Special attention was given to the visit to the library of the Royal Institute and Observatory of the Army in San Fernando (which keeps many precious mathematical and scientific works from the 15th to the 19th century, comprising more than thirty thousand catalogued books). Besides, the participants could attend the show "How the Andalusian horses dance", from the "Royal Andalusian School of Equestrian Art of Jerez" which featured purebred horses. There were also a lunch in the Barrellery and a Gala Dinner in the Clock Museum.

These courses would not have been possible without the generous help of the financing entities: the Ministry of Science and Technology of the Spanish Government, the Regional Ministry of Innovation, Science and Enterprise, the project CONSOLIDER i-Math and the University of Cádiz which, apart from providing installations and economical support for the event, edited the complete works of Antonio Aizpuru, which were given as a present to all the participants. Special thanks are due to the rector of the University of Cádiz, who was supportive and enthusiastic with this initiative from the very first moment.

F. J. Pérez-Fernández                          Puerto Real (Cádiz, Spain)
F. Rambla-Barreno                                   Spring 2011

# ORGANIZING COMMITTEES

## SCIENTIFIC/ORGANIZING COMMITTEE

| | |
|---|---|
| Amin El Kaidi Lhachmi | – Universidad de Almería |
| Juan Carlos Navarro Pascual | – Universidad de Almería |
| Fernando León Saavedra | – Universidad de Cádiz |
| Francisco Javier Pérez Fernández | – Universidad de Cádiz |
| Juan Francisco Mena Jurado | – Universidad de Granada |
| Rafael Payá Albert | – Universidad de Granada |
| Ángel Rodríguez Palacios | – Universidad de Granada |
| María Victoria Velasco Collado | – Universidad de Granada |
| Cándido Piñeiro Gómez | – Universidad de Huelva |
| Ramón J. Rodríguez Álvarez | – Universidad de Huelva |
| Miguel Marano Calzolari | – Universidad de Jaén |
| Francisco Roca Rodríguez | – Universidad de Jaén |
| Daniel Girela Álvarez | – Universidad de Málaga |
| Francisco Javier Martín Reyes | – Universidad de Málaga |
| Antonio Villar Notario | – Universidad Pablo de Olavide |
| Tomás Domínguez Benavides | – Universidad de Sevilla |
| Antonio Fernández Carrión | – Universidad de Sevilla |
| Carlos Pérez Moreno | – Universidad de Sevilla |
| Luis Rodríguez Piazza | – Universidad de Sevilla |

## LOCAL ORGANIZING COMMITTEE

Francisco Benítez Trujillo — Universidad de Cádiz
Antonio Jesús Calderón Martín — Universidad de Cádiz
José Manuel Díaz Moreno — Universidad de Cádiz
Juan Ignacio García García — Universidad de Cádiz
Fernando León Saavedra — Universidad de Cádiz
Francisco Ortegón Gallego — Universidad de Cádiz
Francisco Javier Pérez Fernández — Universidad de Cádiz
Fernando Rambla Barreno — Universidad de Cádiz

## SPONSORS

Ministerio de Ciencia e Innovación
Consejería de Innovación, Ciencia y Empresa de la Junta de Andalucía
Ingenio Mathematica
Universidad de Cádiz
Consejo Social de la Universidad de Cádiz
Vicerrectorado de Investigación de la Universidad de Cádiz
Facultad de Ciencias de la Universidad de Cádiz
Departamento de Matemáticas de la Universidad de Cádiz

# CONTENTS

Preface   v

Organizing committees   ix

In memoriam: Professor Antonio Aizpuru   1
   *F. Javier Pérez-Fernández*

## Part A   Mini-courses   11

Weyl type theorems for bounded linear operators on Banach spaces   13
   *Pietro Aiena*

Finitely additive measures in action   58
   *Joe Diestel and Angela Spalsbury*

Sampling and recovery of bandlimited functions and applications
to signal processing   116
   *Th. Schlumprecht*

## Part B   Plenary speakers   143

Isometric shifts between spaces of continuous functions   145
   *Jesús Araujo*

Uniform algebras of symmetric holomorphic functions   158
   *Richard M. Aron and Pablo Galindo*

Some results on the local theory of normed spaces since 2002 (1997)   165
   *F. J. García-Pacheco*

A survey on linear (additive) preserver problems 174
*Mostafa Mbekhta*

Bounded approximation properties via Banach operator ideals 196
*Eve Oja*

Linear or bilinear mappings between spaces of continuous or
Lipschitz functions 216
*Fernando Rambla-Barreno*

Summability and lineability in the work of Antonio Aizpuru Tomás 226
*Juan B. Seoane-Sepúlveda*

Optimal bounds for the Hardy operator minus the Identity 234
*Javier Soria*

Author Index 247

# In memoriam: Professor Antonio Aizpuru

F. Javier Pérez-Fernández

*Department of Mathematics*
*University of Cádiz, Spain*
*e-mail: javier.perez@uca.es*

> *I want to be with you under the winged souls*
> *of the roses of the cream-coloured almond tree,*
> *for we have many things to talk of,*
> *companion of my soul, my companion.*
>
> (MIGUEL HERNÁNDEZ: UNCEASING LIGHTNING)

## 1. The person

Antonio Aizpuru Tomás was highly involved with all the CIDAMA from the very beginning in 2001. In that year, many researchers in Mathematical Analysis from different Andalusian universities (among them, Antonio) had the idea of starting these courses, oriented towards forming young researchers in Mathematical Analysis in Andalucía, joining efforts from the research projects of our region.

In September 2002 the first edition of the courses is held in Cadiz, under the organization leading of Antonio.

The likable figure of Antonio and this special link to the CIDAMA, which continued in future editions, made the Scientific Committee think of dedicating the fourth edition, to be celebrated in Almería, to the memory of Antonio. Simultaneously in Cádiz we had the intention of organizing a congress in his memory, also in 2009.

Both initiatives joined efforts and it was decided to celebrate the fourth edition as a congress in his memory and held it in Cádiz.

For me, above all, Antonio was a great friend, and independently of any subjectivity he was an extraordinary professor, and excellent mathematician and a wonderful person.

This smiling picture of Antonio, although not suiting him aesthetically, is very illustrative of that likable person and all of us who met him kept an indelible memory:

Fig. 1. Antonio Aizpuru

He passed away aged 53, after a heart attack, on March 1, 2008 in Cadiz. Six days after his passing, F. J. Ruiz Domínguez, a former student of him wrote in a personal website [12] a beautiful text remembering Antonio, which reads as follows:

> *Short, or properly said, squat; pleated trousers with the pockets stained in chalk; dark curly hair; in front of the blackboard, looking at us and smiling, with that smile in all his face, grinning from ear to ear, so much that almost closing his eyes. A smile of a big boy. That is how I remember Antonio Aizpuru....*
>
> *The first [lesson Antonio taught me] was by example: his humanity, good mood, his concerns and devotion to everybody else. Antonio was one of those people making the world a bit better, someone who made you believe the human being is not so bad, that there is good people in the world. And the most important thing is that his example put a seed in all of us that knew him. Perhaps not all soils*

are good for that seed, but in some of us his example delved very deeply.

All that met him are witness of the truthful and exactness of these words.

He was born in September 13th, 1954 in Valencia. As he became an orphan at a very early age, his mother sent him with his brother Gabriel to the "School of Orphans of the Army" (his father was a military man). Madrid is the place where he would meet the love of his life: his wife, Mara. It was in November 10th, 1973, in the "newcomers party", his wife was beginning the degree of Psychology in Madrid's "Complutense" University and he was a student of Mathematics in the same university. Four years later they married in Cádiz.

Therefore, it is fate that brings him to Cádiz, a place he would never want to leave. After four years in Valencia as a school teacher, he returned to Andalusia in 1982, first to Morón and later to San Fernando, in 1983, as full professor of Mathematics in a secondary school.

In that year, by means of the selectivity summits, he met Juan Luis Romero (full professor of Mathematical Analysis in the University of Cádiz) and several more colleagues. He was in a group of secondary school teachers that have decided to go to a Research Seminar with Juan Luis Romero. That was the beginning of several doctoral theses and a university career for some of these teachers, among them Antonio, who I met in 1983 in the house of a common friend who is also a mathematician. The three couples have had dinner together and my friendship with him began.

Those who were his friends knew how much love, passion and dedication he felt for Mathematics, and being this very much, he felt more love and passion for his family. In his own words [6] we can read

> *Mathematics are important, mainly for the people sensitive to beauty. It might be that another human activities are more important: art, poetry, philosophy... But it is sure that for me there are more important things.*     *To my family*

To subsequently add, with his characteristic subtle humor and irony:

> *To the important things: Mara, my daughters and son, my friends, the Spanish potato salads, the Cruzcampo, the Habana club, the federal republic we need, the beach, the tobacco, ...*

And that was how he felt it. The important things were his loved ones and the small things in life, where happiness lies. Likable, close, kind,

*Machado-like* speaking: "a good man", ingenious, funny, deeply democratic and republican, Christian, ... With a very subtle sense of humor. He did not have a visiting card saying he was a full professor. Instead he had one saying:

> Antonio Aizpuru Tomás
> Sophist, rethorical and freethinker

"Freethinker", indeed, the essential part of human condition: the free discernment and the absence of subservience. Faithful to his principles and rigorous in his approaches.

Fig. 2. In his office

Among the many anecdotes that can be cited of that subtle and ironic humor, was the price list for argumentations, according to the category of the event:

- Argumentations on demand:
    - For the proposal (10 €).
    - Against the proposal (10 €).
    - Emphasizing positive and negative aspects of the proposal (18 €).

- Speeches on demand:
  - Of entrance in royal academies (100 €).
  - Of entrance in provincial academies (80 €).
  - Of entrance in athenaeums (60 €).
  - Of entrance in casinos (40 €).
  - Of praise in retirement ceremony (25 €).

## 2. The professor

A born hard-working person. He arrived at the faculty early in the morning, had lunch in little more than half an hour and came back to keep working in his office, which he would leave at sunset. He was completely devoted to teaching and researching. In Fig. 2 we can see him working in his office, which more than an office was another part of his home, full of books and homey effects giving it warmth and having among them a beautiful pedestal, on top of it a picture of his wife.

Mathematics were his other great love. Working as full professor in a secondary school he made his doctoral thesis, entitled "Some classical theorems of measure theory using the Stone spaces of Boolean Algebras", supervised by Juan Luis Romero, and he defended it in 1986. After his thesis he started working in the University of Cádiz, as an associate professor who also worked in a secondary school. Four years later he dedicated exclusively to the university and in 2003 he obtained full professorship.

Apart from a wonderful person, he was an excellent professor, capable of infecting his enthusiasm on mathematics to his students. In the previously cited website [12], his student says:

> *He is likely the professor I've spent more classroom hours with. And what hours!... one never got bored. Antonio filled those hours with his personality, his humor, his presence... I skipped very few of those hours. It was great seeing how he gave life to all those abstract and boring concepts, making them human.*

And the student illustrates it with some anecdotes:

- "I call all the terms of the sequence to my office, and ask them if they live at less than epsilon of the limit",
- "we call the elements $\hat{x}$ 'Chinese' because of the hat they wear, and those who are $x^{**}$ we call them 'captains', because of the two stars",

- "don't enumerate my dense sets, my dense sets are not countable, don't enumerate my dense sets, I'm not separable" [with the rhythm of Sevillian dances],
- and a long, long list.

He always received extraordinary assessments from his students. There are many testimonials speaking of his excellence as a professor, for example the students asked him many years to be their "godfather", and he received each year the prize to the most beloved professor in the student's "San Alberto" party, or that his optative subjects had a high demand, or the numerous students of the degree in Mathematics asking for a last-year grant to be initiated in research[a] by him, or that each year the best students wanted him to be their advisor.

His ex-students, from all years, attended en masse his funeral to bid him the last farewell.

## 3. The researcher

His research work was intense and fruitful. He published, until his passing, fifty-eight papers and three books. Only in 2008, seven of his papers were published. After his passing, between 2008 and 2010, ten papers coauthored by him have been published. He was the advisor of five doctoral theses and when he passed away there were four more theses supervised by him and almost completed. He was the head of several national research projects and also the "alma mater" of our research group *Geometry, operators and series in Banach spaces*.

He worked in the following fields:

- Transitivity of the norm.
- The diameter problem.
- The fixed point property.
- Proving or refuting that in each Banach space there exists a non-trivial vector subspace of norm-attaining functionals.
- Series, summability, matrix convergence and Banach-Lorentz convergence. Natural families and natural Boolean algebras.
- Lineability.
- Statistical convergence.
- Banach-Stone type theorems.

---

[a]Collaboration grant from the national government.

Fig. 3. The day of the dissertation defense of F. J. García Pacheco. From L to R: F. J. Pérez-Fernández, J. C. Navarro, J. L. Romero, R. M. Aron, A. Aizpuru, F. J. García-Pacheco and J. Diestel

- Absolutely valued Banach algebras and absolutely valued Banach spaces.
- Separating mappings.

He devoted a very special attention to the education of young researchers. It was the professional hope of his life to improve mathematical research by educating young researchers and creating in Cádiz a large and sound research group. In Fig. 3 we can see him the day of the dissertation defense of F. J. García-Pacheco and in Fig. 4 the day of the dissertation defense of J. B. Seoane-Sepúlveda.

His work is still bearing fruits after his passing. As we have already indicated, there are still joint works with him being published:

(1) Series and summability in Banach spaces:

- Summability methods using statistical convergence (cf. [3], [4], [7], [11]).
- Matrix methods of summability (cf. [5]).
- Almost summability using Banach-Lorentz convergence (cf. [1], [8]).

Fig. 4. The day of the dissertation defense of J. B. Seoane. From L to R: M. Maestre, J. B. Seoane-Sepúlveda, A. Aizpuru, J. Diestel and F. Rambla

(2) Geometry of Banach spaces. Isometries in Banach spaces. (cf. [6], [9], and [10]).

Apart from the previously mentioned works, on his table remained many other problems he was working on, and they are now being continued by his disciples, in the following topics:

- Banach-Stone theorems on Lipschitz functions.
- Linear and separating mappings between function spaces (scalar and vector-valued case).
- Bilinear and separating mappings between function spaces (scalar case).
- Phillips lemma on effect algebras (with student S. Moreno-Pulido).
- Density by moduli and statistical convergence of simple and double sequences (with student M. C. Listán-García).

## References

1. A. Aizpuru, R, Armario, F. J. García-Pacheco and F. J. Pérez-Fernández. *Banach limits and uniform almost summability*, J. Math. Anal. Appl. **379** (2011), 82–90.

2. A. Aizpuru and F. Rambla. *Diameter preserving linear bijections and $C_0(L)$ spaces*, Bull. Belg. Math. Soc. Simon Stevin, **17** (2010), no. 2, 377–383.
3. A. Aizpuru, M. Nicasio-Llach and A. Sala. *A remark about the statistical Cesro summability and the Orlicz-Pettis theorem.* Acta Math. Hungar. **126** (2010), no. 1-2, 94–98.
4. A. Aizpuru, M. Nicasio-Llach and F. Rambla. *A remark about the Orlicz-Pettis theorem and the statistical convergence.* Acta Math. Sin. (Engl. Ser.) **26** (2010), no. 2, 305–310.
5. A. Aizpuru, C. Pérez-Eslava and J. B. Seoane-Sepúlveda. *Matrix summability methods and weakly unconditionally Cauchy series.* Rocky Mountain J. Math. **39** (2009), no. 2, 367–380.
6. A. Aizpuru. *Apuntes incompletos de Análisis Funcional.* Servicio de Publicaciones de la Universidad de Cádiz (2009).
7. A. Aizpuru and M. Nicasio-Llach. *Spaces of sequences defined by the statistical convergence.* Studia Sci. Math. Hungar. **45** (2008), no. 4, 519–529.
8. A. Aizpuru, R. Armario and F. J. Pérez-Fernández. *Almost Summability and Unconditionally Cauchy Series*, Bull. Belg. Math. Soc. Simon Stevin. **15** (2008), no. 4, 635–644.
9. A. Aizpuru and F. J. García-Pacheco. *A short note about exposed points in real Banach spaces.* Acta Math. Sci. Ser. B Engl. Ed. **28** (2008), no. 4, 797–800.
10. A. Aizpuru and F. J. García-Pacheco. *A note on $L^2$-summand vectors in dual spaces.* Glasg. Math. J. **50** (2008), no. 3, 429–432.
11. A. Aizpuru and M. Nicasio-Llach. *About the statistical uniform convergence.* Bull. Braz. Math. Soc. (N.S.). **39** (2008), no. 2, 173–182.
12. F. J. Ruiz Domínguez, http://wewe0.wordpress.com/2008/03/07/antonio-aizpuru/ (2008).

# PART A

# Mini-courses

# Weyl type theorems for bounded linear operators on Banach spaces

Pietro Aiena*

*Dipartimento di Metodi e Modelli Matematici*
*Facoltà di Ingegneria, Università di Palermo*
*Viale delle Scienze, I-90128 Palermo, Italy*
*e-mail: paiena@unipa.it*

*Keywords*: Localized SVEP, Fredholm theory, Weyl type theorems.

## 1. Introduction

In 1909 H. Weyl [59] studied the spectra of all compact linear perturbations of a self-adjoint operator defined on a Hilbert space and found that their intersection consisted precisely of those points of the spectrum which are not isolated eigenvalues of finite multiplicity. Later, the property established by Weyl for self-adjoint operators has been observed for several other classes of operators, for instance hyponormal operators on Hilbert spaces, Toeplitz operators [33], convolution operators on group algebras [19], and many other classes of operators defined on Banach spaces ([52]). In the literature, a bounded linear operator defined on a Banach space which satisfies this property is said to satisfy *Weyl's theorem*. Weaker variants of Weyl's theorem have been discussed by Harte and Lee [44], while two approximate-point spectrum versions of Weyl's theorem have been introduced by Rakočević [54], *a-Weyl's theorem* ([55], [2]), and the so-called *property (w)* ([54], [17]). In this course we describe all Weyl type theorems, together with some their generalized versions ([25], [27]) and we show the equivalences of these theorems for classes of operators which satisfy certain

---

*The author was supported by ex-60 2008, Fondi di ricerca dell' Universitá di Palermo. These notes form an extended version of the three lectures given in the IV International Course of Mathematical Analysis en Andalucía (Jerez de La Frontera, 8-12 of September 2009). The author thanks the Organizing Scientific Committee for their kind invitation and is very grateful for the warm and friendly atmosphere of the Conference.

"polaroid" conditions on the isolated points of the spectrum ([5], or on the isolated points of the approximate point spectrum. Our main tool is an important property, the so-called single valued extension property (SVEP), introduced by Dunford [41], [42]. The SVEP plays an important role in local spectral theory, see the monograph of Laursen and Neumann [50], and a localized version of SVEP has deep connections with Fredholm theory, see the monograph [1]. In the last part of the course we study the permanence of Weyl type theorems under a quasi-affinity, or more in general, the permanence of Weyl type theorems from an operator $T \in L(X)$ to an operator $S \in L(Y)$, $X$ and $Y$ Banach spaces, in the case that $T$ and $S$ are intertwined asymptotically by an operator $A \in L(X, Y)$.

## 2. Preliminaries

Let $L(X)$ be the algebra of all bounded linear operators acting on an infinite dimensional complex Banach space $X$ and if $T \in L(X)$ let us denote by $\alpha(T)$ the dimension of the kernel ker $T$ and by $\beta(T)$ the codimension of the range $T(X)$. Recall that the operator $T \in L(X)$ is said to be *upper semi-Fredholm*, $T \in \Phi_+(X)$, if $\alpha(T) < \infty$ and the range $T(X)$ is closed, while $T \in L(X)$ is said to be *lower semi-Fredholm*, $T \in \Phi_-(X)$, if $\beta(T) < \infty$. If either $T$ is upper or lower semi-Fredholm then $T$ is said to be a *semi-Fredholm operator*, while if $T$ is both upper and lower semi-Fredholm then $T$ is said to be a *Fredholm operator*. If $T$ is semi-Fredholm then the *index* of $T$ is defined by $\operatorname{ind}(T) := \alpha(T) - \beta(T)$. A bounded operator $T \in L(X)$ is said to be a *Weyl operator*, $T \in W(X)$, if $T$ is a Fredholm operator having index 0. The classes of *upper semi-Weyl* and *lower semi-Weyl* operators are defined, respectively:

$$W_+(X) := \{T \in \Phi_+(X) : \operatorname{ind} T \leq 0\},$$

$$W_-(X) := \{T \in \Phi_-(X) : \operatorname{ind} T \geq 0\}.$$

Clearly, $W(X) = W_+(X) \cap W_-(X)$. These sets of operators define the following spectra: the *Weyl spectrum* defined by

$$\sigma_{\mathrm{w}}(T) := \{\lambda \in \mathbb{C} : \lambda I - T \notin W(X)\},$$

the *upper semi-Weyl spectrum* defined by

$$\sigma_{\mathrm{uw}}(T) := \{\lambda \in \mathbb{C} : \lambda I - T \notin W_+(X)\},$$

and the *lower semi-Weyl spectrum* defined by

$$\sigma_{\mathrm{lw}}(T) := \{\lambda \in \mathbb{C} : \lambda I - T \notin W_-(X)\}.$$

The *ascent* of an operator $T \in L(X)$ is defined as the smallest non-negative integer $p := p(T)$ such that ker $T^p$ = ker $T^{p+1}$. If such integer does not exist we put $p(T) = \infty$. Analogously, the *descent* of $T$ is defined as the smallest non-negative integer $q := q(T)$ such that $T^q(X) = T^{q+1}(X)$, and if such integer does not exist we put $q(T) = \infty$. It is well-known that if $p(T)$ and $q(T)$ are both finite then $p(T) = q(T)$, see [1, Theorem 3.3]. Moreover, if $\lambda \in \mathbb{C}$ then $0 < p(\lambda I - T) = q(\lambda I - T) < \infty$ if and only if $\lambda$ is a pole of the resolvent of $T$. In this case $\lambda$ is an eigenvalue of $T$ and an isolated point of the spectrum $\sigma(T)$, see [45, Prop. 50.2]. A bounded operator $T \in L(X)$ is said to be *Browder* (resp. *upper semi-Browder, lower semi-Browder*) if $T \in \Phi(X)$ and $p(T) = q(T) < \infty$ (resp. $T \in \Phi_+(X)$ and $p(T) < \infty$, $T \in \Phi_-(X)$ and $q(T) < \infty$). Denote by $B(X)$, $B_+(X)$ and $B_-(X)$ the classes of Browder operators, upper semi-Browder operators and lower semi-Browder operators, respectively. Clearly, $B(X) \subseteq W(X)$, $B_+(X) \subseteq W_+(X)$ and $B_-(X) \subseteq W_-(X)$. Of course, also Browder operators define some spectra: the *Browder spectrum* defined by

$$\sigma_b(T) := \{\lambda \in \mathbb{C} : \lambda I - T \notin B(X)\},$$

the *upper semi-Browder spectrum* defined by

$$\sigma_{ub}(T) := \{\lambda \in \mathbb{C} : \lambda I - T \notin B_+(X)\},$$

and *lower semi-Browder spectrum* defined by

$$\sigma_{lb}(T) := \{\lambda \in \mathbb{C} : \lambda I - T \notin B_-(X)\}.$$

Clearly, $\sigma_w(T) \subseteq \sigma_b(T)$, $\sigma_{uw}(T) \subseteq \sigma_{ub}(T)$ and $\sigma_{lw}(T) \subseteq \sigma_{lb}(T)$. Moreover, from classical Fredholm theory there is a perfect duality: if $T^*$ denotes the dual of $T$ we have $\sigma_{ub}(T) = \sigma_{lb}(T^*)$ and $\sigma_{ub}(T^*) = \sigma_{lb}(T)$ and analogous equalities we have for Weyl spectra.

The classical *approximate point spectrum* is defined by

$$\sigma_a(T) := \{\lambda \in \mathbb{C} : \lambda I - T \text{ is not bounded below}\},$$

while the *surjectivity spectrum* of $T \in L(X)$ is defined by

$$\sigma_s(T) := \{\lambda \in \mathbb{C} : \lambda I - T \text{ is not surjective}\}.$$

We have, see [1, Theorem 3.65, part (v)],

$$\sigma_{ub}(T) = \sigma_{uw}(T) \cup \operatorname{acc} \sigma_a(T), \tag{1}$$

$$\sigma_{lb}(T) = \sigma_{lw}(T) \cup \operatorname{acc} \sigma_s(T), \tag{2}$$

and
$$\sigma_{\mathrm{b}}(T) = \sigma_{\mathrm{w}}(T) \cup \mathrm{acc}\,\sigma(T), \qquad (3)$$
where with acc $K$ we denote the set of accumulation points of the set $K \subseteq \mathbb{C}$, see Theorem 3.65 of [1].

The single valued extension property was introduced in the early years of local spectral theory by Dunford [41], [42] and plays an important role in spectral theory, see the recent monographs by Laursen and Neumann [50], or by Aiena [1]. We shall consider the following local version of this property

**Definition 2.1.** Let $X$ be a complex Banach space and $T \in L(X)$. The operator $T$ is said to have *the single valued extension property* at $\lambda_0 \in \mathbb{C}$ (abbreviated SVEP at $\lambda_0$), if for every open disc $U$ of $\lambda_0$, the only analytic function $f : U \to X$ which satisfies the equation $(\lambda I - T)f(\lambda) = 0$ for all $\lambda \in U$ is the function $f \equiv 0$.
An operator $T \in L(X)$ is said to have SVEP if $T$ has SVEP at every point $\lambda \in \mathbb{C}$.

Evidently, an operator $T \in L(X)$ has SVEP at every point of the resolvent $\rho(T) := \mathbb{C} \setminus \sigma(T)$. Moreover, the identity theorem for analytic functions implies that $T \in L(X)$ has SVEP at every point of the boundary $\partial \sigma(T)$ of the spectrum. In particular, both $T$ and the dual $T^*$ have SVEP at every isolated point of $\sigma(T)$.
An important example of operator which has not SVEP is the left shift on $\ell^2(\mathbb{N})$, while the right shift has SVEP. Note that
$$p(\lambda I - T) < \infty \Rightarrow T \text{ has SVEP at } \lambda,$$
and dually
$$q(\lambda I - T) < \infty \Rightarrow T^* \text{ has SVEP at } \lambda,$$
see [1, Theorem 3.8].

Let us consider the *quasi-nilpotent part* of $T$, i.e. is the set
$$H_0(T) := \{x \in X : \lim_{n \to \infty} \|T^n x\|^{\frac{1}{n}} = 0\}.$$

It is easily seen that $\ker (T^m) \subseteq H_0(T)$ for every $m \in \mathbb{N}$. Moreover, $T$ is quasi-nilpotent if and only if $H_0(T) = X$, see [1, Theorem 1.68]. If $T \in L(X)$, the *analytic core* $K(T)$ is the set of all $x \in X$ such that there exists a constant $c > 0$ and a sequence of elements $x_n \in X$ such that $x_0 = x, T x_n = x_{n-1}$, and $\|x_n\| \leq c^n \|x\|$ for all $n \in \mathbb{N}$, see [1] for informations on the subspaces $H_0(T)$, $K(T)$. The subspaces $H_0(T)$ and $K(T)$ may be not closed. We have

$$H_0(\lambda I - T) \text{ closed} \Rightarrow T \text{ has SVEP at } \lambda,$$

see [1].

In the following theorem we collect some characterizations of localized SVEP for semi-Fredholm operators.

**Theorem 2.1 ([1, Chapter 3]).** *Suppose that $\lambda_0 I - T$ is semi-Fredholm. Then the following statements are equivalent:*

*(i) $T$ has SVEP at $\lambda_0$;*

*(ii) $p(\lambda_0 I - T) < \infty$;*

*(iii) $H_0(\lambda_0 I - T)$ is closed.*

*(iv) $H_0(\lambda_0 I - T)$ is finite-dimensional.*

*Dually, the following statements are equivalent:*

*(v) $T^*$ has SVEP at $\lambda_0$;*

*(vi) $q(\lambda_0 I - T) < \infty$.*

*(vii) $K(\lambda I - T)$ is finite-codimensional.*

**Remark 2.1.** The equivalences established in Theorem 2.1 have been proved also in the more general framework of operators of Kato type or quasi-Fredholm operators [3]. We do not give the definitions of these concepts, but the reader can find details on the operators of Kato type in [1, Chapter 3], while quasi-Fredholm operators on Hilbert spaces have been introduced by Labrousse [47] and studied, for Banach space operators, in several papers, for instance in [22], [23].

## 3. Browder theorems

Let write iso $K$ for the set of all isolated points of $K \subseteq \mathbb{C}$. Following Harte and W. Y. Lee [44] we shall say that $T$ satisfies *Browder's theorem* if

$$\sigma_w(T) = \sigma_b(T),$$

or equivalently, by (3), if

$$\lambda I - T \text{ Weyl} \Rightarrow \lambda \in \text{iso}\,\sigma(T). \tag{4}$$

For a bounded operator $T \in L(X)$ define

$$\pi_{00}(T) := \{\lambda \in \text{iso }\sigma(T) : 0 < \alpha(\lambda I - T) < \infty\},$$

and

$$p_{00}(T) := \sigma(T) \setminus \sigma_b(T) = \{\lambda \in \sigma(T) : \lambda I - T \in \mathcal{B}(X)\},$$

the set of all *Riesz points* in $\sigma(T)$. Clearly, $p_{00}(T)$ is the set of poles of the resolvent having finite rank, i.e. $0 < \alpha(\lambda I - T) = \beta(\lambda I - T) < \infty$. Consequently, every point of $p_{00}(T)$ is an isolated point of the spectrum and an eigenvalue of $T$. Finally, let us consider the following set:

$$\Delta(T) := \sigma(T) \setminus \sigma_w(T).$$

If $\lambda \in \Delta(T)$ then $\lambda I - T \in W(X)$ and since $\lambda \in \sigma(T)$ then $\alpha(\lambda I - T) = \beta(\lambda I - T) > 0$, so we can write

$$\Delta(T) = \{\lambda \in \mathbb{C} : \lambda I - T \in W(X), 0 < \alpha(\lambda I - T)\}.$$

The points of $\Delta(T)$ are called *generalized Riesz points*.

The following result shows that Browder's theorem is equivalent to the localized SVEP at the points of the complement in $\mathbb{C}$ of $\sigma_w(T)$.

**Theorem 3.1 ([7]).** *For an operator $T \in L(X)$ the following statements are equivalent:*

*(i)* $p_{00}(T) = \Delta(T)$;

*(ii)* $T$ *satisfies Browder's theorem;*

*(iii)* $T^*$ *satisfies Browder's theorem;*

*(iv)* $T$ *has SVEP at every* $\lambda \notin \sigma_w(T)$;

*(v)* $T^*$ *has SVEP at every* $\lambda \notin \sigma_w(T)$.

By Theorem 3.1 the SVEP for $T$ or $T^*$ entails Browder's theorem, but the following example shows that SVEP for $T$ or $T^*$ is not necessary for Browder's theorem.

**Example 3.1.** Let $T := L \oplus L^* \oplus Q$, where $L$ is the unilateral left shift on $\ell^2(\mathbb{N})$, defined by

$$L(x_1, x_2, \ldots) := (x_2, x_3, \cdots), \quad (x_n) \in \ell^2(\mathbb{N}),$$

and $Q$ is any quasi-nilpotent operator on $\ell^2(\mathbb{N})$. $L$ does not have SVEP, see [1, p. 71], so also $T$ and $T^*$ do not have SVEP, see Theorem 2.9 of [1]. On the other hand, we have $\sigma_b(T) = \sigma_w(T) = \mathbf{D}$, where $\mathbf{D}$ is the closed unit disc in $\mathbb{C}$, so that Browder' theorem holds for $T$.

For a bounded operator $T \in L(X)$ define
$$\pi_{00}^a(T) := \{\lambda \in \text{iso } \sigma_a(T) : 0 < \alpha(\lambda I - T) < \infty\},$$
and
$$p_{00}^a(T) := \sigma_a(T) \setminus \sigma_{\text{ub}}(T) = \{\lambda \in \sigma_a(T) : \lambda I - T \in \mathcal{B}_+(X)\}.$$
Clearly, for every $T \in L(X)$ we have the implications:
$$p_{00}(T) \subseteq \pi_{00}(T) \subseteq \pi_{00}^a(T) \quad \text{and} \quad p_{00}^a(T) \subseteq \pi_{00}^a(T). \tag{5}$$
In particular, every $\lambda \in p_{00}^a(T)$ is an isolated point of $\sigma_a(T)$. Finally, let us consider the following set:
$$\Delta_a(T) := \sigma_a(T) \setminus \sigma_{\text{uw}}(T).$$
Since $\lambda I - T \in W_a(X)$ implies that $(\lambda I - T)(X)$ is closed, we can write
$$\Delta_a(T) = \{\lambda \in \mathbb{C} : \lambda I - T \in W_a(X), 0 < \alpha(\lambda I - T)\}.$$
It should be noted that the set $\Delta_a(T)$ may be empty. This is, for instance, the case of a right shift on $\ell^2(\mathbb{N})$.

We shall say that $T$ satisfies *a-Browder's theorem* if
$$\sigma_{\text{uw}}(T) = \sigma_{\text{ub}}(T),$$
or equivalently, by (1), if
$$\lambda I - T \text{ upper semi-Weyl} \Rightarrow \lambda \in \text{iso } \sigma_a(T).$$

We now establish a first simple characterization of operators satisfying *a*-Browder's theorem.

**Theorem 3.2.** *For a bounded operator* $T \in L(X)$, *a-Browder's theorem holds for* $T$ *if and only if* $p_{00}^a(T) = \Delta_a(T)$. *In particular, a-Browder's theorem holds whenever* $\Delta_a(T) = \emptyset$.

We give now a precise description of operators satisfying *a*-Browder's theorem in terms of SVEP at certain sets.

**Theorem 3.3** ([11]). *If* $T \in L(X)$ *the following statements hold:*

*(i)* $T$ *satisfies a-Browder's theorem if and only if* $T$ *has SVEP at every* $\lambda \notin \sigma_{\text{uw}}(T)$.

*(ii)* $T^*$ *satisfies a-Browder's theorem if and only if* $T^*$ *has SVEP at every* $\lambda \notin \sigma_{\text{lw}}(T)$.

*(iii) If $T$ has SVEP at every $\lambda \notin \sigma_{\mathrm{lw}}(T)$ then a-Browder's theorem holds for $T^*$.*

*(iv) If $T^*$ has SVEP at every $\lambda \notin \sigma_{\mathrm{uw}}(T)$ then a-Browder's theorem holds for $T$.*

An obvious consequence of Theorem 3.2 is that if either $T$ or $T^*$ has SVEP then a-Browder's theorem holds for both $T$ and $T^*$. Furthermore, since $\sigma_{\mathrm{uw}}(T) \subseteq \sigma_{\mathrm{w}}(T)$, a-Browder's theorem for $T$ entails that Browder's theorem holds for $T$.

**Example 3.2.** The following example shows that the reverse of the assertions (iii) and (iv) of Theorem 3.1 generally do not hold. Let $1 \le p \le \infty$ be given, and let $\omega := (\omega_n)$ be a bounded sequence of strictly positive real numbers. The corresponding *unilateral weighted right shift* on $\ell^p(\mathbb{N})$ is defined by

$$Tx := \sum_{n=1}^{\infty} \omega_n x_n e_{n+1} \quad \text{for all } x = (x_n) \in \ell^p(\mathbb{N}),$$

where $(e_n)$ is the standard basis of $\ell^p(\mathbb{N})$. In this case the *spectral radius* of $T$ is given by

$$r(T) = \lim_{n\to\infty} \sup_{k\in\mathbb{N}} (\omega_k \cdots \omega_{k+n-1})^{1/n}.$$

Define

$$i(T) := \lim_{n\to\infty} \inf_{k\in\mathbb{N}} (\omega_k \cdots \omega_{k+n-1})^{1/n},$$

and

$$c(T) := \lim_{n\to\infty} \inf_{k\in\mathbb{N}} (\omega_1 \cdots \omega_n)^{1/n}.$$

We have $i(T) \le c(T) \le r(T)$, and as it has been observed by Shields [57] for every triple of real number $0 \le i \le c \le r$ it is possible to find a weighted right shift on $\ell^p(\mathbb{N})$ such that

$$i(T) = i, \quad c(T) = c, \quad r(T) = r.$$

Suppose now that $0 < i(T) \le c(T) \le r(T)$. Every unilateral weighted right shift has SVEP, since it has no eigenvalues, so $T$ satisfies a-Browder's theorem. On the other hand, the dual $T^*$ of $T$ does not have SVEP at 0, see Theorem 2.88 of [1]. Moreover, by Theorem 1.6.15 of [50] we have that

$$\sigma_a(T) = \{\lambda \in \mathbb{C} : i(T) \le |\lambda| \le r(T)\},$$

and since $i(T)$ is strictly greater than 0 then $0 \notin \sigma_a(T)$. Since the inclusion $\sigma_{wa}(T) \subseteq \sigma_a(T)$ holds for every operator, we conclude that $0 \notin \sigma_{wa}(T)$. This example shows that the assertion (iv) of Theorem 3.1 cannot be reversed.

To show that the converse of the assertion (iii) of Theorem 3.1 in general does not hold, note first that the dual $T^*$ of $T$ is the *unilateral weighted left shift* on $\ell^q(\mathbb{N})$ given by

$$T^*x = (\omega_n x_{n+1}) \quad \text{for all } x = (x_n) \in \ell^q(\mathbb{N}),$$

where, as usual, $1/p + 1/q = 1$ and $\ell^q(\mathbb{N})$ is canonically identified with the dual of $\ell^p(\mathbb{N})$. Now, let $S := T^*$. Then $S^* = T$ has SVEP, so a-Browder's theorem holds for $S$, whilst $S$ does not have SVEP at 0. On the other hand, $0 \notin \sigma_{wa}(T) = \sigma_{ws}(T^*) = \sigma_{ws}(S)$.

We recall that *reduced minimum modulus* of a non-zero operator $T$ is defined by

$$\gamma(T) := \inf_{x \notin \ker T} \frac{\|Tx\|}{\operatorname{dist}(x, \ker T)}.$$

It is well-known that $T(X)$ is closed if and only if $\gamma(T) > 0$. Let $M$, $N$ denote two closed linear subspaces of a Banach space $X$ and define

$$\delta(M, N) := \sup\{\operatorname{dist}(u, N) : u \in M, \|u\| = 1\},$$

in the case $M \neq \{0\}$, otherwise we put $\delta(\{0\}, N) = 0$ for any subspace $N$. The *gap* between $M$ and $N$ is defined by

$$\widehat{\delta}(M, N) := \max\{\delta(M, N), \delta(N, M)\}.$$

Note that the function $\widehat{\delta}$ is a metric on the set of all linear closed subspaces of $X$ and the convergence $M_n \to M$ is obviously defined by $\widehat{\delta}(M_n, M) \to 0$ as $n \to \infty$.

**Theorem 3.4 ([7]).** *For an operator $T \in L(X)$ the following statements are equivalent:*

(i) $T$ *satisfies Browder's theorem;*

(ii) $\Delta(T) \subseteq \operatorname{iso} \sigma(T)$;

(iii) $\Delta(T) \subseteq \partial \sigma(T)$, $\partial \sigma T$) *the topological boundary of $\sigma T$);*

(iv) $\operatorname{int} \Delta(T) = \emptyset$.

(v) *the mapping $\lambda \to \ker(\lambda I - T)$ is not continuous at every $\lambda \in \Delta(T)$ in the gap metric;*

(vi) *the mapping $\lambda \to \gamma(\lambda I - T)$ is not continuous at every $\lambda \in \Delta(T)$;*

*(vii) the mapping* $\lambda \to (\lambda I - T)(X)$ *is not continuous at every* $\lambda \in \Delta(T)$ *in the gap metric.*

Analogously, we have:

**Theorem 3.5 ([11]).** *For a bounded operator* $T \in L(X)$ *the following statements are equivalent:*

*(i)* $T$ *satisfies a-Browder's theorem;*

*(ii)* $\Delta_a(T) \subseteq iso\,\sigma_a(T)$;

*(iii)* $\Delta_a(T) \subseteq \partial\sigma_a(T)$, $\partial\sigma_a(T)$ *the topological boundary of* $\sigma_a(T)$;

*(iv) the mapping* $\lambda \to \ker(\lambda I - T)$ *is not continuous at every* $\lambda \in \Delta_a(T)$ *in the gap metric;*

*(v) the mapping* $\lambda \to \gamma(\lambda I - T)$ *is not continuous at every* $\lambda \in \Delta_a(T)$;

*(vi) the mapping* $\lambda \to (\lambda I - T)(X)$ *is not continuous at every* $\lambda \in \Delta_a(T)$ *in the gap metric.*

We now establish some characterization of operators satisfying Browder's theorem in terms of the quasi-nilpotent parts $H_0(\lambda I - T)$.

**Theorem 3.6 ([7]).** *For a bounded operator* $T \in L(X)$ *Browder's theorem holds precisely when one of the following statements holds;*

*(i)* $H_0(\lambda I - T)$ *is finite-dimensional for every* $\lambda \in \Delta(T)$;

*(ii)* $H_0(\lambda I - T)$ *is closed for all* $\lambda \in \Delta(T)$;

*(iii)* $K(\lambda I - T)$ *is finite-codimensional for all* $\lambda \in \Delta(T)$.

Analogously, we have:

**Theorem 3.7 ([11]).** *For a bounded operator* $T \in L(X)$ *the following statements are equivalent:*

*(i) a-Browder's theorem holds for* $T$.

*(ii)* $H_0(\lambda I - T)$ *is finite-dimensional for every* $\lambda \in \Delta_a(T)$.

*(iii)* $H_0(\lambda I - T)$ *is closed for every* $\lambda \in \Delta_a(T)$.

A bounded operator $T \in L(X)$ is said to be *semi-regular* if it has closed range and
$$\ker T^n \subseteq T(X) \quad \text{for all } n \in \mathbb{N}.$$
It is easily seen that $T \in L(X)$ is semi-regular precisely when $T(X)$ is closed and $\mathcal{N}^\infty(T) \subseteq T^\infty(X)$, where $\mathcal{N}^\infty(T) := \bigcup_{n=1}^\infty \ker T^n$ denotes the

*hyper-kernel* of $T$ and $T^\infty(X) := \bigcap_{n=1}^{\infty} T^n(X)$ denotes the *hyper-range* of $T$. The *Kato spectrum* is defined by

$$\sigma_k(T) := \{\lambda \in \mathbb{C} : \lambda I - T \text{ is not semi-regular}\}.$$

Note that $\sigma_k(T)$ is a non-empty compact subset of $\mathbb{C}$, since it contains the boundary of the spectrum, see [1, Theorem 1.75].

Define

$$\sigma_1(T) := \sigma_w(T) \cup \sigma_k(T).$$

Denote by $\mathcal{H}(\sigma(T))$ the set of all analytic functions defined on a neighborhood of $\sigma(T)$ and let $f(T)$ be defined by means of the classical functional calculus. It should be noted that in general the spectral mapping theorem does not hold for $\sigma_1(T)$. In fact we have the following result:

**Theorem 3.8 ([29]).** *Suppose that $T \in L(X)$. For every $f \in \mathcal{H}(\sigma(T))$ we have*

$$\sigma_1(f(T)) \subseteq f(\sigma_1(T)).$$

*The equality $f(\sigma_1(T)) = \sigma_1(f(T))$ holds for every $f \in \mathcal{H}(\sigma(T))$ precisely when the spectral mapping theorem holds for $\sigma_w(T)$, i.e.,*

$$f(\sigma_w(T)) = \sigma_w(f(T)) \quad \text{for all } f \in \mathcal{H}(\sigma(T)).$$

Note that the spectral mapping theorem for $\sigma_w(T)$ holds if either $T$ or $T^*$ satisfies SVEP. Let $\sigma_f(T)$ denote the *Fredholm spectrum* of $T$.

**Theorem 3.9 ([29]).** *The spectral mapping theorem holds for $\sigma_1(T)$ precisely when $\operatorname{ind}(\lambda I - T) \cdot \operatorname{ind}(\mu I - T) \geq 0$ for each pair $\lambda, \mu \notin \sigma_f(T)$.*

In general, Browder's theorem for $T$ does not entail Browder's theorem for $f(T)$. However, we have the following result.

**Theorem 3.10.** *Suppose that both $T \in L(X)$ and $S \in L(X)$ satisfy Browder's theorem, $f \in \mathcal{H}(\sigma(T))$ and $p$ a polynomial. Then we have:*

*(i) [29] Browder's theorem holds for $f(T)$ if and only if $f(\sigma_1(T)) = \sigma_1(f(T))$.*

*(ii) [29] Browder's theorem holds for $T \oplus S$ if and only if $\sigma_1(T) \cup \sigma_1(S) = \sigma_1(T \oplus S)$.*

*(iii) [44] Browder's theorem holds for $p(T)$ if and only if $p(\sigma_w(T)) \subseteq \sigma_w(p(T))$.*

(iv) [44] Browder's theorem holds for $T \oplus S$ if and only if $\sigma_w(T) \cup \sigma_w(S) \subseteq \sigma_w(T \oplus S)$.

Browder's theorem survives under perturbation of compact operators $K$ commuting with $T$. In fact, we have

$$\sigma_w(T+K) = \sigma_w(T) \quad \text{and} \quad \sigma_b(T+K) = \sigma_b(T); \qquad (6)$$

the first equality is a standard result from Fredholm theory, while the second equality is due to V. Rakočević [56]. It is not difficult to extend this result to Riesz operators commuting with $T$ (recall that $K \in L(X)$ is said to be a *Riesz operator* if $\lambda I - K \in \Phi(X)$ for all $\lambda \in \mathbb{C} \setminus \{0\}$). Indeed, the equalities (6) hold also in the case where $K$ is Riesz [56]. An analogous result holds if we assume that $K$ is a commuting quasi-nilpotent operator, since quasi-nilpotent operators are Riesz. These results may fail if $K$ is not assumed to commute, see [44, Example 12]. In [44] it is also shown that Browder's theorem holds for a Hilbert space operator $T \in L(H)$ if $T$ is reduced by its finite dimensional eigenspaces.

Browder's theorem entails the continuity of some mappings. To see this, we need some preliminary definitions. Let $(\sigma_n)$ be a sequence of compacts subsets of $\mathbb{C}$ and define canonically its *limit inferior* by

$$\liminf \sigma_n := \{\lambda \in \mathbb{C} : \text{there exists } \lambda_n \in \sigma_n \text{ with } \lambda_n \to \lambda\}.$$

Define the *limit superior* of $(\sigma_n)$ by

$$\limsup \sigma_n := \{\lambda \in \mathbb{C} : \text{there exists } \lambda_{n_k} \in \sigma_{n_k} \text{ with } \lambda_{n_k} \to \lambda\}.$$

A mapping $\varphi$, defined on $L(X)$ whose values are compact subsets of $\mathbb{C}$ is said to be *upper semi-continuous at* $T$ (respectively, *lower semi-continuos a* $T$) provided that if $T_n \to T$, in the norm topology, then $\limsup \varphi(T_n) \subseteq \varphi(T)$ (respectively, $\varphi(T) \subseteq \liminf \varphi(T_n)$). If the map $\varphi$ is both upper and lower semi-continuous then $\varphi$ is said to be *continuos* at $T$. In this case we write $\lim_{n \in \mathbb{N}} \varphi(T_n) = \varphi(T)$. In the following result we consider mappings that associate to an operator its Browder spectrum or its Weyl spectrum.

**Theorem 3.11 ([37]).** *If $T_0 \in L(X)$ then the following assertions hold:*

*(i) Browder's theorem holds for $T_0$ if and only if the map $T \in L(X) \to \sigma_b(T)$ is continuous at $T_0$.*

*(ii) If Browder's theorem holds for $T_0$ then the map $T \in L(X) \to \sigma(T)$ is continuous at $T_0$.*

## 4. Weyl type theorems

In this section we consider some classes of operators which present a very nice spectral structure. This structure is the same that H. Weyl observed for a self-adjoint operator defined on a Hilbert space.

**Definition 4.1.** A bounded operator $T \in L(X)$ is said to satisfy *Weyl's theorem*, in symbol $(W)$, if $\sigma(T) \setminus \sigma_w(T) = \pi_{00}(T)$. $T$ is said to satisfy *a-Weyl's theorem*, in symbol $(aW)$, if $\sigma_a(T) \setminus \sigma_{uw}(T) = \pi_{00}^a(T)$. $T$ is said to satisfy property $(w)$, if $\sigma_a(T) \setminus \sigma_{uw}(T) = \pi_{00}(T)$.

The class of operators which satisfy Weyl's theorem was introduced first by Coburn [33], while $a$-Weyl's theorem was introduced by V. Rakočević [55]. The property $(w)$ was considered first in the short article [54] by Rakočević and more recently studied in [17] and [15].

**Theorem 4.1 ([2 and 17]).** *For a bounded operator $T \in L(X)$ then the following assertions hold:*

*(i) Weyl's theorem holds for $T \Leftrightarrow T$ satisfies Browder's theorem and $\pi_{00}(T) = p_{00}(T)$.*

*(ii) a-Weyl's theorem holds for $T \Leftrightarrow T$ satisfies a-Browder's theorem and $\pi_{00}^a(T) = p_{00}^a(T)$.*

*(iii) Property $(w)$ holds for $T \Leftrightarrow T$ satisfies a-Browder's theorem and $\pi_{00}(T) = p_{00}^a(T)$.*

The following diagram resumes the relationships between Weyl's theorems, $a$-Browder's theorem and property $(w)$.

$$\begin{array}{ccc} \text{Property } (w) & \Rightarrow & a\text{-Browder's theorem} \\ \Downarrow & & \Uparrow \\ \text{Weyl's theorem} & \Leftarrow & a\text{-Weyl's theorem} \end{array}$$

(see [54] and [17]). The converse of the implications above in general are not true. Examples of operators satisfying Weyl's theorem but not property $(w)$ may be found in [17]. Note that property $(w)$ is not intermediate between Weyl's theorem and $a$-Weyl's theorem, see [17] for examples.

Let us give a closer look to the condition $p_{00}(T) = \pi_{00}(T)$ which appears in Theorem 4.1. By definition of these sets, the equality $p_{00}(T) = \pi_{00}(T)$ means that the set $\pi_{00}(T)$ of all isolated points of $\sigma(T)$ which are eigenvalue of finite multiplicity coincides with the set of poles of $T$ having finite rank. Let $\mathcal{P}_0(X)$, $X$ a Banach space, denote the class of all operators $T \in L(X)$

for which there exists $p := p(\lambda) \in \mathbb{N}$ such that

$$H_0(\lambda I - T) = \ker(\lambda I - T)^p \quad \text{for all } \lambda \in \pi_{00}(T). \tag{7}$$

The condition (7), and in general the form that has the quasi-nilpotent part $H_0(\lambda I - T)$ as $\lambda$ ranges in certain subsets of $\mathbb{C}$, has a crucial role for establishing Weyl's theorem for several classes of operators.

**Theorem 4.2.** $T \in \mathcal{P}_0(X)$ if and only if $p_{00}(T) = \pi_{00}(T)$. In particular, if either $T$ or $T^*$ has SVEP then Weyl's theorem holds for $T$ if and only if $T \in \mathcal{P}_0(X)$.

**Proof.** The equivalence $T \in \mathcal{P}_0(X)$ if and only if $p_{00}(T) = \pi_{00}(T)$ has been proved in [14]. The SVEP for $T$ or for $T^*$ implies Browder's theorem for $T$, so the result follows from Theorem 4.1. □

Let $P(X)$ denote the class of operator $T \in L(X)$ for which there exists $p := p(\lambda I - T) \in \mathbb{N}$ such that

$$H_0(\lambda I - T) = \ker(\lambda I - T)^p \quad \text{for all } \lambda \in \operatorname{iso} \sigma(T). \tag{8}$$

Clearly, $P(X) \subseteq \mathcal{P}_0(X)$. The class $P(X)$ may be characterized in a very clear way:

**Theorem 4.3.** $T \in P(X)$ if and only if every isolated point of the spectrum is a pole of the resolvent.

**Proof.** Suppose $T$ satisfies (8) and that $\lambda$ is an isolated point of $\sigma(T)$. Then there exists $p \in \mathbb{N}$ such that $H_0(\lambda I - T) = \ker(\lambda I - T)^p$. Since $\lambda$ is isolated in $\sigma(T)$ then, by [1, Theorem 3.74],

$$X = H_0(\lambda I - T) \oplus K(\lambda I - T) = \ker(\lambda I - T)^p \oplus K(\lambda I - T),$$

from which we obtain

$$(\lambda I - T)^p(X) = (\lambda I - T)^p(K(\lambda I - T)) = K(\lambda I - T),$$

so

$$X = \ker(\lambda I - T)^p \oplus (\lambda I - T)^p(X),$$

which implies, by [1, Theorem 3.6], that $p(\lambda I - T) = q(\lambda I - T) \leq p$, hence $\lambda$ is a pole of the resolvent, so that $T$ is polaroid.

Conversely, suppose that $T$ is polaroid and $\lambda$ is an isolated point of $\sigma(T)$.

Then $\lambda$ is a pole, and if $p$ is its order then $H_0(\lambda I - T) = \ker(\lambda I - T)^p$, see Theorem 3.74 of [1]. □

A bounded operator $T \in L(X)$ is called *polaroid* if $T \in \mathcal{P}(X)$. It is well known that $\sigma(T) = \sigma(T^*)$ and that $\lambda$ is a pole of the resolvent of $T$ if and only if $\lambda$ is a pole of the resolvent of $T^*$, so

$$T \text{ is polaroid} \Leftrightarrow T^* \text{ is polaroid}.$$

**Theorem 4.4.** *If either $T$ or $T^*$ has SVEP and $T$ is polaroid then Weyl's theorem holds for both $T$ and $T^*$.*

**Proof.** The polaroid condition for $T$ entails $p_{00}(T) = \pi_{00}(T)$. Since $T^*$ is polaroid we also have $p_{00}(T^*) = \pi_{00}(T^*)$. If either $T$ or $T^*$ has SVEP then Browder's theorem holds for $T$ and $T^*$. □

In order to give a characterization of the set $p_{00}^a(T)$ we need first to give some preliminary definitions.

**Definition 4.2.** $T \in L(X)$ is said to be *Drazin invertible* (with a finite index) if and only if $p(T) = q(T) < \infty$ and this is equivalent to saying that $T = T_0 \oplus T_1$, where $T_0$ is invertible and $T_1$ is nilpotent, see [48, Corollary 2.2] and [46, Prop. A].

The definition of Drazin invertibility suggests the following definitions:

**Definition 4.3.** $T \in L(X)$ is said to be *left Drazin invertible* if $p := p(T) < \infty$ and $T^{p+1}(X)$ is closed, while $T \in L(X)$ is said to be *right Drazin invertible* if $q := q(T) < \infty$ and $T^q(X)$ is closed. If $\lambda I - T$ is left Drazin invertible and $\lambda \in \sigma_a(T)$ then $\lambda$ is said to be a *left pole* of the resolvent of $T$. If $\lambda I - T$ is right Drazin invertible and $\lambda \in \sigma_s(T)$ then $\lambda$ is said to be a *right pole* of the resolvent of $T$.

Clearly, $T \in L(X)$ is both right and left Drazin invertible if and only if $T$ is Drazin invertible. In fact, if $0 < p := p(T) = q(T)$ then $T^p(X) = T^{p+1}(X)$ is the kernel of the spectral projection associated with the spectral set $\{0\}$, canonically defined by

$$P_\sigma = \frac{1}{2\pi i} \int_\Gamma (\lambda I - T)^{-1} d\lambda,$$

where $\Gamma$ is a curve enclosing $\{0\}$ which separates $\sigma$ from the remaining part of the spectrum. see [45, Prop. 50.2].

Also the equality and $\pi_{00}^a(T) = p_{00}^a(T)$ may be expressed in terms of poles of the resolvent. This is a consequence of the following result.

**Theorem 4.5.** *Let $T \in L(X)$. Then we have:*
  (i) $\lambda \in p_{00}^a(T)$ *if and only if $\lambda$ is a left pole of finite rank.*
  (ii) $\lambda \in p_{00}^a(T^*)$ *if and only if $\lambda$ is a right pole of finite rank.*

**Proof.** (i) Suppose $\lambda$ be a left pole of finite rank. We may assume $\lambda = 0$. Then, $0 \in \sigma_a(T)$, $T$ is left Drazin invertible, so $p(T) < \infty$. The condition of left Drazin invertibility is equivalent to saying that $T$ is upper semi B-Browder, i.e. there exists $n \in \mathbb{N}$ such that $T^n(X)$ is closed and the restriction $T_n := T|T^n(X)$ is upper semi-Browder (see [9] for details), in particular upper semi-Fredholm. Since $\alpha(T) < \infty$ then $\alpha(T^n) < \infty$, hence $T^n \in \Phi_+(X)$, from which obtain that $T \in \Phi_+(X)$. Since $p(T) < \infty$ we then conclude that $T \in B_+(X)$, so $0 \notin \sigma_{\mathrm{ub}}(T)$, and consequently $0 \in \sigma_a(T) \setminus \sigma_{\mathrm{ub}}(T) = p_{00}^a(T)$.

Conversely, assume that $0 \in p_{00}^a(T)$. Then $0 \in \sigma_a(T) \setminus \sigma_{\mathrm{ub}}(T)$, hence $p := p(T) < \infty$ and $T \in \Phi_+(X)$. From classical Fredholm theory we know that $T^n \in \Phi_+(X)$ for all $n \in \mathbb{N}$, so $T^{p+1}(X)$ is closed. Thus $T$ is left Drazin invertible. But $0 \in \sigma_a(T)$, thus $0$ is a left pole having finite rank, since $\alpha(T) < \infty$.

(ii) Suppose $\lambda$ be a right pole of finite rank. We may assume $\lambda = 0$. Then $0 \in \sigma_s(T) = \sigma_a(T^*)$, $T$ is right Drazin invertible, so $q(T) < \infty$. The condition of right Drazin invertibility is equivalent to saying that $T$ is lower semi B-Browder, i.e. there exists $n \in \mathbb{N}$ such that $T^n(X)$ is closed and the restriction $T_n := T|T^n(X)$ is lower semi-Browder (see [9] for details), in particular lower semi-Fredholm. Since $\beta(T) < \infty$ then $\beta(T^n) < \infty$, hence $T^n \in \Phi_-(X)$, from which obtain that $T \in \Phi_-(X)$. Since $q(T) < \infty$ we then conclude that $T \in B_-(X)$, or equivalently $T^* \in B_+(X^*)$, hence $0 \notin \sigma_{\mathrm{ub}^*}(T)$, and this implies that $0 \in \sigma_a(T^*) \setminus \sigma_{\mathrm{ub}}(T^*) = p_{00}^a(T^*)$.

Conversely, assume that $0 \in p_{00}^a(T^*)$. Then $0 \in \sigma_a(T^*) \setminus \sigma_{\mathrm{ub}}(T^*)$, hence $p := p(T^*) < \infty$ and $T^* \in \Phi_+(X^*)$. From classical Fredholm theory we know that $T^{*n} \in \Phi_+(X^*)$ for all $n \in \mathbb{N}$, so $T^{n+1}(X)$ is closed. Thus $T^*$ is left Drazin invertible. But $0 \in \sigma_a(T^*)$, thus $\lambda$ is a left pole of $T^*$, or equivalently, by [18, Theorem 2.8] a right pole of $T$. Finally, since $T^* \in \Phi_+(X^*)$ we have $T \in \Phi_-(X)$, so $\beta(T) < \infty$ and hence $0$ is a right pole of finite rank. □

Consequently, the conditions $\pi_{00}(T) = p_{00}^a(T)$ and $\pi_{00}^a(T) = p_{00}^a(T)$ which appear in Theorem 4.1 means, respectively, that $\pi_{00}(T)$ and $\pi_{00}^a(T)$ consist precisely of the set of all left poles of $T$ having finite rank.

**Lemma 4.1.** *If $T \in L(X)$ then the following statements hold:*

*(i) If $T^*$ has SVEP then $\sigma_w(T) = \sigma_{uw}(T)$ and $\sigma(T) = \sigma_a(T)$.*

*(ii) If $T$ has SVEP then $\sigma_w(T) = \sigma_{lw}(T)$ and $\sigma(T) = \sigma_s(T)$.*

**Proof.** (i) The inclusion $\sigma_{uw}(T) \subseteq \sigma_w(T)$ holds for all $T \in L(X)$. To see the opposite inclusion, suppose that $\lambda \notin \sigma_{uw}(T)$. Then $\lambda I - T$ is upper semi-Fredholm with $\mathrm{ind}\,(\lambda I - T) \leq 0$, and by Theorem 2.1 the SVEP of $T^*$ implies that $q(\lambda I - T) < \infty$. By [1, Theorem 3.4] then $\mathrm{ind}\,(\lambda I - T) \geq 0$, hence $\mathrm{ind}\,(\lambda I - T) = 0$ and consequently $\lambda \notin \sigma_w(T)$. A proof of the equality $\sigma(T) = \sigma_a(T)$ may be found in [1, Corollary 2.45].
(ii) The proof is analogous to part (i). □

As a consequence we obtain:

**Theorem 4.6.** *Let $T \in L(X)$.*

*(i) If $T^*$ has SVEP, then property $(w)$, Weyl's theorem and a-Weyl's theorem for $T$ are equivalent.*

*(ii) If $T$ has SVEP, then property $(w)$, Weyl's theorem and a-Weyl's theorem for $T^*$ are equivalent.*

**Proof.** The assertion (i) is an easy consequence of part (i) of Lemma 4.1. If $T^*$ has SVEP then $\sigma(T) = \sigma_a(T)$ and $\sigma_w(T) = \sigma_{uw}(T)$. Moreover, $\pi_{00}(T) = \pi_{00}^a(T)$. Analogously, (ii) easily follows from part (ii) of Lemma 4.1. □

## 5. Weyl type theorems for left and right polaroid operators

In this section we introduce the generalized Weyl type theorems and we consider the equivalence of them in the case that $T$ is polaroid, or left polaroid. These results are essentially contained in [5]. First we need to introduce a generalization of semi-Fredholm operators due to Berkani ([22], [24]). For every $T \in L(X)$ and a nonnegative integer $n$ let us denote by $T_{[n]}$ the restriction of $T$ to $T^n(X)$ viewed as a map from the space $T^n(X)$ into itself (we set $T_{[0]} = T$). $T \in L(X)$ is said to be *semi B-Fredholm* (resp. *B-Fredholm, upper semi B-Fredholm, lower semi B-Fredholm,*) if for some integer $n \geq 0$ the range $T^n(X)$ is closed and $T_{[n]}$ is a semi-Fredholm operator (resp. Fredholm, upper semi-Fredholm, lower semi-Fredholm). In this case $T_{[m]}$ is a semi-Fredholm operator for all $m \geq n$ ([24]). This enables one to define the index of a semi B-Fredholm as $\mathrm{ind}\,T = \mathrm{ind}\,T_{[n]}$. A bounded operator $T \in L(X)$ is said to be *B-Weyl* (respectively, *upper semi*

*B-Weyl, lower semi B-Weyl*) if for some integer $n \geq 0$ $T^n(X)$ is closed and $T_{[n]}$ is Weyl (respectively, upper semi-Weyl, lower semi-Weyl). The classes of operators previously defined generate the *B-Weyl spectrum* $\sigma_{\text{bw}}(T)$, the *upper B-Weyl spectrum* $\sigma_{\text{usbw}}(T)$, and the *lower B-Weyl spectrum* $\sigma_{\text{lsbw}}(T)$. Analogously, a bounded operator $T \in L(X)$ is said to be *B-Browder* (respectively, (respectively, *upper semi B-Browder, lower semi B-Browder*) if for some integer $n \geq 0$ $T^n(X)$ is closed and $T_{[n]}$ is Browder (respectively, upper semi-Browder, lower semi-Browder). The *B-Browder spectrum* is denoted by $\sigma_{\text{bb}}(T)$, the *upper semi B-Browder spectrum* by $\sigma_{\text{usbb}}(T)$ and the *lower semi B-Browder spectrum* by $\sigma_{\text{lsbb}}(T)$.

Also the concept of Drazin invertibility generate some spectra. Specifically, the *left Drazin spectrum* is defined as

$$\sigma_{\text{ld}}(T) := \{\lambda \in \mathbb{C} : \lambda I - T \text{ is not left Drazin invertible}\},$$

while the *right Drazin spectrum* is defined as

$$\sigma_{\text{rd}}(T) := \{\lambda \in \mathbb{C} : \lambda I - T \text{ is not right Drazin invertible}\},$$

and the *Drazin spectrum* is defined as

$$\sigma_{\text{d}}(T) := \{\lambda \in \mathbb{C} : \lambda I - T \text{ is not Drazin invertible}\}.$$

Obviously, $\sigma_{\text{d}}(T) = \sigma_{\text{ld}}(T) \cup \sigma_{\text{rd}}(T)$.

**Theorem 5.1 ([9]).** *For every $T \in L(X)$ we have*

$$\sigma_{\text{usbb}}(T) = \sigma_{\text{ld}}(T), \quad \sigma_{\text{lsbb}}(T) = \sigma_{\text{rd}}(T), \quad \sigma_{\text{bb}}(T) = \sigma_{\text{d}}(T).$$

**Lemma 5.1.** *If $T \in L(X)$ and $p = p(T) < \infty$ then the following statements are equivalent:*

*(i) There exists $n \geq p+1$ such that $T^n(X)$ is closed;*

*(ii) $T^n(X)$ is closed for all $n \geq p$.*

**Proof.** Define $c'_i(T) := \dim(\ker T^i / \ker T^{i+1})$. Clearly, $p = p(T) < \infty$ entails that $c'_i(T) = 0$ for all $i \geq p$, so $k_i(T) := c'_i(T) - c'_{i+1}(T) = 0$ for all $i \geq p$. The equivalence then easily follows from [51, Lemma 12]. □

In the sequel by $T'$ we shall denote the dual of $T \in L(X)$. Denote by $M^\perp$ the *annihilator* of $M \subseteq X$, while by $^\perp N$ we denote the *pre-annihilator* of $N \subseteq X'$.

**Theorem 5.2.** *For every $T \in L(X)$ the following equivalences hold:*

*(i) $T$ is left Drazin invertible $\Leftrightarrow T^*$ is right Drazin invertible.*

*(ii)* $T$ is right Drazin invertible $\Leftrightarrow T^*$ is left Drazin invertible.

*(iii)* $T$ is Drazin invertible if and only if $T^*$ is Drazin invertible.

**Proof.** (i) Suppose that $T$ is left Drazin invertible. Then $p := p(T) < \infty$ and $T^{p+1}(X)$ is closed, and hence also $T^{*p+1}(X')$ is closed. By Lemma 5.1 $T^p(X)$ is closed, and consequently also $T^{*p}(X^*)$ is closed. From the equality ker $T^p$ = ker $T^{p+1}$ and from the classical closed range theorem we then deduce that

$$T^{*p}(X^*) = [\ker T^p]^\perp = [\ker T^{p+1}]^\perp = T^{*p+1}(X^*).$$

This shows that $T^*$ has finite descent $q := q(T^*) \leq p$ and since $T^{*q}(X^*) = T^{*p}(X^*)$ is closed it then follows that $T^*$ is right Drazin invertible.

Conversely, suppose that $T^*$ is right Drazin invertible. Then $q := q(T^*) < \infty$ and $T^{*q}(X^*)$ is closed. From the equality $T^{*q}(X^*) = T^{*q+1}(X^*)$ and from the closed range theorem we then obtain:

$$\ker T^q =^\perp [T^{*q}(X^*)] =^\perp [T^{*q+1}(X^*)] = \ker T^{q+1},$$

and hence $p := p(T) \leq q$. Since $T^{*q+1}(X^*)$ is closed then also $T^{q+1}(X)$ is closed and by Lemma 5.1 it then follows that $T^{p+1}(X)$ is closed, so $T$ is a left Drazin invertible.

(ii) This may be proved in a similar way of part (i).

(iii) Obviously, since both left and right Drazin invertibility entails Drazin invertibility. □

Every left or right Drazin invertible operator is quasi-Fredholm, so Remark 2.1 applies to these operators. The concept of polaroid operators may be split as follows:

**Definition 5.1.** A bounded operator $T \in L(X)$ is said to be *left polaroid* if every isolated point of $\sigma_a(T)$ is a left pole of the resolvent of $T$, while $T \in L(X)$ is said to be *right polaroid* if every isolated point of $\sigma_s(T)$ is a right pole of the resolvent of $T$.

**Theorem 5.3.** *If $T \in L(X)$ is both left and right polaroid then $T$ is polaroid.*

**Proof.** If iso $\sigma(T) = \emptyset$ there is nothing to prove. Suppose then $\lambda \in$ iso $\sigma(T) \neq \emptyset$. Since the boundary the spectrum is contained in $\sigma_a(T)$, see [1, Theorem 2.42], then $\lambda \in$ iso $\sigma_a(T)$, so $\lambda$ is a left pole and hence $p(\lambda I - T) < \infty$. On the other hand, $\lambda \in \sigma_s(T)$, (otherwise we have

$0 = q(\lambda I - T) = p(\lambda I - T)$ and hence $\lambda \notin \sigma(T)$). Therefore, $\lambda \in \operatorname{iso}\sigma_s(T)$ and since $T$ is right polaroid then $q(\lambda I - T) < \infty$, from which conclude that $\lambda$ is a pole of the resolvent of $T$. □

The following example shows that the converse of Theorem 5.3, in general, does not hold.

**Example 5.1.** Let $R$ denote the right shift on $\ell^2(\mathbb{N})$ defined by
$$R(x_1, x_2, \ldots) := (0, x_1, x_2, \ldots) \quad (x_n) \in \ell^2(\mathbb{N}),$$
and let $Q$ be the weighted left shift defined by
$$Q((x_1, x_2, \ldots)) := (x_2/2, x_3/3, \ldots) \quad (x_n) \in \ell^2(\mathbb{N}).$$
$Q$ is a quasi-nilpotent operator, $\sigma(R) = D(0,1)$, where $D(0,1)$ denotes the closed unit disc of $\mathbb{C}$, and $\sigma_a(R) = \Gamma$, where $\Gamma$ is the unit circle of $\mathbb{C}$. Moreover, if $e_n := (0, ..., 0, 1, 0...)$, where 1 is the $n$-th term, then $e_{n+1} \in \ker Q^{n+1}$ while $e_{n+1} \notin \ker Q^n$ for every $n \in \mathbb{N}$, so $p(Q) = \infty$.

Define $T := R \oplus Q$ on $X := \ell^2(\mathbb{N}) \oplus \ell^2(\mathbb{N})$. Clearly, $\sigma(T) = D(0,1)$, and $\sigma_a(T) = \Gamma \cup \{0\}$. We have $p(T) = p(R) + p(Q) = \infty$, so 0 is not a left pole. Therefore, $T$ is polaroid, since iso $\sigma(T) = \emptyset$, but not left polaroid. It is easily seen that the dual $T^*$ is polaroid but not right polaroid, since $q(T^*) = \infty$.

The concept of left and right polaroid are dual each other:

**Theorem 5.4.** *If $T \in L(X)$ then the following equivalences hold:*
*(i) $T$ is left polaroid if and only if $T^*$ is right polaroid.*
*(ii) $T$ is right polaroid if and only if $T^*$ is left polaroid.*

**Proof.** (i) Suppose that $T$ is left polaroid. If iso $\sigma_s(T^*) = \emptyset$ there is nothing to prove. Suppose that $\lambda \in \operatorname{iso}\sigma_s(T^*)$. Then $\lambda$ is an isolated point of $\sigma_a(T)$, hence, by assumption, $\lambda I - T$ is left Drazin invertible. By Theorem 5.2 then $\lambda I - T^*$ is right Drazin invertible, so $\lambda$ is a right pole of the resolvent of $T^*$.

Conversely, suppose $T^*$ right polaroid and let $\lambda$ be an isolated point of $\sigma_a(T)$. Then $\lambda \in \operatorname{iso}\sigma_s(T^*)$ and hence is a right pole of the resolvent of $T^*$. Therefore, $\lambda I - T^*$ is right Drazin invertible so, by Theorem 5.2, $\lambda I - T$ is left Drazin invertible. Thus, $\lambda$ is a left pole of the resolvent of $T$.

(ii) The proof is similar to that of part (i). □

**Definition 5.2.** A bounded operator $T \in L(X)$ is said to be *a-polaroid* if every $\lambda \in \operatorname{iso}\sigma_a(T)$ is a pole of the resolvent of $T$.

Trivially,
$$T \text{ a-polaroid} \Rightarrow T \text{ left polaroid.} \tag{9}$$

Moreover, iso$(T) \subseteq \sigma_a(T)$ for every $T \in L(X)$, since the boundary of $\sigma(T)$ is contained in $\sigma_a(T)$, from which we easily obtain:

$$T \text{ a-polaroid} \Rightarrow T \text{ polaroid.} \tag{10}$$

The following example provides an operator that is left polaroid but not a-polaroid.

**Example 5.2.** Let $R \in \ell^2(\mathbb{N})$ be the unilateral right shift defined as
$$R(x_1, x_2, \dots) := (0, x_1, x_2, \cdots) \quad \text{for all } (x_n) \in \ell^2(\mathbb{N}),$$
and
$$U(x_1, x_2, \dots) := (0, x_2, x_3, \cdots) \quad \text{for all } (x_n) \in \ell^2(\mathbb{N}).$$

If $T := R \oplus U$ then $\sigma(T) = D(0,1)$, so iso $\sigma(T) = \emptyset$. Moreover, $\sigma_a(T) = \Gamma \cup \{0\}$, $\Gamma$ the unit circle, so iso $\sigma_a(T) = \{0\}$. Since $R$ is injective and $p(U) = 1$ it then follows that $p(T) = p(R) + p(U) = 1$. Furthermore, $T \in \Phi_+(X)$ and hence $T^2 \in \Phi_+(X)$, so that $T^2(X)$ is closed. Therefore 0 is a left pole and hence $T$ is left polaroid. On the other hand $q(R) = \infty$, so that $q(T) = q(R) + q(U) = \infty$, so $T$ is not a-polaroid. Note that $T$ is also polaroid.

In the case of Hilbert space operators $T \in L(H)$ instead of the dual $T^*$ it is more appropriate to consider the Hilbert adjoint $T'$. By means of the classical Fréchet-Riesz representation theorem we know that if $U$ is the conjugate-linear isometry that associates to each $y \in H$ the linear form $x \to \langle x, y \rangle$ then

$$\overline{\lambda}I - T' = (\lambda I - T)' = U^{-1}(\lambda I - T)^* U. \tag{11}$$

This obviously implies that

$$\sigma_a(T') = \overline{\sigma_a(T^*)} \quad \text{and} \quad \sigma_s(T') = \overline{\sigma_s(T^*)}.$$

Taking into account these equalities it is not difficult to extend the result of Theorem 5.4 to the Hilbert adjoint $T'$ of $T \in L(H)$.

**Theorem 5.5.** *If $T \in L(H)$, $H$ a Hilbert space, then the following equivalences hold:*

*(i) $T$ is left polaroid if and only if $T'$ is right polaroid.*

*(ii) $T$ is right polaroid if and only if $T'$ is left polaroid.*

*(iii) $T$ is polaroid if and only if $T'$ is polaroid.*

An obvious consequence of Theorem 5.5 and Theorem 5.4 is that for a Hilbert space operator $T$ the property of being $T^*$ and $T'$ left polaroid (respectively, right polaroid, polaroid) are equivalent.

**Theorem 5.6.** *Suppose that $T \in L(X)$. Then the following assertions hold:*
*(i) If $T^*$ has SVEP then the properties of being polaroid, a-polaroid and left polaroid for $T$ are all equivalent.*
*(ii) If $T$ has SVEP then the properties of being polaroid, a-polaroid and left polaroid for $T^*$ are all equivalent.*

**Proof.** (i) Suppose that $T^*$ has SVEP. Then $\sigma(T) = \sigma_a(T)$, see Corollary 2.44 of [1], hence $T$ is polaroid precisely when $T$ is a-polaroid. By (9) the a-polaroid condition for $T$ entails that $T$ is left polaroid. If $T$ is left polaroid and $\lambda \in \text{iso}\,\sigma(T)$ then $\lambda \in \text{iso}\,\sigma_a(T)$, again because the boundary of the spectrum is contained in $\sigma_a(T)$, hence $\lambda$ is a left pole, so that then $p(\lambda I - T) < \infty$. On the other hand, since $\lambda I - T$ is left Drazin invertible, in particular quasi-Fredholm the SVEP of $T^*$ at $\lambda$ entails that $q(\lambda I - T) < \infty$, see [3, Theorem 2.11], and consequently $\lambda$ is a pole of the resolvent of $T$. Therefore $T$ is polaroid.

(ii) If $T$ has SVEP, by Corollary 2.44 of [1] then $\sigma(T) = \sigma_s(T)$ and hence $\sigma(T^*) = \sigma_a(T^*)$, so $T^*$ is polaroid if and only if $T^*$ is a-polaroid. Again from (9) the a-polaroid condition for $T^*$ entails that $T^*$ is left polaroid. Suppose now that $T^*$ is left polaroid. If $\lambda \in \text{iso}\,\sigma(T) = \text{iso}\,\sigma(T^*) = \text{iso}\,\sigma_a(T^*)$ then $\lambda I - T^*$ is left Drazin invertible, hence $\lambda I - T$ is right Drazin invertible, so that $q(\lambda I - T) < \infty$. On the other hand, $\lambda I - T$ is quasi-Fredholm so,by [3, Theorem 2.7], the SVEP for $T$ at $\lambda$ entails, that $p(\lambda I - T) < \infty$. Hence $\lambda$ is a pole of $T$ so that $T$ is polaroid, or equivalently $T^*$ is polaroid. □

**Remark 5.1.** It is easily seen from the equality (11) and from the definition of SVEP that, for a Hilbert space operator $T \in L(H)$, the SVEP for $T^*$ is equivalent to the SVEP for $T'$. Therefore, in Theorem 5.6 the statements (i) and (ii) remain valid if we replace the SVEP for $T^*$ with the SVEP for $T'$.

If $T \in L(X)$ define
$$E(T) := \{\lambda \in \text{iso}\,\sigma(T) : 0 < \alpha(\lambda I - T)\},$$
and
$$E^a(T) := \{\lambda \in \text{iso}\,\sigma_a(T) : 0 < \alpha(\lambda I - T)\}.$$

Evidently, $E(T) \subseteq E^a(T)$ for every $T \in L(X)$. For a bounded operator $T \in L(X)$, let $\Pi(T)$ denote the set of all poles of the resolvent of $T$ and define

$$\Pi_{00}^a(T) := \sigma_a(T) \backslash \sigma_{\text{ubb}}(T) = \{\lambda \in \sigma_a(T) : \lambda I - T \text{ is upper semi } B\text{-Browder}\}.$$

Since $\sigma_{\text{ubb}}(T) = \sigma_{\text{ld}}(T)$, by Theorem 5.1, it is clear that $\Pi_{00}^a(T)$ is the set of all left poles of the resolvent of $T$, hence coincides with $p_{00}^a(T)$. It is not difficult to prove:

$$\Pi_{00}^a(T) \subseteq E^a(T) \quad \text{for all } T \in L(X).$$

**Definition 5.3.** $T \in L(X)$ is said to satisfy *generalized Browder's theorem* if $\sigma_{\text{bb}}(T) = \sigma_{\text{bw}}(T)$, while $T \in L(X)$ is said to satisfy *generalized a-Browder's theorem* if $\sigma_{\text{ubb}}(T) = \sigma_{\text{ubw}}(T)$.

**Theorem 5.7 ([9 and 21]).** *For every $T \in L(X)$ the following equivalences hold:*

*(i) $\sigma_{\text{w}}(T) = \sigma_{\text{b}}(T) \Leftrightarrow \sigma_{\text{bw}}(T) = \sigma_{\text{bb}}(T)$.*

*(ii) $\sigma_{\text{uw}}(T) = \sigma_{\text{ub}}(T) \Leftrightarrow \sigma_{\text{ubw}}(T) = \sigma_{\text{ubb}}(T)$.*

*Consequently, Browder's Theorem and generalized Browder's theorem are equivalent, and analogously a-Browder's theorem and generalized a-Browder's theorem are equivalent.*

**Definition 5.4.** A bounded operator $T \in L(X)$ is said to satisfy *generalized Weyl's theorem*, in symbol $(gW)$, if $\sigma(T) \setminus \sigma_{\text{bw}}(T) = E(T)$. $T \in L(X)$ is said to satisfy *generalized a-Weyl's theorem*, in symbol $(gaW)$, if $\sigma_a(T) \setminus \sigma_{\text{ubw}}(T) = E^a(T)$. $T \in L(X)$ is said to satisfy *generalized property (w)*, in symbol $(gw)$, if $\sigma_a(T) \setminus \sigma_{\text{ubw}}(T) = E(T)$.

Define

$$\Delta_1(T) := \Delta(T) \cup E(T) \quad \text{and} \quad \Delta_1^a(T) := \Delta^a(T) \cup E^a(T).$$

Clearly,

$$\Delta_1(T) = \Delta(T) \cup E(T) \subseteq \Delta^a(T) \cup E^a(T) = \Delta_1^a(T).$$

**Theorem 5.8 ([12]).** *For a bounded operator $T \in L(X)$ the following statements are equivalent:*

*(i) $T$ satisfies generalized Weyl's theorem;*

(ii) $T$ satisfied generalized Browder's theorem and $E(T) = \Pi_{00}(T)$;

(iii) For every $\lambda \in \Delta_1(T)$ there exists $p := p(\lambda) \in \mathbb{N}$ such that $H_0(\lambda I - T) = \ker(\lambda I - T)^p$.

**Theorem 5.9 ([16]).** *For a bounded operator $T \in L(X)$ the following statements are equivalent:*

(i) $T$ satisfies generalized a-Weyl's theorem;

(ii) $T$ satisfies generalized a-Browder's theorem and $E^a(T) = \Pi^a(T)$;

(iii) For every $\lambda \in \Delta_1^a(T)$ there exists $p := p(\lambda) \in \mathbb{N}$ such that $H_0(\lambda I - T) = \ker(\lambda I - T)^p$ and $(\lambda I - T)^n(X)$ is closed for all $n \geq p$.

In particular, every left polaroid operator which has SVEP satisfies $(gaW)$.

Since $\Delta_1(T) \subseteq \Delta_1^a(T)$ we then have:

generalized a-Weyl's theorem for $T \Rightarrow$ generalized Weyl's theorem for $T$.

We also have:

generalized property $(w)$ for $T \Rightarrow$ generalized Weyl's theorem for $T$,

and

generalized property $(w)$ for $T \Rightarrow$ property $(w)$ for $T$,

see [28, Theorem 2.3], [17] and [27]. Generalized property $(w)$ and generalized a-Weyl's theorem are also independent, see [28]. In the following diagrams we resume the relationships between all Weyl's type theorems:

$$(gw) \Rightarrow (w) \Rightarrow (W)$$

$$(gaW) \Rightarrow (aW) \Rightarrow (W).$$

Furthermore,

$$(gw) \Rightarrow (gW) \Rightarrow (W)$$

$$(gaW) \Rightarrow (gW) \Rightarrow (W).$$

The converse of all these implications in general does not hold. In the sequel we shall need the following simple result:

**Lemma 5.2.** *Let $T \in L(X)$. Then we have:*

*(i) $T$ is upper semi B-Fredholm and $\alpha(T) < \infty$ if and only if $T \in \Phi_+(X)$.*

*(ii) $T$ is lower semi B-Fredholm and $\beta(T) < \infty$ if and only $T \in \Phi_-(X)$.*

**Proof.** (i) If $T$ is upper semi B-Fredholm then there exists $n \in \mathbb{N}$ such that $T^n(X)$ is closed and $T_{[n]}$ is upper semi-Fredholm. Since $\alpha(T) < \infty$ then $\alpha(T^n) < \infty$, hence $T^n$ is upper semi Fredholm. From the classical Fredholm theory then also $T$ is upper semi-Fredholm. The converse is obvious. Part (ii) may be proved proved in a similar way. $\square$

We now show the equivalence of some Weyl type theorems for left polaroid or right polaroid operators:

**Theorem 5.10.** *Let $T \in L(X)$. If $T$ is left-polaroid then $(aW)$ and $(gaW)$ for $T$ are equivalent. If $T$ is right-polaroid then $(aW)$ and $(gaW)$ for $T^*$ are equivalent*

**Proof.** (i) As already observed generalized property $(gaW)$ entails property $(aW)$ without any assumption on $T$.

Suppose now that property $(aW)$ holds for $T$, i.e. $\sigma_a(T) \setminus \sigma_{\mathrm{uw}}(T) = \pi_{00}^a(T)$. We have to prove that $\sigma_a(T) \setminus \sigma_{\mathrm{ubw}}(T) = E^a(T)$. We show first that the inclusion $\sigma_a(T) \setminus \sigma_{\mathrm{ubw}}(T) \subseteq E^a(T)$ holds without any assumption on $T$.

Let $\lambda \in \sigma_a(T) \setminus \sigma_{\mathrm{ubw}}(T)$. We can suppose that $\lambda = 0$. Therefore, $0 \in \sigma_a(T)$ and $T$ is upper semi B-Fredholm with index less or equal than 0. By [24, Corollary 3.2] then there exists $\varepsilon > 0$ such that
$$\mu I - T \in W_+(X) \quad \text{for all } 0 < |\mu| < \varepsilon.$$
We claim that $T$ has SVEP at every $\mu$. If $\mu \notin \sigma_a(T)$ this is obvious. Suppose that $\mu \in \sigma_a(T)$. Then $\mu \in \sigma_a(T) \setminus \sigma_{\mathrm{uw}}(T) = \pi_{00}^a(T)$, so $\mu$ is an isolated point of $\sigma_a(T)$ and hence $T$ has SVEP at $\mu$. The following argument shows that $T$ has SVEP at 0. Let $f : D_0 \to X$ be an analytic function defined on an open disc $D_0$ centered at 0 for which the equation $(\lambda I - T)f(\lambda) = 0$ for all $\lambda \in D_0$. Take $0 \neq \mu \in D_0$ and let $D_1$ be an open disc centered at $\mu$ contained in $D_0$. The SVEP of $T$ at $\mu$ implies that $f \equiv 0$ on $D_1$ and hence, from the identity theorem for analytic functions, it then follows that $f \equiv 0$ on $D_0$, so $T$ has SVEP at 0. But $T$ is upper semi B-Fredholm, so, by Theorem 2.7 of [3], $0 \in \mathrm{iso}\,\sigma_a(T)$. Suppose now that $\alpha(T) = 0$. By Lemma 5.2 then $T \in \Phi_+(X)$, so $T(X)$ is closed and, consequently, $0 \notin \sigma_a(T)$, a

contradiction. Therefore $\alpha(T) > 0$, from which we conclude $0 \in E^a(T)$ and hence $\sigma_a(T) \setminus \sigma_{\text{ubw}}(T) \subseteq E^a(T)$.

Suppose now that $T$ is left polaroid and let $\lambda \in E^a(T)$. Then $\lambda$ is an isolated point of $\sigma_a(T)$, and hence by the left polaroid condition $\lambda$ is a left pole of $T$. In particular, $\lambda I - T$ is left Drazin invertible or equivalently, by Theorem 5.1, an upper semi $B$-Browder operator. Since $\sigma_{\text{ubw}}(T) \subseteq \sigma_{\text{ubb}}(T)$ we then have $\lambda \in \sigma_a(T) \setminus \sigma_{\text{ubb}}(T) \subseteq \sigma_a(T) \setminus \sigma_{\text{ubw}}(T)$. Therefore, generalized $a$-Weyl's theorem holds for $T$.

The assertion concerning right-polaroid operators is obvious by Theorem 5.4. □

In the sequel we need the following result:

**Lemma 5.3.** *If $T \in L(X)$ is quasi-Fredholm then there exists $\varepsilon > 0$ such that $\mathcal{N}^\infty(\lambda I - T) \subseteq (\lambda I - T)^\infty(X)$ for all $0 < |\lambda| < \varepsilon$. If $T$ is semi $B$-Fredholm then $\lambda I - T$ is semi-regular in a suitable punctured open disc centered at $0$.*

**Proof.** Observe first that if $T$ is quasi-Fredholm of degree $d$ then $T^n(X)$ is closed for all $n \geq d$, so $T^\infty(X)$ is closed. Furthermore, by Theorem 3.4 of [43] the restriction $T|T^\infty(X)$ is onto, so $T(T^\infty(X)) = T^\infty(X)$. Let $T_0 := T|T^\infty(X)$. Clearly, $T_0$ is onto and hence $\lambda I - T$ is onto for all $|\lambda| < \varepsilon$, where $\varepsilon := \gamma(T_0)$ is the minimal modulus of $T_0$, see [1, Lemma 1.30]. Therefore, $(\lambda I - T)(T^\infty(X)) = T^\infty(X)$ for all $|\lambda| < \varepsilon$. Since $T^\infty(X)$ is closed, by [1, Theorem 1.22] it then follows that

$$T^\infty(X) \subseteq K(\lambda I - T) \subseteq (\lambda I - T)^\infty(X) \quad \text{for all } |\lambda| < \varepsilon.$$

By part (ii) of Theorem 1.3 of [1] we have $\mathcal{N}^\infty(\lambda I - T) \subseteq T^\infty(X)$ for all $\lambda \neq 0$, so we conclude that

$$\mathcal{N}^\infty(\lambda I - T) \subseteq (\lambda I - T)^\infty(X) \quad \text{for all } 0 < |\lambda| < \varepsilon,$$

and the first assertion is proved.

To show the second assertion, suppose that $T$ is semi $B$-Fredholm. Then there exists an open disc $D$ centered at $0$ such that $\lambda I - T$ is semi-Fredholm for all $\lambda \in D \setminus \{0\}$ (this follows as a particular case of a result proved in [24, Corollary 3.2] for operators having topological uniform descent for $n \geq d$). Since semi-Fredholm operators have closed range then the last assertion easily follows. □

**Theorem 5.11.** *If $T$ is polaroid then $(W)$, and $(gW)$ for $T$ are equivalent. Analogously, $(W)$, and $(gW)$ for $T^*$ are equivalent.*

**Proof.** (ii) We have only to show that Weyl's theorem entails generalized Weyl's theorem. Suppose first that $\lambda_0 \in E(T)$. Since $T$ is polaroid then $\lambda_0$ is a pole of $T$, hence $0 < p(\lambda_0 I - T) = q(\lambda_0 I - T) < \infty$. Therefore, $\lambda_0 I - T$ is Drazin invertible or equivalently, by Theorem 5.1, $\lambda_0 I - T$ is $B$-Browder and hence $B$-Weyl. Consequently, $\lambda_0 \in \sigma(T) \setminus \sigma_{\text{bw}}(T)$ and hence $E(T) \subseteq \sigma(T) \setminus \sigma_{\text{bw}}(T)$.

Conversely, assume that $\lambda_0 \in \sigma(T) \setminus \sigma_{\text{bw}}(T)$. Then $\lambda_0 I - T$ is $B$-Weyl and hence, again by [24, Corollary 3.2], there exists $\varepsilon > 0$ such that $\lambda I - T$ is Weyl for all $0 < |\lambda - \lambda_0| < \varepsilon$. By Lemma 5.3 we know that $\lambda I - T$ is semi-regular in a punctured open disc centered at $\lambda_0$, so we can assume that

$$\ker(\lambda I - T) \subseteq \mathcal{N}^\infty(\lambda I - T) \subseteq (\lambda I - T)^\infty(X) \quad \text{for all } 0 < |\lambda - \lambda_0| < \varepsilon.$$

As observed above Weyl's theorem for $T$ entails Browder's theorem for $T$, i.e. $\sigma_{\text{w}}(T) = \sigma_{\text{b}}(T)$. Therefore, $\lambda I - T$ is Browder for all $0 < |\lambda - \lambda_0| < \varepsilon$ and, consequently, $p(\lambda I - T) = q(\lambda I - T) < \infty$. By Lemma 3.2 of [1] we then have

$$\ker(\lambda I - T) = \ker(\lambda I - T) \cap (\lambda I - T)^\infty(X) = \{0\},$$

thus $\alpha(\lambda I - T) = 0$ and since $\lambda I - T$ is Weyl we then conclude that also $\beta(\lambda I - T) = 0$, so $\lambda I - T$ is invertible for all $0 < |\lambda - \lambda_0|$ and hence $\lambda_0 \in \text{iso } \sigma(T)$. To show that $\lambda_0 \in E(T)$ it remains to prove that $\alpha(\lambda_0 I - T) > 0$. Suppose that $\alpha(\lambda_0 I - T) = 0$. Since $\lambda_0 I - T$ is $B$-Weyl then, by Lemma 5.2, $\lambda_0 I - T$ is Weyl and since $\alpha(\lambda_0 I - T) = 0$ it then follows that $\lambda_0 I - T$ is invertible, a contradiction since $\lambda_0 \in \sigma(T)$. Therefore, $\lambda_0 \in E(T)$, so generalized Weyl's theorem holds for $T$.

The last assertion is clear: $T^*$ is also polaroid. □

**Corollary 5.1.** *If $T \in L(X)$ is a-polaroid then $(aW)$, $(gaW)$, $(w)$, $(gw)$ for $T$ are equivalent.*

**Proof.** Every $a$-polaroid operator is left polaroid so, by Theorem 5.10, $(aW)$ and $(gaW)$ are equivalent. Property $(w)$ and $(aW)$ are equivalent, since by Lemma 7.5 we have $\pi_{00}(T) = \pi_{00}^a(T)$. By Lemma 7.5 we also have $E(T) = E^a(T)$, from which it easily follows that $(gaW)$ and $(gw)$ are equivalent. □

In the following example we show that the result of Theorem 5.11 does not hold if we replace the condition of being $a$-polaroid by the weaker conditions of being left polaroid or polaroid.

**Example 5.3.** Let $R$ and $U$ be defined as in Example 5.2. As observed before $T := R \oplus U$ is both left polaroid and polaroid. Moreover, $\sigma_a(T) = \Gamma \cup \{0\}$ and iso $\sigma(T) = \pi_{00}(T) = \emptyset$, so $\sigma_a(T) \setminus \sigma_{uw}(T) = \{0\} \neq \pi_{00}(T)$, i.e. $T$ does not satisfy property $(w)$. On the other hand, we have $\pi_{00}^a(T) = \{0\}$, hence $T$ satisfies $a$-Weyl's theorem.

If $T$, or $T^*$, has SVEP we can say much more:

**Theorem 5.12.** *Let* $T \in L(X)$ *be polaroid. Then we have*

*(i) If $T^*$ has SVEP then $(W)$, $(aW)$, $(w)$, $(gW)$, $(gaW)$ and $(gw)$ for $T$ are equivalent and $T$ satisfies all of them. Moreover, $T^*$ satisfies $(gW)$.*

*(ii) If $T$ has SVEP then $(W)$, $(aW)$, $(w)$, $(gW)$, $(gaW)$ and $(gw)$ for $T^*$ are equivalent and $T^*$ satisfies all of them. Moreover, $T$ satisfies $(gW)$.*

**Proof.** (i) $T$ satisfies $(W)$ by Theorem 4.4. The first statement is then proved if we show that $(W)$ is equivalent to each one of the other Weyl type theorems for $T$, generalized or not. Since $T^*$ has SVEP $(W)$ and $(aW)$ for $T$ are equivalent, by part (i) of [17, Theorem 2.16]. By Theorem 5.6 $T$ is $a$-polaroid hence, by Theorem 5.1, $(aW)$, $(gaW)$, $(w)$, $(gw)$ for $T$ are equivalent. Finally, by Theorem 5.11, $(W)$ and $(gW)$ for $T$ are equivalent. By Theorem 4.4 $T^*$ satisfies $(W)$ and since $T^*$ is polaroid then, by Theorem 5.11, $(gW)$ holds for $T^*$.

(ii) $T^*$ satisfies $(W)$ by Theorem 4.4, so it suffices to prove that $(W)$ is equivalent to each one of the other Weyl type theorems, generalized or not, for $T^*$. Since, by Theorem 5.6, $T^*$ is $a$-polaroid from Theorem 5.1 it then follows that $(aW)$, $(gaW)$, $(w)$, $(gw)$ are equivalent for $T^*$. The SVEP for $T$ entails by part (ii) of [17, Theorem 2.16] that $(W)$ and $(aW)$ are equivalent for $T^*$, while $(W)$ and $(gW)$ for $T^*$ are equivalent by Theorem 5.11. By Theorem 4.4 $T$ satisfies $(W)$ and since $T$ is polaroid this is equivalent to $(gW)$ for $T$, always by Theorem 5.11. □

Let $\mathcal{H}(\sigma(T))$ denote the set of all analytic functions defined on an open neighborhood of $\sigma(T)$ define, by the classical functional calculus, $f(T)$ for every $f \in \mathcal{H}(\sigma(T))$.

**Lemma 5.4.** *Suppose that $f \in \mathcal{H}(\sigma(T))$ is non constant on each of the components of its domain. If $T$ is left polaroid (respectively, right polaroid, polaroid), then $f(T)$ is left polaroid (respectively, right polaroid, polaroid).*

**Proof.** Let $\lambda_0 \in \text{iso}\,\sigma_a(f(T))$. We have to show that $\lambda_0$ is a left pole of $f(T)$. Since $\sigma_a(T)$ satisfies the spectral mapping theorem we have $\lambda_0 \in$

iso $f(\sigma_a(T))$. We show that $\lambda_0 \in f(\text{iso}\,\sigma_a(T))$. Let $\mu_0 \in \sigma_a(T)$ be such that $f(\mu_0) = \lambda_0$. Denote by $\Omega$ the open and connected component of the domain of $f$ which contains $\mu_0$. Suppose that $\mu_0$ is not isolated in $\sigma_a(T)$. Then there exists a sequence $(\mu_n) \subset \sigma_a(T)$ of distinct scalars such that $\mu_n \to \mu_0$. Clearly, for $n$ sufficiently large $\mu_n \in \Omega$ and since $K := \{\mu_0, \mu_1, \mu_2, \dots\}$ is compact subset of $\Omega$, the classical principle of isolated zeros of analytic functions says to us that $f$ may assume the value $\lambda_0 = f(\mu_0)$ only a finite number of points of $K$, so for $n$ sufficiently large $f(\mu_n) \neq f(\mu_0) = \lambda_0$, and since $f(\mu_n) \to f(\mu_0) = \lambda_0$ it then follows that $\lambda_0$ is not an isolated point of $f(\sigma_a(T))$, a contradiction. Hence $\lambda_0 = f(\mu_0)$, with $\mu_0 \in \text{iso}\,\sigma_a(T)$. Since $T$ is left polaroid then $\mu_0$ is a left pole of $T$ and by [18,Theorem 2.9] it then follows that $\lambda_0$ is a left pole of $f(T)$, which proves that $f(T)$ is left polaroid.

The proofs for right polaroid and polaroid operators are analogous, just use the spectral mapping theorems for $\sigma_s(T)$ and $\sigma(T)$, respectively, and [18, Theorem 2.9]. □

The result of Theorem 5.12 may be considerably extended as follows

**Theorem 5.13.** *Let $T \in L(X)$ be polaroid an suppose that $f \in \mathcal{H}(\sigma(T))$ is non constant on each of the components of its domain. Then we have*

*(i) If $T^*$ has SVEP then $(W)$, $(aW)$, $(w)$, $(gW)$, $(gaW)$ and $(gw)$ for $f(T)$ are equivalent and $f(T)$ satisfies all of them.*

*(ii) If $T$ has SVEP then $(W)$, $(aW)$, $(w)$, $(gW)$, $(gaW)$ and $(gw)$ for $f(T^*)$ are equivalent and $f(T^*)$ satisfies all of them.*

**Proof.** (i) If $T^*$ has SVEP then $f(T)^* = f(T^*)$ has SVEP, see [1, Theorem 2.40]. Moreover, $T$ is left polaroid by Theorem 5.6, so $f(T)$ is left polaroid by Lemma 5.4. Again by Theorem 5.6, the SVEP of $f(T)^*$ entails that $f(T)$ is polaroid, hence Theorem 5.12 applies to $f(T)$.

(ii) Argue as in the proof of part (i), just replace $T$ with $T^*$. □

**Remark 5.2.** Obviously, in the case of Hilbert space operators, the condition $T^*$ has SVEP in Theorem 5.13 may be replaced by the SVEP of the adjoint $T'$.

**Theorem 5.14.** *Let $T \in L(X)$. Then we have*

*(i) If $T \in L(X)$ is left-polaroid and has SVEP, $f \in \mathcal{H}(\sigma(T))$ is non constant on each of the components of its domain, then $(aW)$ holds for $f(T)$, or equivalently $(gaW)$ holds for $f(T)$.*

*(ii) If $T \in L(X)$ is polaroid and has SVEP, $f \in \mathcal{H}(\sigma(T))$ is non constant on each of the components of its domain, then $(W)$ holds for $f(T)$, or equivalently $(gW)$ holds for $f(T)$.*

**Proof.** (i) If $T$ is left polaroid then $f(T)$ is left polaroid, by Lemma 5.4. By [1, Theorem 2.40] $f(T)$ has SVEP, hence Theorem 4.4 applies to $f(T)$. The equivalence of $(aW)$ and $(gaW)$ follows from Theorem 5.10.

(ii) If $T$ is polaroid then $f(T)$ is polaroid and has SVEP, so Theorem 4.4 applies to $f(T)$. The equivalence of $(W)$ and $(gW)$ follows from Theorem 5.10. □

## 6. Some applications

Weyl type theorems, in their classical and more recently in their generalized form, have been studied by a large number of authors. The results of the previous sections give us an unifying theoretical framework for establishing all Weyl type theorems for a large number of the commonly considered classes of operators. It should be noted that for these classes of operators, Weyl type theorems, or their generalized versions, have been proved in several papers. The classes of polaroid operators introduced in the previous sections are rather large. In the sequel we list some, by no means all, of these classes of operators. In the following $H$ always denotes a Hilbert space.

(a) A bounded operator $T \in L(X)$ is said to belong to the class $H(p)$ if there exists a natural $p := p(\lambda)$ such that:

$$H_0(\lambda I - T) = \ker(\lambda I - T)^p \quad \text{for all } \lambda \in \mathbb{C}. \tag{12}$$

Clearly, every operator $T$ which belongs to the class $H(p)$ has SVEP. Moreover, from (8) it follows that every $H(p)$ operator $T$ is polaroid. Consequently, by Theorem 5.14, $f(T)$ satisfies Weyl's theorem, or equivalently, $(gW)$ for every $f \in \mathcal{H}(\sigma(T))$ which is not constant on each of the components of its domain.

Note that, by [52, Theorem 3.4], $T \in H(p)$ if and only if there exists a $f \in \mathcal{H}(\sigma(T))$, not constant on each of the components of its domain, such that $f(T)$ is $H(p)$ and this is equivalent to saying that $f(T)$ is $H(p)$ for all $f \in \mathcal{H}(\sigma(T))$. Consequently, if $T$ is algebraically $H(p)$ (i.e. there exists a non trivial polynomial $h$ such that $h(T)$ is $H(p)$) then $T$ is $H(p)$, so $(W)$ and $(gW)$ hold for $f(T)$.

The class $H(p)$ has been introduced by Oudghiri in [52] and in [19] this class of operators has been studied for $p := p(\lambda) = 1$ for all $\lambda \in \mathbb{C}$. Property $H(p)$ is satisfied by every generalized scalar operator, and in particular for p-hyponormal, log-hyponormal or M-hyponormal operators on Hilbert spaces, see [52]. Therefore, algebraically $p$-hyponormal or algebraically $M$-hyponormal operators are $H(p)$. $(W)$ or $(gW)$ for $f(T)$, where $T$ is algebraically $M$-hyponormal, have been proved in different papers [35] and [36], respectively. In [30, Theorem 3.3] it is shown that if $T \in L(H)$ is such that $T'$ is $p$-hyponormal or $M$-hyponormal then $(gaW)$ holds for $f(T)$ for all $f \in \mathcal{H}(\sigma(T))$. Since $T'$ is $H(p)$, hence polaroid (or equivalently $T$ is polaroid), then, by Theorem 5.13 and Remark 5.2, all Weyl's theorems (generalized or not) hold (and are equivalent!) for $f(T)$. Finally, Theorem 5.14 subsumes also [31, Theorem 3.4]: if $T$ is *analytically hyponormal* (i.e. there exists $h \in \mathcal{H}(\sigma(T))$ for which $h(T)$ is hyponormal) then Weyl's theorem holds for $f(T)$. Clearly, since analytically hyponormal operators are $H(p)$, then $(W)$ and $(gW)$ are equivalent for $f(T)$,

(b) A bounded operator $T \in L(X)$ on a Banach space $X$ is said to be *paranormal* if

$$\|Tx\|^2 \leq \|T^2x\|\|x\| \quad \text{holds for all } x \in X.$$

Every paranormal operator on a Hilbert space has SVEP [14]. An operator $T \in L(H)$ for which there exists a complex nonconstant polynomial $h$ such that $h(T)$ is paranormal is said to be *algebraically paranormal*. Every algebraic paranormal operator $T$ defined on a Hilbert space is polaroid, see [14]. Moreover, the SVEP for $h(T)$ entails the SVEP for $T$, see [1, Theorem 2.40]. $(gW)$ for $f(T)$, $T$ algebraically paranormal, has been proved in [60, Theorem 3.1] and [35, Theorem 4.14], but this is immediate from Theorem 5.14 and $(gW)$ is equivalent to $(W)$. In [60, Theorem 3.2] it has been also proved that $(gaW)$ holds for $T$ if $T'$ is algebraically paranormal. This result is clear from Theorem 5.12, and in particular $(gaW)$ for $T$ is equivalent to any type of Weyl's theorem, generalized or not. Furthermore, Theorem 5.13 and Remark 5.2 extend to $f(T)$ all Weyl's type theorems. Similar considerations may be done for *analytically paranormal* operators studied in [32]. Analytically paranormal operators $T$ are polaroid and have SVEP, again by [1, Theorem 2.40], so that $(W)$ holds for $f(T)$, for $f \in \mathcal{H}(\sigma(T))$ not constant on each of the components of its domain, and this is equivalent to saying that $(gW)$ holds for $f(T)$. $(gW)$ for $f(T)$, $T$ analytically paranormal, has been proved [32, Theorem 3.1, part (a)]. $(gaW)$ for $f(T^*)$, $T$ analytically paranormal, was proved in [32, Theorem 3.1, part (b)], that is clear by Theorem 5.13 and Remark 5.2 and, again, all Weyl's type theorems, generalized or not, are equivalent for $f(T^*)$. Also the result of [32, Theorem 3.2, part] easily follows from Theorem 5.13 and Remark 5.2: If $T' \in L(H)$

is analytically paranormal then all Weyl type theorems, generalized or not, are satisfied by $f(T)$ and all of them are equivalent.

(c) A bounded operator $T \in L(H)$ is a *class A operator* if $|T|^2 \leq |T^2|$, and $T$ is said to be a *quasi-class A operator* if $T'|T|^2 T \leq T'|T^2|T$. If $T \in L(H)$ is *algebraically quasi-class operator* (i.e. there exists a nonconstant polynomial $h$ such that $h(T)$ is a quasi-class A operator), then $T$ is polaroid [20, Lemma 2.3] and since $h(T)$ has SVEP [39, Lemma 1.5], also $T$ has SVEP. Consequently, by Theorem 5.14 $f(T)$ satisfies $(W)$, or equivalently $(gW)$. If $T'$ is an algebraically quasi-class operator then $T'$, or equivalently $T$, is polaroid and the SVEP for $T'$ entails, by Theorem 5.13 and Remark 5.2, that all Weyl's theorems (generalized or not) hold (and are equivalent!) for $f(T)$. Therefore Theorem 5.13 subsumes and extends [20, Theorem 2.4 and Theorem 3.3].

(d) Every multiplier $T$ of a semi-simple commutative Banach algebra $A$, (see [1, Chapter 4] for definitions and details) is $H(1)$, see [19], in particular every convolution $T_\mu$ operator of $L^1(G)$, $L^1(G)$ the group algebra of a locally compact Abelian group $G$ is $H(1)$. Therefore, $T$ is polaroid and has SVEP so that $(W)$, or equivalently $(gW)$ holds for $f(T)$. If $A$ is regular and Tauberian, this is the case for instance of the group algebra $L^1(G)$, $G$ a compact Abelian group, then $\sigma(T) = \sigma_a(T)$ for every multiplier $T$, see Corollary 5.88 of [1]. Therefore, if $A$ is regular and Tauberian every multiplier $T$ is $a$-polaroid. In particular, every convolution operator $T_\mu$ on $L^1(G)$ whenever $G$ is compact, is $a$-polaroid. Consequently, by Theorem 5.12, $(W)$, $(aW)$, $(w)$, $(gW)$, $(gaW)$ and $(gw)$ for $f(T^*)$ are equivalent and $f(T^*)$ satisfies all of them.

Another example of $a$-polaroid operator is given by a multiplier $T$ of a Banach algebra $A$ with an orthogonal basis. In fact also in this case $\sigma(T) = \sigma_a(T)$, see Theorem 4.46 of [1], so that Theorem 5.12 applies to $T$.

## 7. Polaroid operators under quasi-affinity

In this section we study the preservation of Weyl type theorems between two operators $T \in L(X)$, $S \in L(Y)$ in the case that these operators are intertwined by a quasi-affinity $A \in L(X, Y)$, or in the more general case that $T$ and $S$ are asymptotically intertwined by $A$. Most of these results may be found in [6] and [13].

**Lemma 7.1.** *Suppose that $S \in L(Y)$ has SVEP at $\lambda_0$, and let $T \in L(X)$ be such that there exists $A \in L(X, Y)$ injective for which $SA = AT$. Then $T$ has the SVEP at $\lambda_0$.*

**Proof.** Let $U \subseteq \mathbb{C}$ be an open neighborhood of $\lambda_0$ and $f : U \to X$ be an analytic function such that $(\lambda I - T)f(\lambda) = 0$, for all $\lambda \in U$. Then

$$(\lambda I - S)Af(\lambda) = A(\lambda I - T)f(\lambda) = 0 \quad \text{for all } \lambda \in U.$$

Since $S$ has SVEP at $\lambda_0$, then $Af(\lambda) = 0$ and since $A$ is injective, then $f(\lambda) = 0$ for all $\lambda \in U$. Therefore $T$ has the SVEP at $\lambda_0$. □

**Definition 7.1.** The operator $A \in L(X, Y)$ between the Banach spaces $X$ and $Y$ is a *quasi-affinity* if it has a trivial kernel and dense range. We say that $T \in L(X)$ is a *quasi-affine transform* of $S \in L(Y)$, and we write $T \prec S$, if there is a quasi-affinity $A \in L(X, Y)$ that intertwines $T$ and $S$, i.e. $SA = AT$. If there exists two quasi-affinities $A \in L(X,Y)$, $B \in L(Y,X)$ for which $SA = AT$ and $BS = TB$ then we say that $S$ and $T$ are *quasi-similar*.

In the sequel for every $T$ we denote with $P_T(\lambda)$ the spectral projection associated with $T \in L(X)$ and the singleton se $\{\lambda\}$.

**Lemma 7.2.** *Suppose that $T \in L(X)$ and $S \in L(Y)$ are intertwined by $A \in L(X,Y)$. If $\lambda \in \mathrm{iso}\,\sigma(T) \cap \mathrm{iso}\,\sigma(S)$ then the spectral projections $P_T(\lambda)$ and $P_S(\lambda)$ are also intertwined by $A$, i.e. $P_S(\lambda)A = AP_T(\lambda)$.*

**Proof.** If $T$ and $S$ are intertwined by $A \in L(X,Y)$ we have $(\mu I - S)A = A(\mu I - T)$ for all $\mu \in \mathbb{C}$. Suppose that $\mu$ belongs to the resolvent of $T$ as well to the resolvent of $S$. Then $A = (\mu I - S)^{-1}A(\mu I - T)$ and this implies that $A(\mu I - T)^{-1} = (\mu I - S)^{-1}A$, from which it easily follows that

$$P_S(\lambda)A = \left(\frac{1}{2\pi i}\int_\Gamma (\mu I - S)^{-1}d\mu\right)A = \frac{1}{2\pi i}\int_\Gamma (\mu I - S)^{-1}A\,d\mu$$
$$= \frac{1}{2\pi i}\int_\Gamma A(\mu I - T)^{-1}d\mu = AP_T(\lambda).$$
□

If $T \prec S$ a classical result due to Rosenblum shows that $\sigma(S)$ and $\sigma(T)$ must overlap, see [43]. But also the stronger condition of quasi-similarity is, in general, not sufficient to preserve the spectrum. This happens only in some special cases, for instance if $T$ and $S$ are quasi-similar hyponormal operators, or whenever $T$ and $S$ have totally disconnected spectra, see [43, Corollary 2.5]. Therefore, it is not quite surprising that, if $T \prec S$, the preservation of "good" spectral properties from $S$ to $T$ requires that some spectral inclusions are satisfied. In the next theorem we give a sufficient condition for which the polaroid condition is preserved under an injective map that intertwines $T$ and $S$.

**Theorem 7.1.** *Suppose that $T \in L(X)$, $S \in L(Y)$ are intertwined by an injective map $A \in L(X,Y)$. If $S$ is polaroid and $\operatorname{iso}\sigma(T) \subseteq \operatorname{iso}\sigma(S)$ then $T$ is polaroid.*

**Proof.** If $\sigma(T)$ has no isolated point then $T$ is polaroid and hence there is nothing to prove. Suppose that $\operatorname{iso}\sigma(T) \neq \emptyset$ and let $\lambda \in \operatorname{iso}\sigma(T)$. Then $\lambda \in \operatorname{iso}\sigma(S)$, hence $\lambda$ is a pole of the resolvent of $S$. Let $P_T(\lambda)$ and $P_S(\lambda)$ be the spectral projections associated to $T$ and $S$ with respect the spectral subset $\{\lambda\}$, respectively. As we have seen before, $P_T(\lambda)$ and $P_S(\lambda)$ are intertwined by $A$, i.e. $P_S(\lambda)A = AP_T(\lambda)$. Since $\lambda$ is a pole of the resolvent of $S$ then $p := p(\lambda I - S) = q(\lambda I - S) < \infty$ and $\ker(\lambda I - S)^p$ coincides with the range of $P_S(\lambda)$, see [1, Theorem 3.74]. Therefore, $(\lambda I - S)^p P_S(\lambda) = 0$, and consequently

$$0 = (\lambda I - T)^p P_S(\lambda) A = (\lambda I - T)^p A P_T(\lambda) = A(\lambda I - T)^p P_T(\lambda).$$

Since $A$ is injective then $(\lambda I - T)^p P_T(\lambda) = 0$. Now, the range of the spectral projection $P_T(\lambda)$ coincides with the quasi-nilpotent part $H_0(\lambda I - T)$, see [1, Theorem 3.74], so

$$H_0(\lambda I - T) = P_T(\lambda)(X) \subseteq \ker(\lambda I - T)^p.$$

The opposite inclusion also holds, since $\ker(\lambda I - T)^n \subseteq H_0(\lambda I - T)$ for all natural $n \in \mathbb{N}$. Therefore, $H_0(\lambda I - T) = \ker(\lambda I - T)^p$ for every $\lambda \in \operatorname{iso}\sigma(T)$ and by Theorem 4.3 it then follows that $T$ is polaroid. □

**Corollary 7.1.** *Let $T \in L(X)$, $S \in L(Y)$ be intertwined by an injective map $A \in L(X,Y)$. Suppose that $S$ is polaroid, has SVEP and $\operatorname{iso}\sigma(T) \subseteq \operatorname{iso}\sigma(S)$. Then we have:*

*(i) $f(T)$ satisfies (W) or equivalently (gW) for all $f \in \mathcal{H}(\sigma(T))$.*

*(ii) (W), (aW), (w), (gW), (gaW) and (gw) for $f(T^*) = f(T)^*$ are equivalent and $f(T^*)$ satisfies all of them.*

**Proof.** (i) $T$ has SVEP by Lemma 7.1, hence $f(T)$ has SVEP for all $f \in \mathcal{H}(\sigma(T))$, see [1, Theorem 2.40]. Furthermore, $f(T)$ is polaroid by Theorem 5.4. By Theorem 5.14 then $f(T)$ satisfies Weyl's theorem, or equivalently, by Theorem 5.10, generalized Weyl's theorem.

(ii) As observed above $f(T)$ has SVEP for all $f \in \mathcal{H}(\sigma(T))$ and $f(T)$ is polaroid by Theorem 5.4. Therefore, $f(T)$ satisfies (W) and (W), (aW), (w), (gW), (gaW), (gw) are equivalent for $f(T^*) = f(T)^*$, by Theorem 5.13. □

In general a polaroid operator has not SVEP. A trivial example is the left shift $T$ on $\ell^2(\mathbb{N})$. This operator is polaroid, since $\sigma(T)$ is the unit disc of $\mathbb{C}$, so iso $\sigma(T) = \emptyset$ and it is well known that $T$ fails SVEP at 0.

In the sequel by a part of an operator $T \in L(X)$ we means the restriction of $T$ to a closed $T$-invariant subspace.

**Definition 7.2.** A bounded operator $T \in L(X)$ is said to be *hereditarily polaroid* if every part of $T$ is polaroid.

It is easily seen that the property of being hereditarily polaroid is similarity invariant, but is not preserved by a quasi-affinity. Since every hereditarily polaroid operator has SVEP, see [38, Theorem 2.8], we readily obtain:

**Corollary 7.2.** *Let $T \in L(X)$, $S \in L(Y)$ be intertwined by an injective map $A \in L(X,Y)$. Suppose that $S$ is hereditarily polaroid, and iso $\sigma(T) \subseteq$ iso $\sigma(S)$. Then $f(T)$ satisfies $(W)$ or equivalently $(gW)$ for all $f \in \mathcal{H}(\sigma(T))$, while $(W)$ and $(W)$, $(aW)$, $(w)$, $(gW)$, $(gaW)$, $(gw)$ are equivalent and hold for $f(T^*)$.*

An interesting class of hereditarily polaroid operators is given by the $H(p)$-operators. In fact, $H(p)$-operators are polaroid and if $T$ is $H(p)$ then the every part of $T$ is $H(p)$ [52, Lemma 3.2], so every $H(p)$ is hereditarily polaroid. Other examples of hereditarily polaroid operators are given by the *completely hereditarily normaloid* operators on Banach spaces. In particular, all paranormal operators on Hilbert spaces and the class of $(p,k)$-*quasihyponormal* operators on Hilbert spaces are hereditarily polaroid, see for details [38]. Also the *algebraically quasi-class A operators* on a Hilbert space considered in [20], are hereditarily polaroid. In fact, every part of an algebraically quasi-class A operator $T$ is algebraically quasi-class A and every algebraically quasi-class A operator is polaroid [20, Lemma 2.3].

The next result shows that hereditarily operators are transformed, always under the assumption iso $\sigma(T) \subseteq$ iso $\sigma(S)$, by quasi-affinities into $a$-polaroid operators.

**Theorem 7.2.** *Suppose that $T \in L(X)$, $S \in L(Y)$ are intertwined by an injective map $A \in L(X,Y)$. If $S$ is hereditarily polaroid and iso $\sigma(T) \subseteq$ iso $\sigma(S)$ then $T^*$ is a-polaroid.*

**Proof.** By Theorem 7.1 $T$ is polaroid, and hence also $T^*$ is polaroid. As observed above, $S$ has SVEP, so $T$ has SVEP by Lemma 7.2. The SVEP for $T$ by [1, Corollary 2.45] entails that $\sigma(T^*) = \sigma(T) = \sigma_s(T) = \sigma_a(T^*)$, and this trivially implies that $T^*$ is $a$-polaroid. □

**Theorem 7.3.** *Let $T \in L(X)$, $S \in L(Y)$ be intertwined by an injective map $A \in L(X,Y)$ and suppose that $\operatorname{iso}\sigma_a(T) \subseteq \operatorname{iso}\sigma_a(S)$. If $S$ is left polaroid then $T$ is polaroid.*

**Proof.** If $\operatorname{iso}\sigma_a(T) = \emptyset$ there is nothing to prove. In fact $\sigma_a(T)$ contains the boundary of $\sigma(T)$, in particular the isolated points of the spectrum, hence $\operatorname{iso}\sigma(T) \subseteq \operatorname{iso}\sigma_a(T) = \emptyset$, so $T$ is polaroid.
Therefore, we can assume $\operatorname{iso}\sigma_a(T) \neq \emptyset$. We first show that
$$A(H_0(\lambda I - T)) \subseteq H_0(\lambda I - S).$$
Let $x \in H_0(\lambda I - T)$. Then
$$\lim_{n\to\infty} \|(\lambda I - S)^n A x\|^{1/n} = \lim_{n\to\infty} \|A(\lambda I - T)^n x\|^{1/n}$$
$$\leq \lim_{n\to\infty} \|(\lambda I - T)^n x\|^{1/n} = 0,$$
thus $Ax \in H_0(\lambda I - S)$ and hence $A(H_0(\lambda I - T)) \subseteq H_0(\lambda I - S)$.

Now, let $\lambda \in \operatorname{iso}\sigma(T)$. It is well known that $\sigma_a(T)$ contains the boundary of $\sigma(T)$, so $\lambda \in \operatorname{iso}\sigma_a(T) \subseteq \operatorname{iso}\sigma_a(S)$. Therefore, $\lambda$ is a left pole of the resolvent of $S$ so, by Theorem 2.7 of [3], there exists a positive natural $\nu$ such that $H_0(\lambda I - S) = \ker(\lambda I - S)^\nu$. Consequently,
$$A(H_0(\lambda I - T)) \subseteq H_0(\lambda I - S) = \ker(\lambda I - S)^\nu,$$
so, if $x \in H_0(\lambda I - T)$ then
$$A(\lambda I - T)^\nu x = (\lambda I - S)^\nu (Ax) = 0.$$
Since $A$ is injective then $(\lambda I - T)^\nu x = 0$ and hence $H_0(\lambda I - T) \subseteq \ker(\lambda I - T)^\nu$. The opposite inclusion is still true, so that $H_0(\lambda I - T) = \ker(\lambda I - T)^\nu$ for every $\lambda \in \operatorname{iso}\sigma(T)$. By Theorem 4.3 we then conclude that $T$ is polaroid. □

**Corollary 7.3.** *Let $T \in L(X)$, $S \in L(Y)$ be intertwined by an injective map $A \in L(X,Y)$. Suppose that $S$ is left polaroid, has SVEP and $\operatorname{iso}\sigma_a(T) \subseteq \operatorname{iso}\sigma_a(S)$. Then we have:*

*(i) $f(T)$ satisfies $(W)$ or equivalently $(gW)$ for all $f \in \mathcal{H}(\sigma(T))$.*

*(ii) $(W)$, $(aW)$, $(w)$, $(gW)$, $(gaW)$ and $(gw)$ for $f(T^*) = f(T)^*$ are equivalent and $f(T^*)$ satisfies all of them.*

**Proof.** By Theorem 7.3 $T$ is polaroid. Since $T$ has SVEP, then we can use the same argument of the proof of Corollary 7.1. □

An important subspace in local spectral theory is given by the *glocal spectral subspace* $\mathcal{X}_T(F)$ associated with a closed subset $F \subseteq \mathbb{C}$. This is defined, for an arbitrary operator $T \in L(X)$ and a closed subset $F$ of $\mathbb{C}$, as the set of all $x \in X$ for which there exists an analytic function $f : \mathbb{C}\setminus F \to X$ which satisfies the identity

$$(\lambda I - T)f(\lambda) = x \quad \text{for all } \lambda \in \mathbb{C} \setminus F.$$

It is known that $H_0(\lambda I - T) = \mathcal{X}_T(\{\lambda\})$ [1, Theorem 2.20]. Recall that a bounded operator $T \in L(X)$, $X$ a Banach space, is said to have *Dunford's property (C)*, shortly property $(C)$, if, for each closed set $F \subseteq \mathbb{C}$, $\chi_T(F)$ is closed. It is well-known that Dunford property $(C)$ implies SVEP.

Other important notions from local spectral theory are defined as follows. Let $U$ be an open subset of $\mathbb{C}$ and denote by $\mathcal{H}(U, X)$ the Fréchet space of all analytic functions $f : U \to X$ with respect the pointwise vector space operations and the topology of locally uniform convergence. $T \in L(X)$ has the *Bishop's property* $(\beta)$ if, for every open $U \subseteq \mathbb{C}$ and every sequence $(f_n) \subseteq \mathcal{H}(U, X)$ for which $(\lambda I - T)f_n(\lambda)$ converges to 0 uniformly on every compact subset of $U$, then also $f_n \to 0$ in $\mathcal{H}(U, X)$. Important classes of operators with property $(\beta)$ are the subnormal operators (i.e. restrictions of normal operator to closed invariant subspaces). Note that

$$\text{property } (\beta) \Rightarrow \text{property } (C) \Rightarrow \text{SVEP},$$

see [50, Proposition 1.2.19].
A bounded operator $T \in L(X)$ has the *decomposition property* $(\delta)$ if $X = \mathcal{X}_T(\overline{U}) + \mathcal{X}_T(\overline{V})$ for every open cover $\{U, V\}$ of $\mathbb{C}$. *Decomposable* operators may be defined in several ways, for instance as the union of the property $(\beta)$ and the property $(\delta)$, see [50, Theorem 2.5.19] for relevant definitions. Note that property $(\delta)$ implies SVEP for $T^*$. In fact $T$ has property $(\delta)$ if and only if $T^*$ has property $(\beta)$, see see [50, Theorem 2.5.19]. Every generalized scalar operator is decomposable, see [50] for definitions and details.

Under the stronger conditions of quasi-similarity and property $(\beta)$, in Theorem 7.1 the assumption on the isolated points of the spectra of $T$ and $S$ may be omitted:

**Theorem 7.4.** *Let $T \in L(X)$, $S \in L(Y)$ be quasi-similar.*

*(i) If $S$ is a polaroid operator which has property $(\beta)$ then $T$ is polaroid, while $T^*$ is a-polaroid.*

(ii) If $S \in L(H)$, $H$ a Hilbert space, is a polaroid operator for which property $(C)$ holds then $T$ is polaroid, while $T^*$ is a-polaroid.

Consequently, under the assumptions (i) or (ii) on $S$, $f(T)$ satisfies $(gW)$ for all $f \in \mathcal{H}(\sigma(T))$, while $(gW)$, $(gaW)$ and $(gw)$ hold $f(T^*) = f(T)^*$.

**Proof.** (i) By a result of Putinar [53] we have $\sigma(S) = \sigma(T)$, hence iso $\sigma(T) \subseteq$ iso $\sigma(S)$, so by Theorem 7.1 $T$ is polaroid, or equivalently $T^*$ is polaroid. Now, property $(\beta)$ implies that $S$ has SVEP and hence, by Lemma 7.2, also $T$ has SVEP. The SVEP for $T$, always by [1, Corollary 2.45], entails that $\sigma(T^*) = \sigma_{\mathrm{a}}(T^*)$, and hence $T^*$ is a-polaroid.

(ii) Also in this case, by a result of Stampfli [58], we have $\sigma(S) = \sigma(T)$, and property $(C)$ entails SVEP, so the assertion follows by using the same argument of part (i).

The last assertion is clear from Theorem 7.1. □

It is well known that hyponormal operators on Hilbert spaces have property $(\beta)$. Theorem 7.4 then applies to these operators, since they are $H(1)$ and hence polaroid. Another class of polaroid operators to which Theorem 7.4 applies is the class of all $p_* - QH$ operators studied in [40]. These operators are $H(1)$ (are also paranormal) and have property $(\beta)$, see [40, Theorem 2.12 and THeorem 2.2].

**Corollary 7.4.** *If $T \in L(X)$, $S \in L(Y)$ are quasi-similar and $S$ is generalized scalar, then $(gW)$ holds for $f(T)$, while $(gW)$, $(gaW)$ and $(gw)$ hold $f(T^*) = f(T)^*$.*

**Proof.** If $S$ is generalized scalar then $S$ is decomposable and hence has property $(\beta)$. Furthermore, $S$ is $H(p)$, see Example 3 in [52], hence polaroid. The assertion then follows from Theorem 7.4. □

A very weak notion of intertwining which dates back to Foiaş, cf. [34, Chapter 4] and [50, Chapter 3]. If $T \in L(X)$ and $S \in L(Y)$ the *commutator* $C(S,T)$ is the mapping on $L(X,Y)$ defined by $C(S,T)(A) := SA - AT$ for all $A \in L(X,Y)$. An operator $A \in L(X,Y)$ is said to *intertwine $T$ and $S$ asymptotically* if

$$\lim_{n \to \infty} \|C(S,T)^n(A)\|^{1/n} = 0, \tag{13}$$

where by induction it is easily to show the binomial identity

$$C(S,T)^n(A) = \sum_{k=0}^{n} \binom{n}{k} (-1)^k S^{n-k} A T^k.$$

Evidently, this notion is a generalization of the intertwining condition $C(S,T)(A) = 0$ which appears in the definition of $T \prec S$. This notion is also a generalization of the higher order intertwining condition:

$$C(S,T)^n(A) = 0 \quad \text{for some } n \in \mathbb{N}.$$

Note that $T$ and $S$ are generalized scalar then the condition (13) holds if and only if $C(S,T)^n(A) = 0$ for some $n \in \mathbb{N}$, see [34, Theorem 4.4.5]. If the pairs $(S,T)$ and $(T,S)$ are both asymptotically intertwined by some quasi-affinity then $T$ and $S$ are said to be *asymptotically quasi-similar*. We recall that if a pair $(S,T)$ is asymptotically intertwined by $A \in L(X,Y)$ then

$$A(\mathcal{X}_T(F) ) \subseteq \mathcal{Y}_S(F) \quad \text{for all closed sets } F \subseteq \mathbb{C}, \tag{14}$$

see Corollary 3.4.5 of [50].

A very particular case of asymptotically quasi-similar operators: $T, S \in L(X)$ are said to be *quasi-nilpotent equivalent* if each of the pairs $(S,T)$ and $(T,S)$ are asymptotically intertwined by the identity operator $I$ on $X$. Note that any quasi-nilpotent operator and the 0 operator are quasi-nilpotent equivalent.

**Example 7.1.** The polaroid condition is not transmitted whenever $S$ and $T$ are asymptotically intertwined by a quasi-affinity, even in the case that the inclusion $\operatorname{iso}\sigma(T) \subseteq \operatorname{iso}\sigma(S)$ is satisfied. For instance, if $T \in L(\ell^2(\mathbb{N}))$ is defined by

$$T(x_1, x_2, \ldots) = (\frac{x_2}{2}, \frac{x_3}{3}, \ldots) \quad \text{for all } (x_n) \in \ell^2(\mathbb{N}).$$

If $S := 0$ then $S$ is polaroid, while the quasi-nilpotent operator $T$ is not polaroid. $T$ and $S$ are, as observed above, quasi-nilpotent equivalent.

A very particular case of quasi-nilpotent equivalence is given if $C(S,T)^n(I) = 0$ for some $n \in \mathbb{N}$. If $T$ and $S$ commutes then $C(S,T)^n(I) = (S-T)^n = 0$. In this case $T$ and $S$ differ from a commuting nilpotent operator $N$ and, without any condition, if $S$ is polaroid then $T$ is also polaroid, see Theorem 2.10 of [15].

Let $E_\infty(T)$ denote the set of all $\lambda \in \operatorname{iso}\sigma(T)$ we have finite multiplicity.

**Theorem 7.5.** *Let $T \in L(X)$ and $S \in L(Y)$. Suppose that $S$ is polaroid, $T$ has SVEP and that the pair $(S,T)$ is asymptotically intertwined by an injective map $A \in L(X,Y)$.*

(i) If $E(T) \subseteq E_\infty(S)$ then $T$ satisfies generalized Weyl's theorem.

(ii) If $\pi_{00}(T) \subseteq \pi_{00}(S)$ then $T$ satisfies Weyl's theorem.

**Proof.** (i) If $\lambda \in E(T)$ then $\lambda \in \text{iso } \sigma(T)$. Since $S$ is polaroid it follows that $\lambda$ is a pole of the resolvent of $S$, or equivalently $p(\lambda I - S) = q(\lambda I - S) < \infty$. If we set $p := p(\lambda I - S) = q(\lambda I - S)$ then $H_0(\lambda I - S) = \ker(\lambda I - S)^p$, see [1, Theorem 3.74]. Since $\lambda \in E_\infty(S)$ we have $\alpha(\lambda I - S) < \infty$, so $\alpha((\lambda I - S)^p) < \infty$, thus $H_0(\lambda I - S)$ is finite dimensional. By (14) we have

$$A(H_0(\lambda I - T) = A(\mathcal{X}_T(\{\lambda\}) \subseteq \mathcal{Y}_S(\{\lambda\}) = H_0(\lambda I - S),$$

and since $A$ is injective it then follows that $H_0(\lambda I - T)$ is finite-dimensional. From the inclusion $\ker(\lambda I - T)^n \subseteq H_0(\lambda I - T)$ for all $n \in \mathbb{N}$ it then easily follows that $p(\lambda I - T) < \infty$. But $\lambda$ is an isolated point of $\sigma(T)$, so the decomposition $X = H_0(\lambda I - T) \oplus K(\lambda I - T)$ holds, consequently $K(\lambda I - T)$ is finite co-dimensional, and since $K(\lambda I - T) \subseteq (\lambda I - T)(X)$ we then conclude that $\beta(\lambda I - T) < \infty$. Therefore, $\lambda I - T$ is Fredholm. But $\lambda$ is an isolated point of $\sigma(T^*) = \sigma(T)$, so $T^*$ has SVEP at $\lambda$ and, since $\lambda I - T$ is Fredholm, this implies that $q(\lambda I - T) < \infty$. Therefore, $\lambda$ is a pole of the resolvent of $T$. This shows that $E(T) \subseteq \Pi(T)$ and hence $E(T) = \Pi(T)$. Since $T$ has SVEP then, by Theorem 5.8, generalized Weyl's theorem holds for $T$.

(ii) If $\lambda \in \pi_{00}(T)$ then $\lambda \in \pi_{00}(S)$, so $\lambda$ is a pole of $S$ and proceeding as in part (i) we obtain that $\lambda I - T$ is Browder, i.e, $\lambda \in p_{00}(T)$. Therefore, $p_{00}(T) = \pi_{00}(T)$ and since $T$ has SVEP it then follows that $T$ satisfies Weyl's theorem. □

Clearly, in the case that there exists an injective operator $A \in L(X, Y)$ such $C(S, T)^n(A) = 0$ for some $n \in \mathbb{N}$, the assumption that $T$ has SVEP in Theorem 7.5 may be replaced by the assumption that $S$ has SVEP. In fact, it is easy to show that under this condition the SVEP carries over from $S$ to $T$. In particular, the results of Theorem 7.5 apply whenever $S$ is hereditarily polaroid.

A bounded operator $T \in L(X)$ is said to be *isoloid* if every isolated point of $\sigma(T)$ is an eigenvalue. Trivially every polaroid operator is isoloid.

**Theorem 7.6.** *Let $T \in L(X)$ and $S \in L(S)$ be asymptotically intertwined by a quasi-affinity $A \in L(X, Y)$. Assume that $S$ is isoloid and $\Pi(S) \subseteq \Pi(T)$.*

*(i) If $S$ has property $(C)$, and if either $T$ or $T^*$ has SVEP, then generalized Weyl's theorem for $S$ implies that $T$ satisfies generalized Weyl's theorem.*

*(ii) If $T$ and $S$ are asymptotically quasi-similar and both $T$ and $S$ have one of the properties $(\delta)$ or $(C)$, then generalized Weyl's theorem for $S$ implies that $T$ satisfies generalized Weyl's theorem.*

**Proof.** (i) If $T$ or $T^*$ has SVEP then Browder's theorem (or equivalently, generalized Browder's theorem) holds for $T$, see [12, Corollary 3.3]. By Theorem 5.8 it then suffices to prove $E(T) = \Pi(T)$. Let $\lambda \in E(T)$. Then $\lambda$ is an eigenvalue of $T$ and an isolated point of $\sigma(T)$. Since $A$ is injective Corollary 3.5.8 of [50] entails that $\lambda$ belongs to $\sigma(S)$ (more precisely, $\lambda$ belongs to the approximate-point spectrum of $S$). Since $A$ has dense range, property $(C)$ for $S$ entails, by [49, Theorem 4.1], that $\sigma(S) \subseteq \sigma(T)$. This implies that $\lambda$ is an isolated point of $\sigma(S)$, hence, by the isoloid condition, $\lambda$ is an eigenvalue of $S$, i.e. $\lambda \in E(S)$. Since $S$ satisfies generalized Weyl's theorem we have $E(S) = \Pi(S) \subseteq \Pi(T)$, so $\lambda \in \Pi(T)$.

(ii) Note that $T$ or $T^*$ has SVEP, so Browder's theorem holds for $T$. By [50, Corollary 3.5.16] we know that $\sigma(T) = \sigma(S)$. If $\lambda \in E(T)$ then $\lambda$ is an isolated point of $\sigma(S)$, hence by the isoloid condition $\lambda \in E(S) = \Pi(S) \subseteq \Pi(T)$. □

**Remark 7.1.** Without assuming the inclusion $\Pi(S) \subseteq \Pi(T)$, the results of Theorem 7.6 do not hold. For instance, the operators $S$ and $T$ of Example 7.1 do not satisfy this inclusion, since we have $\Pi(T) = \emptyset$ while $\Pi(S) = \{0\}$. Clearly, $S$ is isoloid, and also all the other assumptions on $T$ and $S$ are satisfied, since $T$ and $S$ are decomposable.

Note that if $T$ is polaroid then the assumption $\Pi(S) \subseteq \Pi(T)$ in part (ii) of Theorem 7.6 is redundant. In fact, from the equality $\sigma(S) = \sigma(T)$ it follows that every $\lambda \in \Pi(S)$ is an isolated point of $\sigma(T)$, and hence a pole of $T$.

**Remark 7.2.** It is easily to see that under the stronger assumption that $C(S,T)^n(A) = 0$ for some $n \in \mathbb{N}$, the isoloid condition on $S$ in the statement of Theorem 7.6 is not required. To see this, observe first that every eigenvalue of $T$ is an eigenvalue of $S$. In fact, if $x \in X$ is an eigenvector for the eigenvalue $\lambda$ of $T$, then

$$0 = C(S,T)^n(A)x = C(S - \lambda I, T - \lambda I)^n(A)x = (S - \lambda I)^n Ax,$$

and since $A$ is injective then $Ax \neq 0$, so $(S - \lambda I)^n$ is not injective. This implies that also $S - \lambda I$ is not injective, so $\lambda$ is an eigenvalue of $S$. Now, we have seen in the proof of part (i) or part (ii) of Theorem 5.10 that if $\lambda \in E(T)$ then $\lambda$ is an isolated point of $\sigma(S)$. Moreover, $\lambda$ is an eigenvalue of $T$ and hence an eigenvalue of $S$, so $\lambda \in E(S) == \Pi(S) \subseteq \Pi(T)$.

Note that in Theorem 7.6 if we assume $S$ decomposable, always under the assumption $C(S,T)^n(A) = 0$, then the assumption that $T$ has SVEP can be dropped. In fact, $S$ has SVEP and it can shown that the SVEP carries from $S$ to $T$.

**Corollary 7.5.** *Suppose that $T \in L(X)$ and $S \in L(S)$ are decomposable operators which are asymptotically intertwined by a quasi-affinity. If $S$ is isoloid and $\Pi(S) \subseteq \Pi(T)$ then generalized Weyl's theorem for $S$ implies generalized Weyl's theorem for $T$.*

**Proof.**
$S$ has $(\delta)$, both $T$ and $T^*$ have SVEP. Hence Theorem 5.10 applies. □

**Theorem 7.7.** *Suppose that $T, S \in L(X)$ are quasi-nilpotent equivalent, $S$ a polaroid operator which has SVEP and $\Pi(S) \subseteq \Pi(T)$. Then generalized Weyl's theorem for $S$ implies that generalized Weyl's theorem holds for $T$.*

**Proof.** Observe first that quasi-nilpotent equivalence preserves SVEP and quasi-nilpotent equivalent operators have the same spectra [34, Theorem 2.2 and Theorem 2.3]. Moreover, by [49, Proposition 2.2] (taking $A = I$) we have $\mathcal{X}_T(F) \subseteq \mathcal{X}_S(F)$ for every closed subset $F$ of $\mathbb{C}$. Evidently, the opposite inclusion holds for symmetry. In particular, taking $F = \{\lambda\}$ we conclude that $H_0(T - \lambda I) = H_0(S - \lambda I)$ holds for every $\lambda \in \mathbb{C}$.

Now, let $\lambda \in E(T)$. Then $\lambda$ is an isolated point of $\sigma(T) = \sigma(S)$ and $T - \lambda I$ is not injective. Since $N(T - \lambda I) \subseteq H_0(T - \lambda I)$ it then follows that $H_0(S - \lambda I) = H_0(T - \lambda I) \neq \{0\}$. Now, since $S$ is polaroid then there exists $p := p(\lambda) \in \mathbb{N}$ such that $H_0(S - \lambda I) = N(S - \lambda I)^p$, see [10, Theorem 2.9]. If were $S - \lambda I$ injective then $(S - \lambda I)^p$ would be also injective and hence $H_0(S - \lambda I) = \{0\}$, which is impossible. Therefore, $\lambda \in E(S)$, i.e. $E(T) \subseteq E(S)$ and by symmetry we conclude that $E(T) = E(S)$. Since $S$ satisfies generalized Weyl's theorem we have $E(S) = \Pi(S)$ and hence $E(T) \subseteq \Pi(T)$, from which it follows that generalized Weyl's theorem holds for $T$. □

The operators $T$ and $S$ of Example 7.1 show the that without the assumption $\Pi(S) \subseteq \Pi(T)$, the result of Theorem 7.7 in general does not hold.

Two operators $T \in L(X)$, $S \in L(Y)$ are said to be *asymptotically similar* if there exists a bijection $A \in L(X, Y)$ such that $A$ intertwines $S$ and $T$ asymptotically and its inverse $A^{-1}$ intertwines $T$ and $S$ asymptotically. This is a slightly generalization of quasi-nilpotent equivalence, see [34] for details.

**Theorem 7.8.** *Suppose that $T \in L(X)$ and $T \in L(Y)$ are asymptotically similar, $S$ a polaroid which has SVEP and $\Pi(S) \subseteq \Pi(T)$. Then generalized Weyl's theorem for $S$ holds implies generalized Weyl's theorem holds for $T$.*

**Proof.** As noted before, asymptotic quasi-similarity, hence asymptotic similarity, preserves the spectrum and SVEP, see [49, Theorem 3.5]. Therefore, $T$ satisfies generalized Browder's theorem. To show the equality $E(T) = \Pi(T)$ proceed as in the proof of Theorem 7.7, taking into account that $A(H_0(T - \lambda I)) = H_0(S - \lambda I)$, always by [49, Proposition 2.2]. Therefore Generalized Weyl's theorem for $S$ then implies $E(T) = E(S) = \Pi(S) \subseteq \Pi(T)$. □

## References

1. P. Aiena *Fredholm and local spectral theory, with application to multipliers.* Kluwer Acad. Publishers (2004).
2. P. Aiena. *Classes of Operators Satisfying a-Weyl's theorem.* Studia Math. **169** (2005), 105-122.
3. P. Aiena *Quasi Fredholm operators and localized SVEP,* Acta Sci. Math. (Szeged), **73** (2007), 251-263.
4. P. Aiena. *Property (w) and perturbations II,* J. Math. Anal. and Appl. **342**, (2008), 830-837.
5. P. Aiena, E. Aponte, E. Balzan *Weyl type theorems for left and right polaroid operators,* Int. Equa. Oper. Theory. **66** (2010), no. 1, 1-20.
6. P. Aiena, M. Berkani *Generalized Weyl's theorem and quasi-affinity,* Studia Math. **198** (2010), no. 2, 105-120.
7. P. Aiena, M. T. Biondi: *Browder's theorem through localized SVEP,* Mediterr. J. Math. **2** (2005), 137-151.
8. P. Aiena, M. T. Biondi *Property (w) and perturbations* J. Math. Anal. Appl. **336** (2007), 683-692.
9. P. Aiena, M. T. Biondi, C. Carpintero *On Drazin invertibility,* Proc. Amer. Math. Soc. **136**, (2008), 2839-2848.
10. P. Aiena, M. T. Biondi, F. Villafãne *Property (w) and perturbations III* J. Math. Anal. Appl. **353** (2009), 205-214.
11. P. Aiena, C. Carpintero, E. Rosas. *Some characterization of operators satisfying a-Browder theorem.* J. Math. Anal. Appl. **311**, (2005), 530-544.
12. P. Aiena, O. Garcia *Generalized Browder's theorem and SVEP* Mediterranean Jour. of Math. **4**, (2007), 215-228.
13. P. Aiena, M. Chō, M. Gonzalez *Polaroid type operators under quasi-affinities* (2010), to appear in J. Math. Anal. and Appl.
14. P. Aiena, J. R. Guillen *Weyl's theorem for perturbations of paranormal operators,* Proc. Amer. Math. Soc. **35**, (2007), 2433-2442.
15. P. Aiena, J. Guillen, P. Peña *Property (w) for perturbation of polaroid operators.* Linear Alg. and Appl. **424** (2008), 1791-1802.

16. P. Aiena, T. L. Miller *On generalized a-Browder's theorem.* Studia Math. **180**, (3), (2007), 285-300.
17. P. Aiena, P. Peña *A variation on Weyl's theorem.* J. Math. Anal. Appl. **324** (2006), 566-579.
18. P. Aiena, J. E. Sanabria *On left and right poles of the resolvent.* Acta Sci. Math. **74** (2008),669-687.
19. P. Aiena, F. Villafañe *Weyl's theorem for some classes of operators.* Int. Equa. Oper. Theory **53**, (2005), 453-466.
20. J. An, Y. M. Han *Weyl's theorem for algebraically Quasi-class A operators.* Int. Equa. Oper. Theory **62**, (2008), 1-10.
21. M. Amouch, H. Zguitti *On the equivalence of Browder's and generalized Browder's theorem.* Glasgow Math. Jour. **48**, (2006), 179-185.
22. M. Berkani *On a class of quasi-Fredholm operators.* Int. Equa. Oper. Theory **34** (1), (1999), 244-249.
23. M. Berkani *Restriction of an operator to the range of its powers,* Studia Math. **140** (2), (2000), 163-175.
24. M. Berkani, M. Sarih *On semi B-Fredholm operators,* Glasgow Math. J. **43** (2001), 457-465.
25. *M. Berkani: Index of B-Fredholm operators and Generalization of a Weyl's Theorem,* Proc. Amer. Math. Soc. **130** (2002), no. 6, 1717-1723.
26. *M. Berkani: B-Weyl spectrum and poles of the resolvent,* J. Math. Anal. Appl. **272** (2002), no. 2, 596-603.
27. *M. Berkani, J. J. Koliha* : *Weyl type theorems for bounded linear operators,* Acta Sc. Math. (Szeged) **69** (2003), no. 1-2, 359-376.
28. M. Berkani, M. Amouch *On the property (gw).* Mediterr. J. Math. 5 (2008), no. 3, 371-378.
29. X. Cao, M. Guo, B. Meng. *A note on Weyl's theorem.* **133**, 10, (2003), Proc. Amer. Math. Soc. 2977-2984.
30. X. Cao, M. Guo, B. Meng *Weyl type theorems for p-hyponormal and M-hyponormal operators,* Studia Math. **163** (2) (2004), 177-187.
31. X. Cao *Weyl's type theorem for analytically hyponormal operators,* Linear Alg. and Appl. **405** (2005), 229-238.
32. X. Cao *Topological uniform descent and Weyl's type theorem,* Linear Alg. and Appl. **420** (2007), 175-182.
33. L. A. Coburn *Weyl's theorem for nonnormal operators.* Michigan Math. J. **20** (1970), 529-544.
34. I. Colojoară, C. Foiaş *Theory of generalized spectral operators.* Gordon and Breach, New York, 1968.
35. R. E. Curto, Y. M. Han *Generalized Browder's and Weyl's theorems for Banach spaces operators,* J. Math. Anal. Appl. **2** (2007), 1424-1442.
36. R. E. Curto, Y. M. Han *Weyl's theorem for algebraically paranormal operators,* Integ. Equa. Oper. Theory **50**, (2004), No.2, 169-196.
37. S. V. Djordjević, Y. M. Han. *Browder's theorem and spectral continuity.* Glasgow Math. J. **42**, (2000), 479-486.
38. B.P. Duggal, *Hereditarily polaroid operators, SVEP and Weyl's theorem* J. Math. Anal. Appl. **340**, (2008), 366-373.

39. B.P. Duggal, I. H. Jeon, I. H. Kim *On Weyl's theorem for quasi-class A operators* J. Korean Math. Soc. **43**, (2006), 899-909.
40. B.P. Duggal, I. H. Jeon *On p-quasi hyponormal operators* Linear Algebra and its Applications. **422**, (2007), 331-340.
41. N. Dunford *Spectral theory I. Resolution of the identity.* Pacific J. Math. **2** (1952), 559-614.
42. N. Dunford *Spectral operators.* Pacific J. Math. **4** (1954), 321-54.
43. S. Grabiner *Spectral consequences of the existence of interwining operators*, Comment. Math. Pace Mat. **22** (1980-81), 227-238.
44. R. Harte, Woo Young Lee *Another note on Weyl's theorem.* Trans. Amer. Math. Soc. **349** (1997), 2115-2124
45. H. Heuser *Functional Analysis*, Marcel Dekker, New York 1982.
46. J. J. Koliha *Isolated spectral points* Proc. Amer. Math. Soc. **124** (1996), 3417-3424.
47. J.P. Labrousse *Les opérateurs quasi-Fredholm.*, Rend. Circ. Mat. Palermo, XXIX **2**, (1980) 161-258.
48. D. C. Lay *Spectral analysis using ascent, descent, nullity and defect.* Math. Ann. **184** (1970), 197-214.
49. K. B. Laursen, M. M. Neumann: *Asymptotic intertwining and spectral inclusions on Banach spaces.*, Czechoslovak Math. J. **43** (118) (1993), no. 3, 483-497.
50. K. B. Laursen, M. M. Neumann *Introduction to local spectral theory.* Clarendon Press, Oxford 2000.
51. M. Mbekhta, V. Müller *On the axiomatic theory of the spectrum II.* Studia Math. **119** (1996), 129-147.
52. M. Oudghiri *Weyl's and Browder's theorem for operators satysfying the SVEP.* Studia Math. **163**, 1, (2004), 85-101.
53. M. Putinar *Quasi-similarity of tuples with Bishop's property* ($\beta$). Int. Eq. Op. Th. **15** (1992), 1047-1052.
54. V. Rakočević *On a class of operators.* Mat. Vesnik **37** (1985), 423-426.
55. V. Rakočević *Operators obeying a-Weyl's theorem.* Rev. Roumaine Math. Pures Appl. **34** (1989), no. 10, 915-919.
56. V. Rakočević. *Semi-Browder operators and perturbations.* Studia Math. **122** (1996), 131-137.
57. A. L. Shields *Weighted shift operators and analytic function theory* Topics in Operator theory, Mathematical Survey N. **105** (1974) (ed. C. Pearcy), 49-128. Am. Math. Soc. Providence, RI.
58. G. J. Stampfli *Quasi-similarity of operators.* Proc. Roy. Irish. Acad. **81**, (1981), 109-119.
59. H. Weyl *Uber beschrankte quadratiche Formen, deren Differenz vollsteig ist.* Rend. Circ. Mat. Palermo **27**, (1909), 373-392.
60. H. Zguitti *A note on generalized Weyl's theorem*, J. Math. Anal. Appl. **316** (1) (2006), 373-381.

# Finitely additive measures in action

Joe Diestel

*Department of Mathematical Sciences*
*Kent State University, Kent, Ohio, USA*
*e-mail: j_diestel@hotmail.com*

Angela Spalsbury

*Department of Mathematics and Statistics*
*Youngstown State University, Youngstown, Ohio, USA*
*e-mail: aspalsbury@yahoo.com*

## 1. Introduction

Naturally the study of finitely additive measures predates measure theory; perhaps the first object of interest in the theory was *Jordan content*. Let $B$ be a bounded subset of $\mathbb{R}$ and define the outer content $j^*$, and inner content $j_*$ of $B$ by

$$j^*(B) = \inf \left\{ \sum_{j \leq n} l(I_j) \right\},$$

where the infimum is taken over all collections $\{I_{,1}, \ldots, I_n\}$ of pairwise disjoint intervals such that $B \subseteq I_1 \cup \cdots \cup I_n$ and

$$j_*(B) = \sup \left\{ \sum_{j \leq n} l(I_j) \right\},$$

where the supremum is taken over all collections $\{I_{,1}, \ldots, I_n\}$ of pairwise disjoint intervals such that $I_1 \cup \cdots \cup I_n \subseteq B$.

The set $B$ is **Jordan measurable** if

$$j_*(B) = j^*(B).$$

Of course, it's critical here to realize that our infs and sups are taken over *finite* collections of families of pairwise disjoint intervals. As a consequence one finds that *the collection $\mathcal{J}[a,b]$ of Jordan measurable subsets of the interval $[a,b]$ is an algebra of sets.* More to the point, $E \subseteq [a,b]$ is in $\mathcal{J}[a,b]$ precisely when $\chi_E$ is Riemann integrable.

Hindsight being what it is, one might easily conclude that $\mathcal{J}[a,b]$ was a noble effort destined to fail. This is not quite the case, as we will see.

The advent of Lebesgue measure accompanied as it was by many fundamental advances, turned *much* of the attention away from finitely additive phenomena. Much but not all.

Before discussing some of the earliest encounters with finitely additive measure theory in geometry, to be followed by applications in Banach space theory, let us make a few points about the differences between finitely additive and countably additive measures, differences which we posit emphasize their similarities.

Though the natural domain of finitely additive measures is an algebra and although it's so that many finitely additive measures (even on a $\sigma$-algebra) are unbounded, we will discuss bounded additive measures on sigma-algebras, at least to begin.

Let $S$ be a set and $\Sigma$ be a $\sigma$-field of subsets of $S$. Let $\mathrm{ba}(\Sigma)$ denote the linear space of all bounded (finitely) additive scalar-valued measures $\mu$ defined on $\Sigma$. Then $\mathrm{ba}(\Sigma)$ is a Banach spaces with either of two equivalent norms. The first norm of $\mu \in \mathrm{ba}(\Sigma)$:

$$||\mu||_\infty = \sup\{|\mu(E)| : E \in \Sigma\};$$

the second, the variation norm

$$||\mu||_1 = \sup\{\sum_i |\mu(E_i)|\},$$

where the supremum is taken over all families $\{E_i\}$ of pairwise disjoint members of $\Sigma$. It's well-known that

$$||\mu||_\infty \leq ||\mu||_1 \leq 4||\mu||_\infty.$$

The use of the variation norm in the study of $\mathrm{ba}(\Sigma)$ is motivated in large part by its importance in the study of the space $\mathrm{ca}(\Sigma)$ of countably additive members of $\mathrm{ba}(\Sigma)$ but also to some extent to the following classical result.

**Theorem 1.1 (Hildebrandt, Fichtenholtz-Kantorovich).**
$(ba(\Sigma), \|\ \|_1)$ *is isometrically isomorphic to the dual* $\mathcal{B}(\Sigma)^*$ *of the Banach space* $\mathcal{B}(\Sigma)$ *of all bounded scalar-valued* $\Sigma$-*measurable functions defined on* $S$.

The action of $\mu \in ba(\Sigma)$ on $f \in \mathcal{B}(\Sigma)$ is given by integration. The integral $\int f \, d\mu$ is defined first for $f$ a simple $\Sigma$-measurable function; it's then noted that for such $f$'s,

$$\left| \int f \, d\mu \right| \leq \|f\|_\infty \|\mu\|_1,$$

and so $\int d\mu$ can be uniquely extended to the completion of the space of simple $\Sigma$-measurable functions in a linear manner. Finally the very definition of $\mathcal{B}(\Sigma)$ says it *is* the completion of the space of simple $\Sigma$-measurable functions in the supremum norm.

The integral so generated indicates the simplicity of integrating bounded measurable functions vis-a-vis bounded additive measures. This simplicity comes at a price, of course: The price is that we have few limit theorems. The Bounded Convergence, Monotone Convergence, and Dominated Convergence Theorems require countable additivity to run properly; on the other hand, it is the sheer simplicity that allows for the quick and efficient use of finitely additive measures.

Regarding the Convergence theorems, we rush to mention the following remarkable result of Banach which we'll prove later, a result which is (to us) surprising and often overlooked.

**Theorem 1.2 (Banach).** *Let* $\mu \in ba(\Sigma)$. *Suppose* $(f_n) \subseteq \mathcal{B}(\Sigma)$ *is uniformly bounded and*

$$\lim_n \liminf_k |f_n(s_k)| = 0$$

*for each sequence* $(s_k)$ *in* $S$. *Then*

$$\lim_n \int f_n \, d\mu = 0.$$

Being a non-countably additive member of $ba(\Sigma)$ carries with it a certain stigma which is somewhat unwarranted. Members of $ba(\Sigma)$ want to be countably additive and that they fail is, in a sense, not their fault.

To be sure, if $\mu \in \text{ba}(\Sigma)$ and $(E_n)$ is a pairwise disjoint sequence of members of $\Sigma$ then $\sum_n |\mu(E_n)| < \infty$; so $\mu$ add up the sets $E_n$ *unconditionally*. It's just that $\sum_n \mu(E_n)$ need not be $\mu(\cup_n E_n)$.

This is an old shtick, one worth repeating though. The default is in the underlying $\sigma$-field. In fact, if we recall the representation theorem of M.H.Stone then we will realize that there is a zero-dimensional compact Hausdorff space $K$ where the field $\mathcal{F}$ of clopen sets is isomorphic (as a Boolean algebra of sets) to $\Sigma$; if $\mu \in \text{ba}(\Sigma)$ then $\mu$ has a trace $\mu'$ that lives on $\mathcal{F}$, namely $\mu'(F) = \mu(E)$ where $E$ and $F$ are in correspondence under the Stone isomorphism. Note that $\mu'$ is bounded (it has the same bounded range as $\mu$) and finitely additive ($\mathcal{F}$ and $\Sigma$ are isomorphic as Boolean algebras). But here's the catch: if $(F_n)$ is a sequence of pairwise disjoint (clopen) members of $\mathcal{F}$ whose union $\cup_n F_n \in \mathcal{F}$ then $\mu'(\cup_n F_n) = \sum_n \mu'(E_n)$; in fact, $\cup_n F_n \in \mathcal{F}$ means that $\cup_n F_n$ is clopen, so closed, so compact, while each $F_n$ is clopen and, as result, the ostensibly infinite union of is actually a finite union and $\mu'$ knows what to do with finite unions (as did $\mu$!). Of course, $\mu'$ being countably additive on a field that's a base for $K$'s topology has a unique countably additive regular extension to the Borel $\sigma$-field of subsets of $K$.

There's more that can (and should) be said here. Let $\mu \in \text{ba}(\Sigma)$ and suppose $(E_n)$ is a sequence of pairwise disjoint members of $\Sigma$. Then there's a subsequence $(E'_n)$ of $(E_n)$ such that $\mu$ is countably additive on the $\sigma$-field $\Sigma_0$ generated by $(E'_n)$. Now each $\mu \in \text{ba}(\Sigma)$ can be written as a difference of non-negative members of $\text{ba}(\Sigma)$– the function $|\mu|(E) = \sup\{\sum |\mu(A_i)|\}$, where the sup is taken over all pairwise disjoint sequences from $\Sigma$ with $A_i \subseteq E$ is a member of $\text{ba}(\Sigma)$ so

$$\mu = \frac{|\mu| - \mu}{2} - \frac{|\mu| + \mu}{2}$$

does the trick; so it's enough to show this for $\mu \in \text{ba}(\Sigma)$ that is non-negative. In this case, it's easy to see that

$$\lim_k \mu\left(\bigcup_{j=k}^{\infty} E_j\right) = 0;$$

after all, $\sum_n \mu(E_n) < \infty$. If we pick $m_j \nearrow m$ in such a way that $\mu(\cup_{k=m_j}^{\infty} E_k) < \frac{1}{2^j}$, then $\mu$ is countably additive on the $\sigma$-field generated by the sequence $(E'_j = E_{m_j})$. This stunningly easy observation was made by Lech Drewnowski who used it prove a remarkable variety of limit theorems for sequences in ba($\Sigma$), including the Vitali-Hahn-Saks theorem and Nikodym's Convergence theorem. For the sake of some semblance of completeness, we state these results in their finite additive clothing. (A fine place to see these ideas is Bob Huff's 1973 Penn State lectures on Vector Measures.)

A few definitions are needed. Let $\mu, \lambda \in$ ba($\Sigma$) with $\mu \geq 0$. We say that $\lambda$ is **absolutely continuous** with respect to $\mu$ if given an $\epsilon > 0$ there is a $\delta > 0$ so that $|\lambda(E)| \leq \epsilon$ where $\mu(E) \leq \delta$. Let $\Lambda \subseteq$ ba($\Sigma$); we say that $\Lambda$ is **uniformly absolutely continuous with respect to** $\mu$ (greater than or equal to zero and a member of ba($\Sigma$)) if given $\epsilon > 0$ there is a $\delta > 0$ such that $|\lambda(E)| \leq \epsilon$ whenever $\mu(E) \leq \delta$ for all $\lambda \in \Lambda$. Again for $\Lambda \subseteq$ ba($\Sigma$) we say that $\Lambda$ is **uniformly additive** if given an pairwise disjoint sequence $(E_n)$ in $\Sigma$ we have

$$\lim_n \lambda(E_n) = 0$$

uniformly for $\lambda \in \Lambda$.

**Theorem 1.3 (Vitali-Hahn-Saks Theorem).** *Let $(\lambda_n) \subseteq$ ba($\Sigma$) and suppose each $\lambda_n$ is absolutely continuous with respect to the non-negative member $\mu$ of ba($\Sigma$). If $\lim_n \lambda_n(E)$ exists for each $E \in \Sigma$ then $\{\lambda_n : n \in \mathbb{N}\}$ is uniformly continuous with respect to $\mu$.*

**Theorem 1.4 (Nikodym's Convergence Theorem).**
*Let $(\lambda_n) \subseteq$ ba($\Sigma$) and suppose $\lim_n \lambda_n(E)$ exists for each $E \in \Sigma$. Then $\{\lambda_n : n \in \mathbb{N}\}$ is uniformly additive.*

A word of caution here. In Nikodym's Convergence theorem nothing is assumed about the uniform boundedness of the sequence $(\lambda_n)$; *nothing need be assumed*. This is so thanks a remarkable result of Grothendieck, generalizing an equally remarkable result of Nikodym.

**Theorem 1.5 (The Nikodym-Grothendieck Boundedness Theorem).** *Let $\phi \neq \Lambda \subseteq$ ba($\Sigma$) and suppose that for each $E \in \Sigma$ we have*

$$\sup_\Lambda |\lambda(E)| < \infty.$$

*Then $\Lambda$ is a bounded set in ba($\Sigma$).*

Grothendieck's discovery of this theorem went largely unnoticed as he stated it as an exercise (!) with hints, in a lecture series presented in Sao

Paulo. Such is the fate of a number of results in finitely additive measure theory; relative anonymity.

As mentioned above, the natural domain for a (bounded) finitely additive measure is an algebra of sets and there is a long list of open questions revolving around the validity of results like the Vitali-Hahn-Saks thereom and the Nikodym-Grothendieck theorem for bounded additive measures on an algebra of sets.

We mention but two such problems.

- For which algebras of sets is the Nikodym Convergence theorem so?
- For which algebras of sets is the Nikodym-Grothendieck Boundedness theorem so?

Important examples of non-sigma complete affirmative responses to each of the questions exist; also the final answers to these questions are different. Indeed the algebra $\mathcal{J}[a,b]$ of Jordan measurable subsets of $[a,b]$ enjoys the fruits of the Nikodym-Grothendieck Boundedness theorem but fails the test for Nikodym's Convergence theorem.

## 2. Bounded, additive measures

The goal of this chapter is to identify the dual $C(K)^*$ of the Banach space $C(K)$ of all continuous scalar-valued functions on the compact Hausdorff space $K$.

The first section is devoted to the study of the space $\mathcal{B}(\mathcal{A})$ of bounded scalar-valued functions that are uniform limits of simple functions modeled on a field $\mathcal{A}$ of subsets of a set $\Omega$; more precisely, we determine the dual $\mathcal{B}(\mathcal{A})^*$ of $\mathcal{B}(\mathcal{A})$ to be the space ba$(\mathcal{A})$ of bounded finitely additive scalar-valued measures defined on $\mathcal{S}$ with $\int f \, d\mu$ being the value of a member $\mu \in \mathcal{B}(\mathcal{A})^*$ at an $f \in \mathcal{B}(\mathcal{A})$. The integral $\int f \, d\mu$ is defined and its basic properties discussed. This duality in hand, we study ba$(\mathcal{A})$ more closely, exposing its vector lattice structure.

In the second section we observe that if $K$ is a compact Hausdorff space and $\mathcal{A}$ is the field generated by $K$'s topology, then each member of $C(K)$ belongs to $\mathcal{B}(\mathcal{A})$. This allows us to hone in on $C(K)^*$: after introducing the notion of regularity, we are able to show that the space rba$(\mathcal{A})$ of regular members of ba$(\mathcal{A})$ is, in fact, the dual of $C(K)^*$.

In the third chapter, we present what is usually understood to be the Riesz Theorem. We start with a few words about outer measure and their role in extending measures from fields to the generated $\sigma$-fields. This is followed by the classical Hahn extension theorem to the effect that any countably additive non-negative extended real-valued measure on a field $S$ of subsets of a given set has a countably additive non-negative extension to the $\sigma$-field generated by $\mathcal{A}$; the extension is unique if the original measure is finite. This is followed by a result of Alexandroff which says that if $\mathcal{A}$ is the Borel field of a compact Hausdorff space, and $\mu$ is a member of rba($\mathcal{A}$), then $\mu$ *is* countably additive on $\mathcal{A}$. This is followed by noting that the Hahn extension of a regular, countably additive member of ba($\mathcal{A}$) to the Borel $\sigma$-field is automatically regular as well.

All we do in this chapter can be gleaned from careful picking and choosing in Dunford-Schwartz.

## 3. The dual of $\mathcal{B}(\mathcal{A})$

### 3.1. Introduction

Let $S$ denote a non-empty set. Let $\mathcal{A}$ be an algebra of subsets of $S$ so if $E, F \in \mathcal{A}$ then $E \cup F, E \cap F$ and $E^c \in \mathcal{A}$.

**Let $\mathcal{S}(\mathcal{A})$ denote the linear space of all simple scalar-valued functions modeled on $\mathcal{A}$,** that is, $f \in \mathcal{S}(\mathcal{A})$ whenever $f$ is of the form

$$f(s) = \sum_{k=1}^{n} a_k \chi_{A_k},$$

where $a_1, \ldots, a_n$ are scalars and $A_1, \ldots, A_n \in \mathcal{A}$. Because $\mathcal{A}$ *is* an algebra, if $f \in \mathcal{S}(\mathcal{A})$ then $f$ can be represented in the form

$$f(s) = \sum_{j=1}^{m} b_j \chi_{B_j}, \qquad (1)$$

where $b_1, \ldots, b_m$ are the scalar values of $f$ and $B_1, \ldots B_m$ are pairwise disjoint members of $\mathcal{A}$ with $S = B_1 \cup \cdots \cup B_m$.

For $f \in \mathcal{S}(\mathcal{A})$ we consider the supremum norm of $f$

$$\|f\|_\infty = \sup\{|f(s)| : s \in S\};$$

$(\mathcal{S}(\mathcal{A}), \|\cdot\|_\infty)$ is a normed linear space with the added properties that $f \cdot g \in \mathcal{S}(\mathcal{A})$ and $|f| \in \mathcal{S}(\mathcal{A})$ if $f, g \in \mathcal{S}(\mathcal{A})$. It is also noteworthy that

$$\|f \cdot g\|_\infty \leq \|f\|_\infty \|g\|_\infty \quad \text{and} \quad \| |f| \|_\infty = \|f\|_\infty.$$

**Let $\mathcal{B}(\mathcal{A})$ denote the uniform closure of $\mathcal{S}(\mathcal{A})$ within $l^\infty(S)$**, which is the Banach space of all bounded scalar-valued functions defined on $S$; again $f \cdot g$ and $|f| \in \mathcal{B}(\mathcal{A})$ if $f, g \in \mathcal{B}(\mathcal{A})$ and

$$\|f \cdot g\|_\infty \leq \|f\|_\infty \|g\|_\infty \quad \text{and} \quad \| |f| \|_\infty = \|f\|_\infty.$$

Our first goal is to identify the dual of $\mathcal{B}(\mathcal{A})$. **We will denote by ba$(\mathcal{A})$ the collection of all bounded finitely additive measures defined on $\mathcal{A}$.** Then ba$(\mathcal{A})$ is a linear space. It is easy to equip ba$(\mathcal{A})$ with a Banach-space norm:

$$\|\mu\| = \|\mu\|_\infty = \sup\{|\mu(A)| : A \in \mathcal{A}\};$$

since $l^\infty(\mathcal{A})$ is a Banach space this norm just relies on the easily established fact that ba$(\mathcal{A})$ is a closed (linear) subspace of $l^\infty(\mathcal{A})$. The uniform norm, while it plays a role in the study of ba$(\mathcal{A})$ is *not* the natural norm for this space. Rather, the variation norm is the one best-equipped to serve ba$(\mathcal{A})$. For any $A \in \mathcal{A}$, we define $|\mu|(A)$, where $\mu \in$ ba$(\mathcal{A})$ by

$$|\mu|(A) = \sup\left\{\sum_{B \in \Pi} |\mu(B)|\right\},$$

where $\Pi = \{B_1, \ldots, B_m\}$ is any finite partition of $A$ into (pairwise disjoint) members of $\mathcal{A}$. Here's an easily proved, yet essential, relationship involving the variation.

**Theorem 3.1.** *For any $A \in \mathcal{A}$, and any $\mu \in$ ba$(\mathcal{A})$,*

$$\sup_{B \in \mathcal{A}, B \subseteq A} |\mu(B)| \leq |\mu|(A) \leq 4 \sup_{B \in \mathcal{A}, B \subseteq A} |\mu(B)|.$$

**Proof.** It's plain that

$$\sup_{B \in \mathcal{A}, B \subseteq A} |\mu(B)| \leq |\mu|(A),$$

so we concentrate on the upper estimate. It, too, is reasonably easy.

Let $\Pi$ be a finite partition of $A \in \mathcal{A}$ into disjoint members of $\mathcal{A}$. If $\mu$ is real-valued (and bounded) then $\Pi = \Pi^+ \cup \Pi^-$, where

$$\Pi^+ = \{B \in \Pi : \mu(B) > 0\}$$

and
$$\Pi^- = \{B \in \Pi : \mu(B) \leq 0\};$$
of course, $\Pi^+$ and $\Pi^-$ are disjoint collections and plainly
$$\sum_{B \in \Pi} |\mu(B)| = \sum_{B \in \Pi^+} \mu(B) - \sum_{B \in \Pi^-} \mu(B)$$
$$= \mu\left(\bigcup_{B \in \Pi^+} B\right) - \mu\left(\bigcup_{B \in \Pi^-} B\right) \quad \text{(after all } \mu \text{ is finitely additive)}$$
$$\leq 2 \sup_{B \in \mathcal{A}, B \subseteq A} |\mu(B)|.$$

If $\mu$ is complex-valued then we write
$$\mu(A) = \text{Re } \mu(A) + i \text{ Im } \mu(A)$$
and take note of the fact that both Re $\mu(A)$ and Im $\mu(A)$ are real-valued bounded finitely additive measures of $A \in \mathcal{A}$; apply what we know appropriately and be done with it. □

Of course, a consequence of these deliberations is the fact that
$$||\mu||_1 = |\mu|(S)$$
is also a Banach space norm of ba($\mathcal{A}$) equivalent to $||\mu||_\infty$.

An added bonus:

**Proposition 3.1.** $|\mu| : \mathcal{A} \to [0, \infty)$ *is also member of ba($\mathcal{A}$) when* $\mu \in$ *ba($\mathcal{A}$).*

**Proof.** In fact, if $A, B \in \mathcal{A}$ and $A \cap B = \emptyset$ then any partition $\Pi$ of $A \cup B$ into finitely many pairwise disjoint members of $\mathcal{A}$ leads to partitions $\Pi_A$ and $\Pi_B$ of $A$ and $B$, respectively, into finitely many pairwise disjoint members of $\mathcal{A}$, namely
$$\Pi_A = \{A \cap F : F \in \Pi\}, \quad \Pi_B = \{B \cap F : F \in \Pi\}.$$
It follows that
$$\sum_{F \in \Pi} |\mu(F)| = \sum_{F \in \Pi} |\mu(A \cap F) + \mu(B \cap F)|$$
$$\leq \sum_{F \in \Pi} |\mu(A \cap F)| + |\mu(B \cap F)|$$
$$= \sum_{E \in \Pi_A} |\mu(E)| + \sum_{E \in \Pi_B} |\mu(E)|$$
$$\leq |\mu|(A) + |\mu|(B);$$

so
$$|\mu|(A \cup B) \le |\mu|(A) + |\mu|(B).$$

On the other hand, if $A, B$ are disjoint members of $\mathcal{A}$ and $\epsilon > 0$ is given then we can pick partitions $\Pi_A$ and $\Pi_B$ of $A$ and $B$ respectively, into finitely many pairwise disjoint members of $\mathcal{A}$ in such a way that

$$|\mu|(A) \le \sum_{F \in \Pi_A} |\mu(F)| + \frac{\epsilon}{2},$$

and

$$|\mu|(B) \le \sum_{F \in \Pi_B} |\mu(F)| + \frac{\epsilon}{2}.$$

Since $\Pi = \Pi_A \cup \Pi_B$ is a partition of $A \cup B$ into finitely many pairwise disjoint members of $\mathcal{A}$,

$$\begin{aligned}
|\mu|(A \cup B) &\ge \sum_{F \in \Pi} |\mu(F)| \\
&= \sum_{F \in \Pi_A} |\mu(F)| + \sum_{F \in \Pi_B} |\mu(F)| \\
&\ge |\mu|(A) - \frac{\epsilon}{2} + |\mu|(B) - \frac{\epsilon}{2} \\
&= |\mu|(A) + |\mu|(B) - \epsilon.
\end{aligned}$$

The usual $\epsilon$psilonics take over to let us say

$$|\mu|(A \cup B) \ge |\mu|(A) + |\mu|(B),$$

and $|\mu|$ is additive. $\square$

## 3.2. The general integral and the dual of $\mathcal{B}(\mathcal{A})$

Wherever measures exist, integrals cannot be far behind. And so it is with $\mathcal{B}(\mathcal{A})$ vis-a-vis ba($\mathcal{A}$). We start with $f \in \mathcal{S}(\mathcal{A})$ and $\mu \in$ ba($\mathcal{A}$). Suppose

$$f = \sum_{j=1}^{m} b_j \chi_{B_j}$$

where $b_1, \ldots, b_m$ are the values of $f$ and $B_1, \ldots, B_m$ constitute a partition of $S$ into finitely many pairwise disjoint members of $\mathcal{A}$; define $\int_A f \, d\mu$ for $A \in \mathcal{A}$ by

$$\int_A f \, d\mu = \int_A \sum_{j=1}^{m} b_j \chi_{B_j} \, d\mu = \sum_{j=1}^{m} b_j \mu(A \cap B_j).$$

Here's what's so:

- $\int_A f \, d\mu$ is well-defined.
- $\int_A f \, d\mu$ is additive in $f$ for each $A \in \mathcal{A}, \mu \in \text{ba}(\mathcal{A})$.
- $\int_A f \, d\mu$ is additive in $\mu$ for each $f \in \mathcal{S}(\mathcal{A}), A \in \mathcal{A}$.
- $\int_A f \, d\mu$ is additive in $A$ for each $f \in \mathcal{S}(\mathcal{A}), \mu \in \text{ba}(\mathcal{A})$.
- $\left|\int_A f \, d\mu\right| \leq \int_A |f| \, d|\mu|$ for each $A \in \mathcal{A}, f \in \mathcal{S}(\mathcal{A})$ and $\mu \in \text{ba}(\mathcal{A})$.

The first four are straightforward but examples of the descriptive value of the word 'tedious.' The last is harmless to prove but with the first four in hand is of some importance: let $f = \sum_{j=1}^{m} b_j \chi_{B_j}$ be expressed as in (1), where $b_1, \ldots b_m$ are the (scalar) values of $f$ and $B_1, \ldots, B_m$ are a decomposition of $S$ into pairwise disjoint members of $\mathcal{A}$. Then

$$\left|\int_A f \, d\mu\right| = \left|\sum_{j=1}^{m} b_j \mu(A \cap B_j)\right|$$

$$\leq \sum_{j=1}^{m} |b_j| |\mu(A \cap B_j)|$$

$$\leq \sum_{j=1}^{m} |b_j| |\mu|(A \cap B_j) = \int_A |f| \, d|\mu|.$$

The upshot of all this is that any $\mu \in \text{ba}(\mathcal{A})$ acts as a bounded a linear functional on $\mathcal{S}(\mathcal{A})$, via the action

$$f \in \mathcal{S}(\mathcal{A}) \to \int_S f \, d\mu.$$

It is worth taking note of the fact that if $A \in \mathcal{A}$ then

$$\mu_A : \mathcal{A} \to \mathbb{K}, \quad \mu_A(B) = \mu(A \cap B)$$

is also in $\text{ba}(\mathcal{A})$ and has $\|\mu_A\|_1 = |\mu|(A)$. What's more

$$\int_A f \, d\mu = \int_S f \, d\mu_A,$$

and so

$$\left|\int_A f \, d\mu\right| \leq \|f\|_\infty |\mu|(A).$$

We're ready to 'compute' $\mathcal{B}(\mathcal{A})^*$. We formulate the end result as follows.

**Theorem 3.2 (Hildebrandt, Fichtenholtz/Kantorovitch).** *There is an isometric isomorphic relationship between $\mathrm{ba}(\mathcal{A})$ and $\mathcal{B}(\mathcal{A})^*$; if $x^* \in \mathcal{B}(\mathcal{A})^*$ then $x^*$ determines a unique $\mu \in \mathrm{ba}(\mathcal{A})$ via the formula*

$$\mu(A) = x^*(\chi_A)$$

*which means that for any $f \in \mathcal{B}(\mathcal{A})$ we have*

$$x^*(f) = \int f \, d\mu.$$

*Under this correspondence, $||x^*|| = ||\mu||_1$.*

Henceforth when we say $\mathcal{B}(\mathcal{A})^* = \mathrm{ba}(\mathcal{A})$, we mean to invoke the above theorem and all it portends.

**Proof.** Starting with $\mu \in \mathrm{ba}(\mathcal{A})$, we use our discussion prior to the theorem to define $\int d\mu$, which we will now see, is a bounded linear functional on $\mathcal{B}(\mathcal{A})$ with norm less than or equal to $||\mu||_1$. To see this note that

$$\left| \int f \, d\mu \right| \leq \int |f| \, d|\mu|$$
$$= \sum_{j=1}^{m} |b_j| |\mu|(B_j)$$
$$\leq \sum_{j=1}^{m} \left( \sup_{1 \leq j \leq m} |b_j| \right) |\mu|(B_j)$$
$$\leq \left( \sup_{1 \leq j \leq n} |b_j| \right) \sum_{j=1}^{m} |\mu|(B_j)$$
$$\leq ||f||_\infty |\mu|(S) = ||f||_\infty ||\mu||_1.$$

So we can extend $\int d\mu$ to a bounded linear functional on the uniform closure of $\mathcal{S}(\mathcal{A})$ in $l^\infty(S)$, that is, to $\mathcal{B}(\mathcal{A})$. The resulting bounded linear functional will still be denoted by $\int d\mu$. Since the (unique) bounded linear extension of $\int d\mu$ has the same norm as the original linear functional we have

$$\left| \int f \, d\mu \right| \leq ||f||_\infty ||\mu||_1$$

for all $f \in \mathcal{B}(\mathcal{A})$ and $\mu \in \mathrm{ba}(\mathcal{A})$. Therefore $\int d\mu$ is a bounded linear functional on $\mathcal{B}(\mathcal{A})$ with norm less than or equal to $||\mu||_1$.

Now if we start with an $x^* \in \mathcal{B}(\mathcal{A})^*$ and define $\mu_{x^*}(A) = x^*(\chi_A)$ then $\mu_{x^*}$ is easily seen to be a member of $\mathrm{ba}(\mathcal{A})$ since

- $|\mu_{x^*}(A)| = |x^*(\chi_A)| \leq ||x^*||$, and
- if $A, B \in \mathcal{B}$ with $A \cap B = \emptyset$ then

$$\mu_{x^*}(A \cup B) = x^*(\chi_{A \cup B}) = x^*(\chi_A + \chi_B) = x^*(\chi_A) + x^*(\chi_B) = \mu_{x^*}(A) + \mu_{x^*}(B).$$

We can find a decomposition $S = B_1 \cup \cdots \cup B_m$ of $S$ into pairwise disjoint members of $\mathcal{A}$ so that if $\epsilon > 0$ is preordained then

$$||\mu_{x^*}||_1 = |\mu_{x^*}|(S) \leq \sum_{j=1}^{m} |\mu_{x^*}(B_j)| + \epsilon.$$

Look at $f \in \mathcal{S}(\mathcal{A})$:

$$f := \sum_{j=1}^{m} (\mathrm{sign}(\mu_{x^*}(B_j)))\chi_{B_j},$$

where $|\mathrm{sign}(\mu(B_j))| = 1$ and serves to make

$$(\mathrm{sign}(\mu_{x^*}(B_j))) \cdot \mu_{x^*}(B_j) = |\mu_{x^*(B_j)}|;$$

then

$$x^*(f) = x^* \left( \sum_{j=1}^{n} \mathrm{sign}(\mu_{x^*}(B_j)) \cdot \chi_{B_j} \right)$$

$$= \sum_{j=1}^{n} \mathrm{sign}(\mu_{x^*}(B_j)) \cdot \mu_{x^*}(B_j)$$

$$= \sum_{j=1}^{m} |\mu_{x^*}(B_j)|$$

$$\geq ||\mu||_1 - \epsilon.$$

But $||f||_\infty = 1$. So

$$||x^*|| \geq |x^*(f)| \geq ||\mu_{x^*}||_1 - \epsilon.$$

Since $\epsilon > 0$ while preordained, was arbitrary so

$$||x^*|| \geq ||\mu_{x^*}||_1. \qquad \square$$

To Summarize: Start with $\mu \in \mathrm{ba}(\mathcal{A})$, generate a linear functional $\int d\mu$ on $\mathcal{B}(\mathcal{A})$ of norm less than or equal to $||\mu||_1$.

Start with $x^* \in \mathcal{B}(\mathcal{A})^*$, generate $\mu_{x^*} \in \text{ba}(\mathcal{A})$ by $\mu_{x^*} = x^*(\chi_A)$ and find $||\mu_{x^*}||_1 \leq ||x^*||$. Check that

$$x^*(f) = \int f \, d\mu_{x^*}.$$

The above correspondence is *more* than simply an identification of $\text{ba}(\mathcal{A})$ with the dual of $\mathcal{B}(\mathcal{A})$ as a Banach space. If $\mu \in \text{ba}(\mathcal{A})$ is real valued then the functional $\int d\mu$ takes real values whenever $f \in \mathcal{B}(\mathcal{A})$ is real valued; moreover if $\mu \in \text{ba}(\mathcal{A})$ is a non-negative real valued measure then $\int d\mu$ take non-negative real values whenever it's acting on a non-negative member $f$ of $\mathcal{B}(\mathcal{A})$. It's here where the special structure of $\text{ba}(\mathcal{A})$ plays a crucial role in understanding $\mathcal{B}(\mathcal{A})^*$. We want to investigate $\text{ba}(\mathcal{A})$ a bit more deeply in order to get a better understanding of this duality $\mathcal{B}(\mathcal{A})^* = \text{ba}(\mathcal{A})$.

Earlier we said that if $\mu \in \text{ba}(\mathcal{A})$ then $|\mu| \in \text{ba}(\mathcal{A})$, too, where for $A \in \mathcal{A}$

$$|\mu|(A) = \sup \left\{ \sum_{j=1}^m |\mu(B_j)| \right\},$$

the supremum being taken over all partitions $B_1, \ldots, B_m$ of $A$ into a finite union of members of $\mathcal{A}$. We saw that

$$|\mu(A)| \leq |\mu|(A)$$

for each $A \in \mathcal{A}$; it follows that

$$\mu^+ = \frac{(|\mu| + \mu)}{2}$$

is, if $\mu$ *is real valued*, a non-negative valued member of $\text{ba}(\mathcal{A})$ as is

$$\mu^- = \frac{(|\mu| - \mu)}{2},$$

again, *in the case that $\mu$ is real valued*. Of course, elementary arithmetic tells us that if *if $\mu$ is real valued then*

$$\mu = \mu^+ - \mu^-,$$

where $\mu^+, \mu^-$ are members of $\text{ba}(\mathcal{A})$ that take only *non-negative real values.*

An alternative description of $\mu^+$ (and so of $\mu^-$) is worth our while to contemplate. We assume $\mu \in \mathrm{ba}(\mathcal{A})$ is real valued and look at the map on $\mathcal{A}$ given by
$$\nu(A) = \sup\{\mu(E) : E \in \mathcal{A}, E \subseteq A\}.$$
Since $\mu(\emptyset) = 0$, we have that $\nu(A) \geq 0$ for each $A \in \mathcal{A}$. Let $A, B \in \mathcal{A}$ with $A \cap B = \emptyset$. Then
$$\begin{aligned}\nu(A \cup B) &= \sup\{\mu(E) : E \in \mathcal{A}, E \subseteq A \cup B\} \\ &= \sup\{\mu((E \cap A) \cup (E \cap B)) : E \in \mathcal{A}, E \subseteq A \cup B\} \\ &\leq \sup\{\mu(E) : E \in \mathcal{A}, E \subseteq A\} + \sup\{\mu(E) : E \in \mathcal{A}, E \subseteq B\} \\ &\quad \text{(since } \mu((E \cap A) \cup (E \cap B)) = \mu(E \cap A) + \mu(E \cap B)) \\ &= \nu(A) + \nu(B);\end{aligned}$$
so $\nu$ is subadditive. More is so: let $\epsilon > 0$ be given and chose $E, F \in \mathcal{A}$ so that $E \subseteq A, F \subseteq B$ and
$$\nu(A) < \mu(E) + \frac{\epsilon}{2}, \nu(B) < \mu(F) + \frac{\epsilon}{2};$$
$E, F$ are disjoint and so
$$\begin{aligned}\nu(A) + \nu(B) &< \mu(E) + \frac{\epsilon}{2} + \mu(F) + \frac{\epsilon}{2} \\ &= \mu(E) + \mu(F) + \epsilon \\ &= \mu(E \cup F) + \epsilon \\ &\leq \nu(A \cup B) + \epsilon.\end{aligned}$$
Since $E \subseteq A, F \subseteq B$,
$$E \cup F \subseteq A \cup B,$$
and since $\epsilon > 0$ was arbitrary,
$$\nu(A) + \nu(B) \leq \nu(A \cup B),$$
and $\nu$ is additive. *Therefore*
$$\nu(A) = \sup\{\mu(E) : E \in \mathcal{A}, E \subseteq A\}$$
*defines a non-negative member of* $\mathrm{ba}(\mathcal{A})$ *for which* $\mu(A) \leq \nu(A)$ *for each* $A \in \mathcal{A}$.

How do $\nu$ and $\mu^+$ compare? Well, take $A \in \mathcal{A}$, let $E \in \mathcal{A}$ with $E \subseteq A$. Then
$$\mu(E) = \mu(A) - \mu(A \backslash E) \leq \mu(A) + |\mu(A \backslash E)| \leq \mu(A) + |\mu|(A \backslash E),$$

and since
$$\mu(E) \leq |\mu(E)| \leq |\mu|(E),$$
we have
$$2\mu(E) \leq \mu(A) + |\mu|(A\setminus E) + |\mu|(E)$$
$$= \mu(A) + |\mu|(A) = (\mu + |\mu|)(A).$$
Thus
$$\nu(A) \leq \left(\frac{\mu + |\mu|}{2}\right)(A) = \mu^+(A).$$

On the other hand, if $A \in \mathcal{A}$ and $\epsilon > 0$ is given then we can find a partition $\Pi = \{B_1, \ldots B_m\}$ of $A$ into pairwise disjoint members of $\mathcal{A}$ so that
$$|\mu|(A) \leq \sum_{j=1}^{m} |\mu(B_j)| + \epsilon.$$

Again we split $\Pi$ into $\Pi^+ \cup \Pi^-$ where
$$\Pi^+ = \{B_j \in \Pi : \mu(B_j) > 0\}, \quad \Pi^- = \Pi\setminus\Pi^+.$$

Then
$$|\mu|(A) \leq \sum_{j\in\Pi^+} \mu(B_j) - \sum_{j\in\Pi^-} \mu(B_j) + \epsilon$$
while
$$\mu(A) = \sum_{j\in\Pi^+} \mu(B_j) + \sum_{j\in\Pi^-} \mu(B_j).$$

Add 'em up:
$$(\mu + |\mu|)(A) \leq 2 \sum_{j\in\Pi^+} \mu(B_j) + \epsilon$$
$$= 2\mu\left(\bigcup_{j\in\Pi^+} B_j\right) + \epsilon$$
$$\leq 2\nu(A) + \epsilon.$$

Since $\epsilon > 0$ was arbitrary,
$$(\mu + |\mu|)(A) \leq 2\nu(A),$$
or
$$\mu^+(A) = \frac{\mu + |\mu|}{2}(A) \leq \nu(A).$$

It follows that $\nu = \mu^+$.

If we order the real valued members of ba($\mathcal{A}$) via $\mu \leq \nu$ when $\mu(A) \leq \nu(A)$ for $A \in \mathcal{A}$ then ba($\mathcal{A}$) is a vector lattice. Here $|\mu|$ serves as the 'absolute value' and $\mu^+$ as the positive part, $\mu^-$ as the negative part of any real valued $\mu \in$ ba($\mathcal{A}$).

It is somewhat informative to know that if $\mu, \nu$ are real-valued members of ba($\mathcal{A}$) then the supremum in ba($\mathcal{A}$) of $\mu$ and $\nu$ is far from the pointwise supremum; in fact

$$(\mu \vee \nu)(A) = \sup\{\mu(E) + \nu(A\setminus E) : E \in \mathcal{A}, E \subseteq A\}$$

serves as the supremum of $\mu$ and $\nu$ in ba($\mathcal{A}$). It is a worthwhile exercise to show that $\mu \vee \nu$, as so defined, is in ba($\mathcal{A}$) and is the tightest fit above both $\mu$ and $\nu$ in ba($\mathcal{A}$).

To summarize the above developments, we formulate the following.

**Theorem 3.3.** *ba($\mathcal{A}$) is a Banach lattice with the variation norm $||\mu||_1$ for $\mu \in$ ba($\mathcal{A}$).*

- *If $\mu \in$ ba($\mathcal{A}$) then $|\mu|(A)$ defined as above serves as $|\mu|$.*
- *If $\mu, \nu \in$ ba($\mathcal{A}$) then $\mu \vee \nu$ is given by*

$$\mu \vee \nu(A) = \sup\{\mu(E) + \nu(A \setminus E) : E \in \mathcal{A}, E \subseteq A\}$$

- $|| |\mu| ||_1 = ||\mu||_1$, *as it should.*
- *Each $\mu \in$ ba($\mathcal{A}$) can be written in the form*

$$\mu = \lambda^+ - \lambda^- + i(\nu^+ - \nu^-)$$

*where $\lambda^+, \lambda^-, \nu^+, \nu^- \in$ ba($\mathcal{A}$) with each $\geq 0$; in this case,*

$$||\mu||_1 = ||\lambda^+||_1 + ||\lambda^-||_1 + ||\nu^+||_1 + ||\nu^-||_1.$$

- *A member of ba($\mathcal{A}$) is non-negative on non-negative members of $\mathcal{BA}$ if and only if $\mu(A) \geq 0$ for each $A \in \mathcal{A}$.*

## 4. The dual of $C(K)$, $K$ a compact, Hausdorff space

The development of integration with respect to members of ba($\mathcal{A}$) has a most important application: the description of $C(K)^*$, where $K$ is a compact, Hausdorff space.

We set the stage. We let $K$ be a compact, Hausdorff space, and $\mathcal{A}$ be the **field** of subsets of $K$ generated by the closed subsets of $K$. Then $C(K)$ is the space of all continuous scalar-valued functions defined on $K$, and, as above, $\mathcal{B}(\mathcal{A})$ is the uniform closure of the space of $\mathcal{S}(\mathcal{A})$ of simple scalar-valued functions on $K$ with sets of constancy members of $\mathcal{A}$.

We start by noticing the following basic fact: *each $x \in C(K)$ is integrable with respect to each $\mu \in \text{ba}(\mathcal{A})$*. Indeed, we can cover $x(K)$ by open sets $G_1, \ldots, G_n$, each with diameter less than $\epsilon$. We disjointify:

$$A_1 = G_1, A_2 = G_2 \backslash G_1, \ldots A_k = G_k \backslash (G_1 \cup \cdots G_{k-1}), \ldots$$

If $A_k \neq \emptyset$ then pick $a_k \in A_k$; if $A_k = \emptyset$, let $a_k = 0$. Now $x^{\leftarrow}(G_k)$ is open so it belongs to $\mathcal{A}$; it follows that each $E_k = x^{\leftarrow}(A_k) \in \mathcal{A}$. The function

$$x_\epsilon = \sum_{k=1}^n a_k \chi_{E_k}$$

belongs to $\mathcal{S}(\mathcal{A})$ and satisfies

$$\|x - x_\epsilon\|_\infty \leq \epsilon,$$

so $x \in \mathcal{B}(\mathcal{A})$.

Notice that for any $\mu \in \text{ba}(\mathcal{A})$, $\int f \, d\mu$ acts as a continuous linear functional on $\mathcal{B}(\mathcal{A})$ which contains $C(K)$ as a closed linear subspace. Hence each $\mu \in \text{ba}(\mathcal{A})$ defines a member of $C(K)^*$:

$$f \to \int f \, d\mu.$$

## 4.1. The dual of $C(K)$ is the space of regular bounded measures on $\mathcal{A}$

**Definition 4.1.** We say that $\mu$ is **regular** if given $A \in \mathcal{A}$ and $\epsilon > 0$ there is a closed set $F \subseteq K$ and an open set $U \subseteq K$ so $F \subseteq A \subseteq U$ and $|\mu(E)| < \epsilon$ whenever $E \subseteq U \backslash F$.

**We denote by rba($\mathcal{A}$) the linear space of regular members of ba($\mathcal{A}$).**

Naturally members of rba($\mathcal{A}$) also determine bounded linear functionals on $C(K)$.

**Theorem 4.1.**

$$C(K)^* = rba(\mathcal{A}).$$

We prove this theorem in Propositions 4.1 and 4.2.

**Proposition 4.1.** *If $\mu \in rba(\mathcal{A})$ then $f \to \int f \, d\mu$ is a bounded linear functional $x^*$ on $C(K)$ for which*

$$||x^*||_{C(K)^*} = |\mu|(K).$$

**Proof.** If we let $\epsilon > 0$ be given then there are disjoint members $E_1, \ldots, E_n$ of $\mathcal{A}$ so that

$$|\mu|(K) \leq \sum |\mu(E_k)| + \epsilon.$$

Let $F_k$ be a closed set so that $F_k \subseteq E_k$ and

$$|\mu|(E_k \setminus F_k) \leq \frac{\epsilon}{n}$$

for $k = 1, \ldots, n$. Now let $G_1, \ldots, G_n$ be open sets that are disjoint and satisfy $F_k \subseteq G_k$ for each $k = 1, \ldots, n$; the normality of $K$ allows us to find the $G_k$'s. Because $\mu \in rba(\mathcal{A})$ we can shrink the $G_k$'s so that

$$|\mu|(G_k \setminus F_k) < \frac{\epsilon}{n},$$

as well. Now let $x_1, \ldots, x_n \in C(K), 0 \leq x_n \leq 1$ with $x_k = 0$ outside of $G_k$ and $x_k = 1$ on $F_k$. Let $a_1, \ldots, a_n \in \mathbb{K}$ satisfy $|a_k| = 1$ and $a_k \mu(E_k) = |\mu(E_k)|$. Put

$$x_0 = \sum_{k=1}^{n} a_k x_k.$$

Note

$$|x^*(x_0) - |\mu|(K)| \leq 3\epsilon.$$

What's more, $||x_0|| = 1$. It follows that

$$||x^*|| = \sup_{x \in B_{C(K)}} |x^*(x)| = |\mu|(K). \qquad \square$$

Our next task is to show that this covers all $x^*$'s in $C(K)^*$. Extend $x^*$ to a member of $\mathcal{B}(\mathcal{A})^*$, still denoted by $x^*$, with the norm of the extension that of $x^*$. We've seen already in Theorem 3.2 that there is a $\lambda \in ba(\mathcal{A})$ such that for any $f \in \mathcal{B}(\mathcal{A})$,

$$x^*(f) = \int f \, d\lambda.$$

**Proposition 4.2.** *Given any $\lambda \in ba(\mathcal{A})$ there is a $\mu \in rba(\mathcal{A})$ such that*

$$\int x \, d\lambda = \int x \, d\mu$$

for each $x \in C(K)$.

Since
$$x^*(x) = \int x\, d\lambda \left( = \int x\, d\mu \right)$$
for each $x \in C(K)$, this will finish the proof of Theorem 4.1 that
$$C(K)^* = \mathrm{rba}(\mathcal{A}).$$

Before we continue with the proof of the proposition, we define

**Definition 4.2.** let $\lambda$ be a real-valued function defined on a field $\mathcal{A}$ of subsets of a set $S$; suppose that $\lambda(\emptyset) = 0$. A member $E$ of $\mathcal{A}$ is called a $\lambda$–**set** if for each $A \in \mathcal{A}$,
$$\lambda(A) = \lambda\left(A \bigcap E\right) + \lambda\left(A \bigcap E^c\right) = \lambda\left(A \bigcap E\right) + \lambda(A\backslash E).$$

**Theorem 4.2.** *If $\lambda : \mathcal{A} \to \mathbb{R}$, $\lambda(\emptyset) = 0$ then the family of $\lambda$–sets is a subfield of $\mathcal{A}$ on which $\lambda$ is finitely additive. Moreover if $E_1, \ldots, E_n$ are pairwise disjoint $\lambda$–subsets of $S$ then for each $A \in \mathcal{A}$*
$$\lambda\left(A \bigcap \left(\bigcup_{k=1}^{n} E_k\right)\right) = \sum_{k=1}^{n} \lambda\left(A \bigcap E_k\right).$$

**Proof.** Plainly, $\emptyset, S$ are $\lambda$–sets. Also $E^c$ is a $\lambda$–set if (and only if) $E$ is. Suppose $E, F$ are $\lambda$–sets. Take $A \in \mathcal{A}$. Since $E$ is a $\lambda$–set
$$\lambda(A \cap F) = \lambda(A \cap F \cap E) + \lambda((A \cap F)\backslash E). \tag{2}$$

Since $F$ is a $\lambda$–set
$$\lambda(A) = \lambda(A \cap F) + \lambda(A \backslash F). \tag{3}$$

Also
$$\lambda(A \cap (E \cap F)^c) = \lambda(A \cap (E \cap F)^c \cap F) + \lambda(A \cap (E \cap F)^c \cap F^c).$$

But
$$(E \cap F)^c \cap F = E^c \cap F, \text{ and } (E \cap F)^c \cap F^c = F^c$$

so
$$\lambda(A \cap (E \cap F)^c)) = \lambda(A \cap F \cap E^c) + \lambda(A \cap F^c). \tag{4}$$

It follows that

$$\lambda(A) = \lambda(A \cap F) + \lambda(A \cap F^c) \text{ (by (3))}$$
$$= \lambda(A \cap (E \cap F)) + \lambda(A \cap F \setminus E) + \lambda(A \cap F^c) \text{ (by (2))}$$
$$= \lambda(A \cap (E \cap F)) + \lambda(A \cap (E \cap F)^c) \text{ (by (4))}.$$

From this we see that $E \cap F$ is also a $\lambda$–set. This shows that the collection of $\lambda$–sets is a subfield of $\mathcal{A}$.

Now suppose $E$ and $F$ are disjoint $\lambda$–sets and $A \in \mathcal{A}$. Since $E$ is a $\lambda$–set
$$\lambda(A \cap (E \cap F)) = \lambda(A \cap (E \cup F) \cap E) + \lambda(A \cap (E \cup F) \cap E^c) = \lambda(A \cap E) + \lambda(A \cap F)$$
because $E, F$ are *disjoint*. □

**Proof.** (of Proposition 4.2) The Jordan decomposition of real valued members of ba($\mathcal{A}$) tells us that $\lambda$ is of the form
$$\lambda = \lambda_1 - \lambda_2 + i\lambda_3 - i\lambda_4$$
where $\lambda_1, \lambda_2, \lambda_3, \lambda_4$ are non-negative real valued members of ba($\mathcal{A}$); hence, we need only show that if $\lambda \in$ ba($\mathcal{A}$) is non-negative real valued then there is a $\mu \in$ rba($\mathcal{A}$) such that
$$\int x \, d\lambda = \int x \, d\mu$$
for each $x \in C(K)$.

Generically: $F$ is a closed subset of $K$, $G$ is open subset of $K$, and $E$ is any subset of $K$. Set
$$\mu_1(F) = \inf\{\lambda(G) : F \subseteq G\}, \quad \mu_2(E) = \sup\{\mu_1(F) : F \subseteq E\}.$$
Both $\mu_1, \mu_2$ take non-negative values and each is ascending on its domain.

Suppose $F_1 \setminus G_1 \subseteq G$. Then $F_1 \subseteq G_1 \cup G$ and so
$$\lambda(G_1 \cup G) \leq \lambda(G_1) + \lambda(G)$$
implies
$$\mu_1(F_1) \leq \lambda(G_1) + \lambda(G).$$
Therefore
$$\mu_1(F_1) \leq \lambda(G_1) + \mu(F_1 \setminus G_1),$$

after all, $F_1\backslash G_1 \subseteq G$, a closed set on $K$. Since $G_1$ is any open set, we let $G_1$ range over the open sets that contain a given $F \cap F_1$ and we see

$$\mu_1(F_1) \leq \mu_1(F \cap F_1) + \mu_1(F_1\backslash G_1).$$

But $F_1\backslash F = F_1\backslash(F_1 \cap F)$ which contains $F_1\backslash G_1$ so

$$\mu_1(F_1) \leq \mu_1(F \cap F_1) + \mu_2(F_1\backslash F).$$

If we now let $E \subseteq K$ and $F_1$ range over the closed sets contained in $E$ the result is

$$\mu_2(E) \leq \mu_2(F \cap E) + \mu_2(E\backslash F), \tag{5}$$

where $E$ and $F$ are unrestricted within their generic classes.

We plan to show the reverse inequality stated in (5) holds as well.

Suppose $F_1$ and $F_2$ are *disjoint* closed sets; envelop $F_1$ and $F_2$ in *disjoint* open sets $G_1$ and $G_2$ respectively. Let $G$ contain $F_1 \cup F_2$. Then

$$\lambda(G) \geq \lambda(G \cap G_1) + \lambda(G \cap G_2) \geq \mu_1(F_1) + \mu_1(F_2);$$

hence

$$\mu_1(F_1 \cup F_2) \geq \mu(F_1) + \mu(F_2).$$

Now look to generic $E, F$ and let $F_1$ range over the closed subsets of $E \cap F$ and $F_2$ range over the closed sets contained in $E\backslash F$. The result:

$$\mu_2(E) \geq \mu_1(F_1 \cup F_2) \geq \mu_2(E \cap F) + \mu_2(E\backslash F)$$

and so

$$\mu_2(E) = \mu_2(E \cap F) + \mu_2(E\backslash F). \tag{6}$$

Look familiar? It should. Were $\mu_2$ an outer measure, then (6) would say that each closed set $F$ is $\mu_2$−measurable.

Now $\mu_2$ is *not* necessarily an outer measure but it does fit within the above framework. It follows from this that $\mu_2$ is defined on $2^K$ and each closed subset of $K$ is a $\mu_2$ set. Let $\mu = \mu|_{\mathcal{A}}$. Then $\mu \in \text{ba}(\mathcal{A})$. More is so: $\mu \in \text{rba}(\mathcal{A})$.

Indeed for any closed set $F \subseteq K$,

$$\mu_1(F) = \mu_2(F) = \mu(F),$$

as a careful inspection of the respective definitions will clearly expose.

From Theorem 4.2 of our study of ba($\mathcal{A}$) we can glean the following: if $A \in \mathcal{A}, F \subseteq A \subseteq G$ (generically) then

$$\sup\{\mu(B) : B \in \mathcal{A}, B \subseteq A\} \leq |\mu|(G \backslash F) \leq 4\sup\{\mu(B) : B \in \mathcal{A}, B \subseteq A\}.$$

So to show that $\mu$ is regular it is enough to realize that by $\mu_2$'s very definition (and the coincidence of $\mu$ with $\mu_2, \mu_1$ for closed sets $F \subseteq K$) that for any $A \in \mathcal{A}$

$$\mu(A) = \sup_{F' \subseteq A} \mu(F');$$

so $\mu$ is "inner regular." But we're inside a compact Hausdorff space $K$ and

$$\lambda(K) = \mu_1(K) = \mu_2(K) = \mu(K) < \infty.$$

Hence

$$\begin{aligned}
\mu(A) &= \mu(K) - \mu(A^c) \\
&= \mu(K) - \sup_{F \subseteq A^c} \mu(F) \\
&= \mu(K) + \inf_{F \subseteq A^c} \{\mu(F^c) - \mu(K)\} \\
&= \mu(K) - \mu(K) + \inf_{F \subseteq A^c} \mu(F^c) \\
&= \inf_{F \subseteq A^c} \mu(F^c) = \sup_{A \subseteq G} \mu(G),
\end{aligned}$$

and $\mu$ is "outer regular" too. Therefore $\mu \in \text{rba}(\mathcal{A})$.

Finally it remains to show that

$$\int x \, d\lambda = \int x \, d\mu$$

for each and every $x \in C(K)$. But again we need only do this for non-negative real-valued $x$'s. Since each $x \in C(K)$ is bounded, we'll turn our attention to $x \in C(K)$ so that $0 \leq x \leq 1$.

Let $\epsilon > 0$. Partition $K$ into *disjoint* members $E_1, \ldots, E_n$ of $\mathcal{A}$ in such a way that

$$\int x \, d\mu \leq \sum_{j=1}^{n} \left( \inf_{k \in E_j} f(k) \right) \mu(E_j) + \epsilon.$$

Since $\mu$ is regular, we can find closed sets $F_1, \ldots, F_n$ with $F_j \subseteq E_j$ such that

$$\int x \, d\mu \leq \sum_{j=1}^{n} \left( \inf_{k \in E_j} f(k) \right) \mu(F_j) + 2\epsilon.$$

But $K$ is normal and $x \in C(K)$ so we can envelop the $F_j$'s within *disjoint* open sets $G_j$ where $F_j \subseteq G_j$ for $j = 1, \ldots, n$ with

$$\inf_{k \in G_j} x(k) \geq \inf_{k \in E_j} x(k) - \frac{\epsilon}{n(\mu(K))};$$

this being done,

$$\int x \, d\mu \leq \sum_{j=1}^{n} \left( \inf_{k \in G_j} x(k) \right) \mu(G_j) + 3\epsilon.$$

But remember that

$$\mu_1(F) = \mu_2(F) = \mu(F)$$

for closed sets $F \subseteq K$, and if $G$ is open with $F \subseteq G$ then $\mu(F) \leq \lambda(G)$. It follows that $\mu(G) \leq \lambda(G)$ and

$$\sum_{j=1}^{n} \left( \inf_{k \in G_j} x(k) \right) \mu(G_j) \leq \sum_{j=1}^{n} \left( \inf_{k \in G_j} x(k) \right) \lambda(G_j) \leq \int x \, d\lambda.$$

So

$$\int x \, d\mu \leq \sum_{j=1}^{n} \left( \inf_{k \in G_j} x(k) \right) \mu(G_j) + 3\epsilon$$
$$\leq \sum_{j=1}^{n} \left( \inf_{k \in G_j} x(k) \right) \lambda(G_j) + 3\epsilon$$
$$\leq \int x \, d\lambda + 3\epsilon.$$

Since $\epsilon > 0$ was arbitrary, we have that for any $x \in C(K)$, with $0 \leq x \leq 1$,

$$\int x \, d\mu \leq \int x \, d\lambda.$$

If we apply this to $(1 - x)$ and keep in mind that

$$\mu(K) = \lambda(K)$$

then the reverse inequality follows from

$$\int (1 - x) d\mu \leq \int (1 - x) d\lambda,$$

and that is all she wrote! □

## 4.2. The dual of $C(K)$ is the space of regular, countably additive measures on the $\sigma$-field of Borel subsets of $K$

**Definition 4.3.** By an **outer measure** on $S$, we mean a non-negative extended real valued set function $\lambda$ defined on some $\sigma$-field $\Sigma$ of subsets of $S$ satisfying

(1) $\lambda(\emptyset) = 0$
(2) $\lambda(A) \leq \lambda(B)$ whenever $A \subseteq B$
(3) $\lambda\left(\bigcup_n E_n\right) \leq \sum_n \lambda(E_n)$ for any sequence $(E_n)$ of members of $\Sigma$.

Let $\mu \in \text{ba}(\mathcal{A})$; we say that $\mu$ is countably additive on $\mathcal{A}$ if given a sequence $(E_n)$ of pairwise disjoint members of $\mathcal{A}$ with $\bigcup_n E_n \in \mathcal{A}$ we have that

$$\mu\left(\bigcup_n E_n\right) = \sum_n \mu(E_n).$$

The fundamental result regarding outer measure is the following

**Theorem 4.3 (Caratheodory).** *If $\lambda$ is an outer measure then the family of $\lambda$-sets is a $\sigma$-field; $\lambda$ is countably additive thereupon.*

**Proof.** Recall that $E \in \Sigma$ is a $\lambda$-set if for each $A \in \Sigma$,

$$\lambda(A) = \lambda(A \cap E) + \lambda(A \cap E^c).$$

We've seen already that the collection of $\lambda$-sets is a subfield of $\Sigma$; to show that this collection is, in fact, a sub-$\sigma$-field, it is enough to show that if $E$ is the union of a sequence $(E_n)$ of pairwise disjoint $\lambda$-sets then $E$ is a $\lambda$-set.

We saw earlier that if $A \in \Sigma$ then for each $k$

$$\begin{aligned}\lambda(A) &= \lambda\left(A \cap \left(\bigcup_{j=1}^k E_j\right)\right) + \lambda\left(A \cap \left(\bigcup_{j=1}^k E_j\right)^c\right) \\ &= \sum_{j=1}^k \lambda(A \cap E_j) + \lambda\left(A \cap \left(\bigcup_{j=1}^n E_j\right)^c\right) \\ &\geq \sum_{j=1}^k \lambda(A \cap E_j) + \lambda\left(A \cap \left(\bigcup_n E_n\right)^c\right).\end{aligned}$$

From this we see that
$$\lambda(A \cap E) + \lambda(A \cap E^c) \geq \lambda(A)$$
$$\geq \sum_n \lambda(A \cap E_n) + \lambda(A \cap E^c)$$
$$\geq \lambda(A \cap E) + \lambda(A \cap E^c)$$
and $E$ is indeed a $\lambda$-set. If we replace $A$ by $A \cap E$ then
$$\lambda(A \cap E) = \sum_n \lambda(A \cap E_n)$$
results. □

**Lemma 4.1 (Generating outer measures from measures on fields).** *let $\mu$ be a non-negative, countably additive extended real valued function defined on a field $\mathcal{A}$ of subsets of a set $S$. For any $A \subseteq S$ define*
$$\mu^*(A) = \inf \left\{ \sum_n \mu(E_n) \right\}$$
*where the infimum is taken over all sequences $(E_n)$ of members of $\mathcal{A}$ such that $A \subseteq \bigcup_n E_n$. Then $\mu^*$ is an outer measure and on each set of $\mathcal{A}$ is a $\mu^*$-set. Further*
$$\mu^*(A) = \mu(A)$$
*for any $A \in \mathcal{A}$.*

**Proof.** Insofar as $\mu^*$ being an outer measure is concerned, only $\mu^*$'s countable subadditivity needs explication. To this end, let $(E_n)$ be a sequence of subsets of $S$ and suppose $E = \bigcup_n E_n$. Let $\epsilon > 0$. For each $n$, let $(E_{n,m})_m$ be a sequence chosen from $\mathcal{A}$ in such a judicious fashion that
$$E_n \subseteq \bigcup_m E_{n,m}, \quad \sum_m \mu(E_{n,m}) \leq \mu^*(E_n) + \frac{\epsilon}{2^{n+1}}.$$
Of course
$$E \subseteq \bigcup_{n,m} E_{n,m}$$
and so
$$\mu^*(E) \leq \sum_{n,m} \mu(E_{n,m}) \leq \sum_n \mu^*(E_n) + \epsilon.$$
This establishes the countable subadditivity of $\mu^*$. So $\mu^*$ is an outer measure defined on the $\sigma$-field of *all* subsets of $S$.

If $E \in \mathcal{A}$ then $E \subseteq E$ implies that $\mu^*(E) \leq \mu(E)$. On the other hand, if $E \in \mathcal{A}$ and $E \subseteq \bigcup_n E_n$ where each $E_n \in \mathcal{A}$, then the disjointification of $E_n$:

$$D_1 = E_1, D_2 = E_2 \setminus E_1, \ldots, D_j = E_j \setminus \bigcup_{k=1}^{j-1} E_k, \ldots,$$

is a disjoint sequence of members of $\mathcal{A}$ with $\bigcup_n E_n = \bigcup_n D_n$. It follows that

$$\mu(E) = \mu\left(E \cap \left(\bigcup_n D_n\right)\right)$$
$$= \mu\left(\bigcup_n (E \cap D_n)\right)$$
$$= \sum_n \mu(E \cap D_n) \quad (\mu \text{ is countably additive on } \mathcal{A})$$
$$\leq \sum_n \mu(D_n)$$
$$\leq \sum_n \mu(E_n).$$

It follows that $\mu(E) \leq \mu^*(E)$ too. So if $E \in \mathcal{A}$ then $\mu(E) = \mu^*(E)$.

Finally we show that each $E \in \mathcal{A}$ is a $\mu^*$-set. We need to test vis-a-vis arbitrary $A \subseteq S$. Because $\mu^*$ is an outer measure and $A = (A \cap E) \cup (A \cap E^c)$,

$$\mu^*(A \cap E) + \mu^*(A \cap E^c) \leq \mu^*(A).$$

With an eye on proving the reverse inequality, let $\epsilon > 0$ be given. Find $(E_n) \subseteq \mathcal{A}$ so that

$$A \subseteq \bigcup_n E_n, \text{ and } \sum_n \mu(E_n) \leq \mu^*(A) + \epsilon.$$

Now

$$A \cap E \subseteq \bigcup_n (E_n \cap E) \text{ and } A \cap E^c \subseteq \bigcup_n (E^c \cap E_n),$$

so

$$\mu^*(A \cap E) + \mu^*(A \cap E^c) \leq \sum_n \mu(E_n \cap E) + \sum_n (E_n \cap E^c)$$
$$= \sum_n \mu(E_n \cap E) + \mu(E_n \cap E^c)$$
$$= \sum_n \mu(E_n)$$
$$\leq \mu^*(A) + \epsilon.$$

Since $\epsilon > 0$ was unspecified with regards to size, it follows that

$$\mu^*(A \cap E) + \mu^*(A \cap E^c) \leq \mu^*(A).$$

In other words, each $E \in \mathcal{A}$ is a $\mu^*$-set. □

### Theorem 4.4 (Hahn's extension).

*Every countably additive non-negative extended real valued measure on a field $\mathcal{A}$ has a countably additive non-negative extension to the $\sigma$-field generated by $\mathcal{A}$. If $\mu$ is finite then this extension is unique.*

**Proof.** By Lemma 4.1, $\mu$ determines an outer measure $\mu^*$ whose value at a given subset of the universe of $\mathcal{A}$ is given by

$$\mu^*(A) = \inf \sum_n \mu(E_n),$$

where the infimum is taken over all sequences $(E_n)$ of members of $\mathcal{A}$ such that $A \subseteq \bigcup_n E_n$. If $\Sigma$ is the $\sigma$-field of $\mu^*$-sets then $\mu^*$ is countably additive, non-negative on $\Sigma$; what's more, $\Sigma$ contains the $\sigma$-field generated by $\mathcal{A}$ and for $A \in \mathcal{A}$, $\mu^*(A) = \mu(A)$. All this we proved when we saw how to generate an outer measure from $\mu$. So $\mu^*$ restricted to the $\sigma$-field generated by $\mathcal{A}$ is the extension of $\mu$ that we want.

Suppose $\mu$ is finite. Suppose $\mu_2$ is a non-negative countably additive extension of the measure $\mu$ (as defined on $\mathcal{A}$) to the $\sigma$-field generated by $\mathcal{A}$. Take a member $E$ that is in the generated $\sigma$-field. Cover $E$ by a sequence $(E_n)$ of members of $\mathcal{A}$: $E \subseteq \bigcup_n E_n$. Then

$$\mu_2(E) \leq \mu_2 \left( \bigcup_n E_n \right) \leq \sum_n \mu_2(E_n) = \sum_n \mu(E_n).$$

It follows that

$$\mu_2(E) \leq \mu^*(E).$$

Also
$$\mu_2(E^c) \leq \mu^*(E^c).$$
But
$$\mu^*(S) = \mu_2(S) < \infty$$
so
$$\mu_2(E^c) = \mu(S) - \mu(E) \leq \mu^*(E^c) = \mu(S) - \mu^*(E)$$
and equality prevails. □

We denote by ca($\mathcal{A}$) the collection of all countably additive measures defined on $\mathcal{A}$.

**Corollary 4.1.** *Every bounded, scalar-valued countably additive measure on a field $\mathcal{A}$ (of subsets of a set $S$) has a unique scalar-valued, countably additive extension to the $\sigma$–field $\Sigma$ generated by $\mathcal{A}$.*

**Proof.** Write $\mu(A) = \operatorname{Re}\mu(A) + i\operatorname{Im}\mu(A)$. Then $\operatorname{Re}\mu(A), \operatorname{Im}\mu(A) \in \operatorname{ca}(\mathcal{A})$.

The Jordan decomposition theorem says that there are non-negative members of $\rho_1, \rho_2, i_1, i_2 \in \operatorname{ca}(\mathcal{A})$ so that
$$\operatorname{Re}\mu = \rho_1 - \rho_2, \quad \operatorname{Im}\mu = i_1 - i_2.$$
Each of $\rho_1, \rho_2, i_1,$ and $i_2$ are prey for Hahn's extension theorem. The result is the existence of *unique* $\nu_1, \nu_2, \eta_1$ and $\eta_2 \in \operatorname{ca}(\Sigma)$ such that each of $\nu_1, \nu_2, \eta_1, \eta_2$ are non-negative real valued and $\nu_1|_\mathcal{A} = \rho_1, \nu_2|_\mathcal{A} = \rho_2, \eta_1|_\mathcal{A} = i_1, \eta_2|_\mathcal{A} = i_2$. The extension
$$\nu_1 - \nu_2 + i(\eta_1 - \eta_2)$$
is it! □

**Theorem 4.5 (Alexandroff).** *Let $\mu$ be a regular bounded additive scalar-valued measure defined on the Borel field $\mathcal{A}$ (the field generated by the open sets) of the compact Hausdorff space $K$. Then $\mu$ is countably additive on $\mathcal{A}$.*

**Proof.** Let $\epsilon > 0$. Suppose $(E_n)$ is a sequence of pairwise disjoint members of $\mathcal{A}$ with $E = \bigcup_n E_n \in \mathcal{A}$. There is an $F \in \mathcal{A}$, $F$ closed, $F \subseteq E$, with $|\mu|(E \setminus F) < \epsilon$.

Moreover for each $n$ we can find a $G_n$ that is open with $E_n \subseteq G_n$ and
$$|\mu|(G_n \setminus E_n) < \frac{\epsilon}{2^n}.$$

Now
$$F \subseteq E = \bigcup_n E_n \subseteq G_n;$$

so $K$'s compactness, which is inherited by the closed set $F$, ensures us that there is an $N$ so
$$F \subseteq G_1 \cup \cdots \cup G_N.$$

It follows that
$$\sum_n |\mu|(E_n) \geq \sum_n |\mu|(G_n) - \epsilon$$
$$\geq \sum_{m \leq N} |\mu|(G_n) - \epsilon$$
$$\geq |\mu|(F) - \epsilon$$
$$\geq |\mu|(E) - 2\epsilon.$$

So $\epsilon > 0$ being arbitrary soon reveals that
$$\sum_n |\mu|(E_n) \geq |\mu|(E) = |\mu|\left(\bigcup_n E_n\right).$$

On the other hand, for any $M \in \mathbb{N}$,
$$|\mu|(E) = |\mu|\left(\bigcup_n E_n\right) \geq |\mu|\left(\bigcup_{n \leq M} E_n\right) = \sum_{n \leq M} |\mu|(E_n),$$

so
$$|\mu|\left(\bigcup_n (E_n)\right) \geq \sum_n |\mu|(E_n),$$

too. So $|\mu| \in \text{rca}(\mathcal{A})$ and with it, $\mu \in \text{rca}(\mathcal{A})$ as well. $\square$

**Theorem 4.6.** *Let $\mu$ be a regular bounded additive non-negative real valued measure defined on the Borel field of subsets of the compact Hausdorff space $K$. Then $\mu$ has a unique regular countably additive non-negative real valued extension to the Borel $\sigma$–field $\mathcal{B}o(K)$.*

**Proof.** By Alexandroff's Theorem, $\mu$ is countably additive on $\mathcal{A}$. If we define $\mu^*(A)$ for $A \subseteq K$ by

$$\mu^*(A) = \inf\left\{\sum_n \mu(E_n)\right\},$$

where the infimum is taken over all sequences $(E_n)$ of members of $\mathcal{A}$ such that $A \subseteq \bigcup_n E_n$ then $\mu^*$ is an outer measure. Moreover each member of $\mathcal{A}$ is $\mu^*$–measurable and if $E \in \mathcal{A}$ then $\mu^*(E) = \mu(E)$. What's more, the Hahn Extension Theorem says that $\mu^*$ is countably additive on the $\sigma$–field $\mathcal{B}o(K)$ generated by $\mathcal{A}$, that is, the $\sigma$–field of Borel subsets of $K$. In particular, if $E \in \mathcal{B}o(K)$ and $\epsilon > 0$ then we can find $(E_n) \subseteq \mathcal{A}$ so

$$E \subseteq \bigcup_n E_n, \text{ and } \mu^*\left(\bigcup_n E_n \backslash E\right) < \frac{\epsilon}{2}.$$

But $\mu$ is regular so for each $n$ there is an open set $G_n$ so that

$$E_n \subseteq G_n \text{ and } \mu(G_n \backslash E_n) < \frac{\epsilon}{2^{n+1}}.$$

Now $E \subseteq \bigcup_n E_n \subseteq G_n = G$, an open set. Also

$$\mu^*(G\backslash E) \leq \mu^*(G\backslash \bigcup_n E_n) + \mu^*\left(\bigcup_n E_n \backslash E\right)$$

$$= \mu^*\left(\bigcup_n G_n \backslash \bigcup_n E_n\right) + \mu^*\left(\bigcup_n E_n \backslash E\right)$$

$$\leq \mu^*\left(\bigcup_n (G_n \backslash E_n)\right) + \mu^*\left(\bigcup_n E_n \backslash E\right)$$

$$\leq \sum_n \mu^*(G_n \backslash E_n) + \mu^*\left(\bigcup_n E_n \backslash E\right) < \frac{\epsilon}{2} + \frac{\epsilon}{2} = \epsilon.$$

$\mu^*$ is an outer measure on $\mathcal{B}o(K)$; the fact that $\mu^*$ is inner regular follows from this. Uniqueness is gratis by Hahn's Theorem. □

**Addendum** We point out that there are important finitely additive measures that are *not* countably additive. Probably the most famous of these examples are provided by 'Banach limits.' Recall the theorem of Banach:

**Theorem 4.7.** *There exists a linear functional LIM on $B(\mathbb{N})$ such that*

$$\liminf_n x_n \leq LIM(x) \leq \limsup_n x_n$$

for all $x = (x_n) \in B(\mathbb{N})$; if $T : B(\mathbb{N}) \to B(\mathbb{N})$ is the map $Tx = (x_2, x_3, \ldots, x_{n+1}, \ldots)$ where $x = (x_1, \ldots, x_n, \ldots)$ then

$$LIM(x) = LIM(Tx),$$

as well.

Such a functional determines, thanks to the duality $B(\mathbb{N})^* = \text{ba}(2^{\mathbb{N}})$, a member $L \in \text{ba}^+(2^{\mathbb{N}})$. It is plain that $L(\{n\}) = 0$ for each $n \in \mathbb{N}$ and $L(\mathbb{N}) = 1$. Also $L$ is *not* countably additive.

**Remark 4.1.** There are actually lots of LIMs! In fact there are $2^c$ of them.

## 5. Finitely Additive Measures and Geometry

The Problem of Measure: Can one assign to each bounded set $E \subseteq \mathbb{R}^n$ a non-negative measure of $E$, $\mu(E)$, in such a way that

- $\mu([0,1]^n) = 1$,
- $\mu(A) = \mu(B)$, whenever $A$ and $B$ are congruent, and
- $\mu(E_1 \cup \ldots \cup E_n) = \mu(E_1) + \cdots \mu(E_n)$ whenever $E_1, \ldots, E_n$ are pairwise disjoint bounded subsets of $\mathbb{R}^n$?

The answer to this question is yes if $n = 1, 2$ but no for $n \geq 3$.

We will discuss this problem from a modern viewpoint (albeit with somewhat old-fashioned sensitivities). To do this we first will introduce the notion of an amenable group.

**Definition 5.1.** A group $G$ is **amenable** if there is a finitely additive probability $\mu$ on $G$ (actually on $2^G$) that's left invariant, that is, for every $x \in G$ and for every $E \subseteq G$,

$$\mu(xE) = \mu(E).$$

Denote, as usual, by $B(G)$ the Banach space of all bounded real-valued functions on $G$.

**Note.** If $\mu$ is a left invariant finitely additive probability on $G$ then $\int d\mu$ is a well-defined functional on $B(G)$ that satisfies

- $\int f \, d\mu$ is linear in $f$;
- $\int f \, d\mu \geq 0$ if $f \geq 0$;
- $\int 1 \, d\mu = 1$;
- if $x \in G$ and $_x f(g) := f(xg)$ then $\int {_g f} \, d\mu = \int f \, d\mu$.

Consequently

$$\inf\{f(x) : x \in G\} \leq \int f \, d\mu \leq \sup\{f(x) : x \in G\}.$$

**Theorem 5.1.** *If $G$ is amenable then there is a biinvariant finitely additive probability on $G$.*

**Proof.** Let $\mu$ be a left invariant finitely additive probability on $G$; $\mu$'s existence comes from the amenability of $G$. Let $\mu_0$ be the right invariant finitely additive probability

$$\mu_0(E) := \mu(E^{-1}).$$

For any $A \subseteq G$, let

$$f_A(x) := \mu(Ax^{-1}).$$

Then $f_A \in B(G)$ and $|f_A(x)| \leq 1$ for all $x$. Look at $\nu$ defined as

$$\nu(A) := \int f_A \, d\mu_0.$$

Then

$$\nu(G) = \int f_G \, d\mu_0 = \int 1 \, d\mu_0 = 1,$$

so $\nu$ is a probability. (It's plain that $\nu$ is finitely additive and $\nu(A) \geq 0$ for each $A$.) Moreover, for any $x \in G$, and for any $A \subseteq G$

$$f_{xA}(g) = \mu(xAg^{-1}) = \mu(Ag^{-1}) = f_A(g),$$

so $\nu$ is left invariant. Furthermore, denoting by $(f_x)_g$ the function $f(g) = f(gx^{-1})$, we see

$$(f_A)_x(g) = f_A(gx^{-1}) = \mu(A(gx^{-1})^{-1}) = \mu(Axg^{-1}) = f_{Ax}(g);$$

since $\mu_0$ is right invariant,

$$\nu((f_A)_x) = \int (f_A)_x(g) \, d\mu_0(g) = \int f_A(g) \, d\mu_0(g) = \nu(f_A),$$

$\nu$ is also right invariant. □

Amenability is enjoyed by many, but not all, groups. Here's a theorem that tells much of the story.

**Theorem 5.2.** *The following classes consist entirely of amenable groups.*

*(1) All finite groups.*

(2) All abelian groups.
(3) Any subgroup of an amenable group.
(4) $G/N$ if $G$ is amenable and $N$ is an invariant (i.e., normal) subgroup
(5) $G$, if $G$ has an invariant subgroup $N$ such that $N$ and $G/N$ are amenable.
(6) Any direct union of a directed family of amenable groups.

We delay the proof of this theorem until later. Rather we prove the solvability of the groups $G_1$ and $G_2$ of isometries of $\mathbb{R}$ and $\mathbb{R}^2$, then discuss the situation for $\mathbb{R}^n$ for $n \geq 3$.

The group of isometries of $\mathbb{R}^n$ are of particular interest: denote by $G_n$ the group of all isometries of $\mathbb{R}^n$ onto itself. In this section 'We' will show that $G_1$ and $G_2$ are solvable and so they are amenable. It follows that in either case, $\mathbb{R}$ or $\mathbb{R}^2$, there is an invariant finitely additive probability defined on all subsets.

The proof is, in fact, a simple consequence of lining up well-known geometric/algebraic features of life in $\mathbb{R}$ and $\mathbb{R}^2$ in suitable order. Some notation is, of course, necessary to lay the proper ground work.

We let $G\mathcal{L}_n$ denote the group of invertible linear transformations taking $\mathbb{R}^n$ onto $\mathbb{R}^n$; remember that members of $G\mathcal{L}_n$ are determined by the fact that they take independent sets to independent sets.

We let $A_n$ denote the group of affine bijections of $\mathbb{R}^n$ to $\mathbb{R}^n$; members of $A_n$ are determined by their values at any $(n+1)$ points, no three of which are collinear, that is mappings which preserve line segments.

Recall from elementary geometry the following facts.

Fact 5.3. Each $f \in A_n$ can written in the form

$$f = \tau l,$$

where $\tau$ is a translation and $l \in G\mathcal{L}_n$.

In this case, the map $\pi : A_n \to G\mathcal{L}_n$, given by

$$\pi(f) = l$$

is a homomorphism with kernel $T_n$, the subgroup of $A_n$ consisting of translations. Hence $T_n$ is an invariant subgroup of $A_n$ and $A_n/T_n$ is isomorphic to $G\mathcal{L}_n$.

Another enlightening feature of life in the Euclidean world:

**Fact 5.4.** Every isometry of $\mathbb{R}^n$ onto itself is affine.

Linear isometries of $\mathbb{R}^n$ coincide with the isometries of $S^{n-1} = \{x \in \mathbb{R}^n : ||x|| = 1\}$ onto $S^{n-1}$. Hence they are precisely the maps whose action is reflected by an orthogonal matrix. (Recall that an orthogonal matrix $A$ is one for which $A^{-1} = A^t$.) The group

$$\mathcal{O}_n := \mathcal{G}_n \cap GL_n$$

is called the **orthogonal group**.

Of note is the fact that if $A \in \mathcal{O}_n$ then

$$\text{id}|_{\mathbb{R}^n} = AA^{-1} = AA^t;$$

hence

$$1 = \det \text{id}|_{\mathbb{R}^n} = \det(AA^t) = \det A \det A^t = (\det A)^2.$$

Therefore any $A \in \mathcal{O}_n$ has $\det A = \pm 1$.

Looking back at Fact 5.3, the map $\pi$ takes $\mathcal{G}_n$ to $\mathcal{O}_n$; $\mathcal{T}_n$ is an invariant subgroup of $\mathcal{G}_n$ and $\mathcal{G}_n/\mathcal{T}_n$ is isomorphic to $\mathcal{O}_n$.

Once more character needs to be mentioned: the special orthogonal group, denoted $\mathcal{SO}_n$, given by

$$\mathcal{SO}_n = \{A \in \mathcal{O}_n : \det A = 1\}.$$

Members of $\mathcal{SO}_n$ are called **rotations** and so $\mathcal{SO}_n$ can be viewed as the rotation group of $S^{n-1}$.

In $\mathbb{R}$ the only orthogonal maps are $\text{id}|_\mathbb{R}$ and $-\text{id}|_\mathbb{R}$. So

$$\mathcal{SO}_1 = \{\text{id}|_\mathbb{R}\}, \mathcal{O}_1 = \{\text{id}|_\mathbb{R}, -\text{id}|_\mathbb{R}\}, \text{ and } \mathcal{G}_1 = \{x \to a + x : a \in \mathbb{R}\}.$$

Consequently

$$\mathbb{Z}_2 \cong \mathcal{O}_1 \cong \mathcal{G}_1/\mathcal{T}_1,$$

where $\mathcal{T}_1 \cong \mathbb{R}$. Therefore $\mathcal{G}_1$ is solvable since

$$\{\text{id}\} \triangleleft \mathcal{T}_1 \triangleleft \mathcal{G}_1.$$

What about $\mathbb{R}^2$? Let $\rho_\theta$ denote the counterclockwise rotation about the origin in $\mathbb{R}^2$ through $\theta$ radians. Then $\rho_\theta$ is represented by the orthogonal matrix

$$\begin{pmatrix} \cos\theta & -\sin\theta \\ \sin\theta & \cos\theta \end{pmatrix}.$$

Then

$$\mathcal{SO}_2 = \{\rho_\theta : 0 \le \theta < 2\pi\},$$

is an abelian group that's isomorphic to the circle group $\mathbb{T}$. Also (denoting by $\mathcal{SG}_2$ the subgroup of members $f \in G_2$ so that $\pi(f)$ has determinant 1)

$$G_2/\mathcal{SG}_2 \cong \mathbb{Z}_2.$$

Therefore $G_2$ is solvable since

$$\{\mathrm{id}|_{\mathbb{R}^2}\} \triangleleft T_2 \triangleleft \mathcal{SG}_2 \triangleleft G_2.$$

**Corollary 5.1.** *For $n = 1, 2$ there exists a finitely additive probability $\mu$ defined on all subsets of $\mathbb{R}^n$ such that*

$$\mu(E) = \mu(B)$$

*whenever $E$ and $B$ are congruent.*

The soul of the negative solution to the amenability of $G_n$ for $n \ge 3$ is found in the following stunning result, often called **Hausdorff's paradox.** For a beautiful exposition of this and related topics, we refer the reader to the paper of Karl Stromberg in the American Mathematical Monthly titled "The Banach-Tarski Paradox" which appeared in the March 1979 issue, pages 151-161.

**Lemma 5.1.** *The unit sphere $S^2$ of $\mathbb{R}^3$ can be decomposed into the disjoint union*

$$S^2 = Q \cup R \cup S \cup T$$

*of four sets; $Q$ is countable, $R, S, T$ are pairwise congruent and $R$ is congruent to $S \cup T$.*

Now suppose the lemma to be so and assume that $\mu$ is a solution to the problem of measure in $\mathbb{R}^3$. Write $B_0 = B_{\mathbb{R}^3} \setminus \{0\}$ in the form

$$B_0 = Q_0 \cup R_0 \cup S_0 \cup T_0$$

where the subscript '0' indicates that the set is the union of rays emanating from the origin and ending on the sphere at a point of the subscript set. (eg., $Q_0 = \cup_{q \in Q}(0, q]$.)

We first show that $\mu(Q_0) = 0$. Introduce spherical coordinates and assume (a rotation of $S^2$ may be needed) that $0z$ passes through the none of the points of $Q$. A point in $S^2$ is determined by two angles: "latitude" which measures $N - S$ and "longitude" which measures $E - W$.)

The set of longitudes of points of $Q$ is countable as is the set of differences thereof. So we can choose an angle $\lambda$ distinct from all the angles of longitude of $Q$ and different from all the differences. Rotate $0z$ through the angle $\lambda$; $Q$ moves to a (countable) set $Q'$ which is disjoint from $Q$. Again we can find an angle $\lambda'$ distinct from all the points of $Q \cup Q'$ and from all the differences thereof. Rotate $0z$ through the angle $\lambda'$ to get a set $Q''$ which is the image of $Q'$; $Q, Q'$ and $Q''$ are congruent by construction. Continue in this vain.

For any $n$, we get pairwise disjoint sets

$$Q, Q', Q^1, \ldots, Q^{(n-1)}$$

and accompanying them are the pairwise disjoint sets

$$Q_0, Q_0', Q_0^1, \ldots, Q_0^{(n-1)}.$$

Of course all these sets are pairwise congruent as well. Then

$$\begin{aligned}\mu(B_0) &\geq \mu(Q_0 \cup \cdots \cup Q_0^{(n-1)}) \\ &= \mu(Q_0) + \mu(Q_0') + \mu(Q_0'') + \cdots + \mu(Q_0^{(n-1)}) \\ &= n\mu(Q_0).\end{aligned}$$

But $\mu(B_0)$ is finite and we can do the above for each $n \in \mathbb{N}$. It must be that $\mu(Q_0) = 0$.

We also have

$$\begin{aligned}\mu(B_0) &= \mu(Q_0 \cup R_0 \cup S_0 \cup T_0) \\ &= \mu(Q_0) + \mu(R_0) + \mu(S_0) + \mu(T_0) \text{ (because } Q_0, R_0, S_0, T_0 \\ &\quad \text{are pairwise disjoint)} \\ &= \mu(R_0) + \mu(S_0) + \mu(T_0) \text{ (because } \mu(Q_0) = 0) \\ &= 3\mu(R_0) \text{ (because } R_0, S_0, T_0 \text{ are congruent)}.\end{aligned}$$

At the same time
$$\mu(B_0) = \mu(R_0) + \mu(S_0 \cup T_0) = 2\mu(R_0)$$
because $R_0$ and $S_0 \cup T_0$ are congruent.

Hence
$$2\mu(R_0) = 3\mu(R_0) = \mu(B_0).$$
Since $\mu(B_0) > 0$, chaos is the only explanation.

The failure of $G_3$ to be amenable soon says the same for $G_4$. After all, if there *is* a finitely additive $G_4$-invariant probability on $\mathbb{R}^4$, say $\mu$, then $\nu$ defined as
$$\nu(E) = \mu(E \times \mathbb{R})$$
for $E \subseteq \mathbb{R}^3$ is a $G_3$-invariant finitely additive probability on $\mathbb{R}^3$. Plainly we can piggy back this argument to any $\mathbb{R}^n, n \geq 3$.

We've been dealing with finitely additive *bounded* measures, and they are reasonably well-behaved. To be sure, the problem of measure as we've stated it is *not* quite what was classically of interest. Here's what the problem really was: can one extend Lebesgue measure from its domain of Lebesgue measurable subsets of $\mathbb{R}^n$ to a measure defined on all the subsets of $\mathbb{R}^n$ in an additive manner?

To be sure, if $\lambda_n$ is Lebesgue measure on $\mathbb{R}^n$ then $\lambda_n$ is unbounded but countably additive. So to deal with possible extensions some new machinery is called for.

Suppose $\mathcal{A}$ is a Boolean algebra of subsets of a set $S$. A subset $\mathcal{A}_0$ of $\mathcal{A}$ is a subring of $\mathcal{A}$ if $\mathcal{A}_0$ is closed under intersections, contains 0, and is closed under finite unions of symmetric differences.

Our first result tells us that finitely additive extensions from subrings to algebras exist. We then use amenability to find that sometimes $\mathcal{G}$-invariant finitely additive measures can be extended, too, in an invariant manner.

**Theorem 5.5 (Measure Extension Theorem).** *Let $\mathcal{A}_0$ be a subring of the Boolean algebra $\mathcal{A}$ and let $\mu : \mathcal{A}_0 \to [0, \infty]$ be finitely additive with $\mu(0) = 0$. Then there is a finitely additive $\mu : \mathcal{A} \to [0, \infty]$ such that $\mu = \bar{\mu}|_{\mathcal{A}_0}$.*

**Proof.**

<u>Case 1</u> Suppose $\mathcal{A}$ is finite. In this case $\mathcal{A}$ is generated by its atoms, $a_1, \ldots, a_n$ say.

If $\mathcal{A}$ has just one atom then $\mathcal{A} = \{0, 1\}$ and the assertion is trivial. Otherwise, let $b$ be minimal in $\mathcal{A}_0 \setminus \{0\}$. (Of course if $\mathcal{A}_0$ is just $\{0\}$ then $\bar{\mu} \equiv 0$ works.) Let $c = b'$.

Look at $\mathcal{A}_c = \{a \in \mathcal{A} : a \subseteq c\}$, a Boolean algebra with $0, c$ as distinguished elements and of course, $b \notin \mathcal{A}_c$. Consider $\mu|_{\mathcal{A}_0 \cap \mathcal{A}_c}$ : if $a_0$ is an atom of $\mathcal{A}$ and $a \subseteq b$ then $a_0 \notin \mathcal{A}_c$, so $\mathcal{A}_c$ has fewer atoms than does $\mathcal{A}$, and inductively speaking, we can find a finitely additive $\nu : \mathcal{A}_c \to [0, \infty]$ so that

$$\nu|_{\mathcal{A}_0 \cap \mathcal{A}_c} = \mu|_{\mathcal{A}_0 \cap \mathcal{A}_c}.$$

Now define $\bar{\mu}$ at any atom of $\mathcal{A}$ as follows:

- if $a \subseteq c$ then $\bar{\mu}(a) = \nu(a)$
- $\bar{\mu}(a) = \mu(b)$
- if $a \subseteq b$ but $a \neq a_0$ then $\bar{\mu}(a) = 0$.

Notice that $\bar{\mu}$ agrees with $\nu$ on $\mathcal{A}_c$ and $\bar{\mu}$ also agrees with $\mu$ on $\mathcal{A}_0$. Because $b$ is minimal in $\mathcal{A}_0 \setminus \{0\}$, if $d \in \mathcal{A}_0$ then either $d \wedge b = 0$ or $d \geq b$. In the former case, $d \leq c$ so $\bar{\mu}(d) = \nu(d) = \mu(d)$; in the latter case, $d \geq b$ so $d = b \vee (d - b)$ and

$$\bar{\mu}(d) = \bar{\mu}(b) + \bar{\mu}(d - b) = \mu(b) + \nu(d - b) = \mu(b) + \mu(b - d) = \mu(d).$$

<u>Case 2</u> General $\mathcal{A}$.

For any *finite* subalgebra $\mathcal{B}$ of $\mathcal{A}$ let $\mathcal{M}(\mathcal{B})$ be the set

$$\{\nu \in [0, \infty]^{\mathcal{A}} : \nu(0) = 0, \mu|_{\mathcal{B} \cap \mathcal{A}_0} = \nu|_{\mathcal{B} \cap \mathcal{A}_0}, \nu|_{\mathcal{B}} \text{ is finitely additive}\}.$$

Each set $\mathcal{M}(\mathcal{B})$ is non-empty (from Case 1) and closed in $[0, \infty]^{\mathcal{A}}$. Moreover if $\mathcal{B}_1, \ldots, \mathcal{B}_n$ are finite subalgebras of $\mathcal{A}$ then there is a *finite* subalgebra of $\mathcal{B} \cap \mathcal{A}$ so that $\mathcal{B}_1, \ldots, \mathcal{B}_n \subseteq \mathcal{B}$; in other words,

$$\{\mathcal{M}(\mathcal{B}) : \mathcal{B} \text{ is a finite subalgebra of } \mathcal{A}\}$$

enjoys the finite intersection property. So

$$\bigcap \mathcal{M}(\mathcal{B}) \neq \emptyset.$$

Any $\mu \in \bigcap \mathcal{M}(\mathcal{B})$ works. $\square$

**Theorem 5.6 (Invariant Extension Theorem).** *Let $\mathcal{A}_0$ be a subring of the Boolean algebra $\mathcal{A}$, suppose $G$ is an amenable group of automorphisms of $\mathcal{A}$ that takes $\mathcal{A}_0$ into itself. Let $\mu : \mathcal{A}_0 \to [0, \infty]$ be a finitely additive measure with $\mu(0) = 0$, and suppose $\mu$ is $G$-invariant. Then $\mu$ has a $G$-invariant extension $\bar{\mu} : \mathcal{A} \to [0, \infty]$.*

**Proof.** We know that there is a finitely additive extension $\nu : \mathcal{A} \to [0, \infty]$ of $\mu$.

Since $G$ is amenable there is a left invariant finitely additive probability $\theta$ on $2^G$. For $b \in \mathcal{A}$ define $f_\theta : G \to \mathbb{R}$ by

$$f_b(g) = \nu(g^{-1}(b)),$$

and define $\bar{\mu}(b)$ by

$$\bar{\mu}(b) = \int f_b \, d\theta.$$

if $f_b \in \mathcal{B}(G)$ all is well; if $f_b$ is an unbounded then $\bar{\mu}(b) = \infty$. Then $\bar{\mu}$ is $G$-invariant finitely additive and extends $\mu$ to $\mathcal{A}$. □

**Remark 5.1.** The procedure just described would be much less circumspect had we stuck to *bounded* additive measures!

Here is another consequence of the Hahn-Banach theorem particularly tailored to our present needs; it is due to L. Kantorovitch in much more general circumstances.

**Lemma 5.2.** *Let $\mathcal{A}$ be a Boolean subalgebra of the Boolean algebra $\mathcal{B}$ of subsets of a set $S$. Then any $\mu \in ba^+(\mathcal{A})$ has an extension $\bar{\mu} \in ba^+(\mathcal{B})$ such that $||\bar{\mu}|| = ||\mu||$.*

**Proof.**
Keep in mind the duality $\mathcal{B}(\mathcal{A})^* = ba(\mathcal{A})$ established by Hildebrandt/Fichenholtz-Kantorovich. Let $\mu \in ba^+(\mathcal{A}) = \mathcal{B}(\mathcal{A})^{*+}$; define $\rho : \mathcal{B}(\mathcal{B}) \to \mathbb{R}$ by

$$\rho(g) = \inf \left\{ \int x \, d\mu : g \leq x \in \mathcal{B}(\mathcal{A}) \right\}.$$

It is clear (since $1 \in \mathcal{B}(\mathcal{A})$!) that $\rho : \mathcal{B}(\mathcal{B}) \to \mathbb{R}$ is sublinear and $\int x \, d\mu = \rho(x)$ for $x \in \mathcal{B}(\mathcal{A})$. The Hahn-Banach Theorem applies to provide us with a linear extension $F$ of $\int d\mu$ to all of $\mathcal{B}(\mathcal{B})$ such that

$$F(g) \leq \rho(g)$$

for all $g \in \mathcal{B}(\mathcal{B})$. If $g \in \mathcal{B}(\mathcal{B}), g \geq 0$ then $-g \leq 0$ and so

$$-F(g) = F(-g) \leq \rho(-g) \leq \int 0 \, d\mu = 0;$$

it follows that $F(g) \geq 0$. Being a positive linear functional on $\mathcal{B}(\mathcal{B})$ ensures $F \in \mathcal{B}(\mathcal{B})^*$, and of course $F(q) \geq 0$ whenever $q \geq 0$ so $F$ corresponds to $\bar{\mu} \in \text{ba}^+(\mathcal{B})$. It's plain that $\bar{\mu}$ extends to $\mu$ and that $||\bar{\mu}|| \leq ||\mu||$ since $F \leq \rho$. □

**Corollary 5.2.** *Suppose $\mathcal{A}$ is a Boolean subalgebra of the Boolean algebra $\mathcal{B}$ of subsets of the set $S$. Suppose $G$ is an amenable group of automorphisms of $\mathcal{B}$ onto itself. Suppose $\mu \in \text{ba}^+(\mathcal{A})$ is a $G$–invariant probability (so $A \in \mathcal{A}$ implies $gA \in \mathcal{A}$ and $\mu(gA) = \mu(A)$ for each $g \in G$). Then there is a $G$–invariant $\bar{\mu} \in \text{ba}^+(\mathcal{B})$ so that $\bar{\mu}|_\mathcal{A} = \mu$, with $\bar{\mu}$ a probability as well.*

**Proof.** By the Extension theorem just proved, there is a $\nu \in \text{ba}^+(\mathcal{B})$ such that $\nu|_\mathcal{A} = \mu$ and $||\nu|| = ||\mu|| = 1$. Let $\theta$ be a biinvariant finitely additive probability on $G$. For $B \in \mathcal{B}$, define $f_B$ by $f_B : G \to \mathbb{R}$

$$f_B(g) = \nu(g^{-1}B)$$

so $f_B \in \mathcal{B}(G)$. Define $\bar{\mu}$ by

$$\bar{\mu}(B) = \int f_B \, d\theta.$$

Then $\bar{\mu}$ does the trick. □

As a matter of fact, finitely additive measures with non-negative extended real values are often of interest, particularly with respect to the problem of measure. For this reason, the question of extending such measures is also of considerable importance. The basic Extension Theorem is a bit different in detail in this situation but, as you'll see, the existence of invariant extensions follows the same path as following in the case of dealing with members of $\text{ba}(\mathcal{A})$ – averaging works!

<u>Indication of the proof of Theorem 5.2</u>

(b) If $G$ is a finite group then there is a natural measure that does the tricks:

$$\mu(E) = \frac{\#E}{\#G}$$

for any $E \subseteq G$.

(c) Suppose $\mu$ is a left invariant finitely additive probability on $G$, and $H$ is a subgroup of $G$. Pick one element from each coset $Hg$ of $H$ and put your choices into a basket $B$. Define $\nu(A)$, for $A \subseteq H$ by

$$\nu(A) = \mu\left(\bigcup_{g \in B} Ag\right).$$

It is easy to verify that if $A_1 \cap A_2 \neq \emptyset$ then

$$\left(\bigcup_{g \in B} A_1 g\right) \cap \left(\bigcup_{g \in B} A_2 g\right) = \emptyset;$$

so $\nu$ is finitely additive. It's plainly a probability. Naturally is $A \subseteq H$ and $h \in H$ then

$$\nu(hA) = \mu\left(\bigcup_{g \in B} hAg\right) = \mu\left(h\bigcup_{g \in B} Ag\right)$$

$$= \mu\left(\bigcup_{g \in B} Ag\right) = \nu(A).$$

(d) Suppose $N$ is an invariant subgroup of the amenable group $G$, and suppose that $\mu$ is a left invariant finitely additive probability on $G$. Define $\nu$ on $G/N$ by

$$\nu(A) = \mu(\pi_N^{\leftarrow}(A))$$

where $\pi_N : G \to G/N$ is the canonical quotient homomorphism. Clearly $\nu$ is a finitely additive probability on $G/N$. Moreover for any $g \in G$ and $A \subseteq G/N$,

$$\nu(\pi_N(g)A) = \mu(\pi_N^{\leftarrow}\pi_n(g) \cdot A)$$
$$= \mu(g \cdot \pi_N^{\leftarrow}(A)) = \mu(\pi_N^{\leftarrow}(A)) = \nu(A).$$

(e) Suppose $N$ is an invariant subgroup of the group $G$ and both $N$ and $G/N$ are amenable. Let $\mu_N$ and $\mu_{G/N}$ be the left invariant finitely additive probabilities on their respective domains. Let $A \subseteq G$. Define $f_A : G \to \mathbb{R}$ by

$$f_a(g) = \mu_N(N \cap g^{-1}A);$$

if $g_1 N = g_2 N$ then $g_2^{-1} g_1 \in N$ and so, letting $h = g_2^{-1} g_1 \in N$,

$$\begin{aligned} f_A(g_2) &= \mu_N(N \cap g_2^{-1} A) = \mu_N(N \cap h g_1^{-1}) \\ &= \mu_N(h(N \cap g_1^{-1} A)) \\ &= \mu_N(N \cap g_1^{-1} A) = f_A(g_1), \end{aligned}$$

and so $f_A$ 'lifts' to $G/N$. The lifting is of course, a bounded and real-valued function $F_A$ on $G/N$. Define

$$\mu(A) = \int F_A \, d\mu_{G/N}.$$

We see easily that $\mu$ is a finitely additive probability on $G$, and since

$$f_{gA} = {}_g(f_A) = f_A(g^{-1} \cdot),$$

it follows that

$$F_{gA} = {}_g(F_A) = F_A(g^{-1} \cdot)$$

so that

$$\mu(gA) = \int F_{gA} \, d\mu_{G/N} = \int_g (F_A) \, d\mu_{G/N} = \int F_A \, d\mu_{G/N} = \mu(A).$$

(f) Suppose

$$G = \bigcup_{i \in I} G_i,$$

where each $G_i$ is amenable (and comes equipped with a left invariant finitely additive probability $\mu_i$) and for any $i, j \in I$ there is $h \in I$ so $G_i$ and $G_j$ are each subgroups of $G_h$.

Look at $K = [0,1]^{2^G}$, a compact Hausdorff space. For each $i \in I$, let

$$\mathcal{M}_i = \{\mu \in \mathrm{ba}^+(G) : \mu(G) = 1, \mu \text{ is left invariant}\}.$$

Then $\mathcal{M}_i$ is *pointwise closed* as a subset of $K$! Think about that!! Moreover $\mathcal{M}_i \neq \emptyset$ for each $i \in I$:

$$\mu(A) = \mu_i(A \cap G_i)$$

defines a member of $\mathcal{M}_i$. If $i, j \in I$ and $G_i, G_j \subseteq G_k$ then $\mathcal{M}_i \cap \mathcal{M}_j$ contains $\mathcal{M}_k$. Hence $\{\mathcal{M}_i : i \in I\}$ has the finite intersection property:

$$\bigcap_{i \in} \mathcal{M}_i \neq \emptyset.$$

Any $\mu \in \bigcap_{i \in I} \mathcal{M}_i$ bears witness to $G$'s amenability.

Now for (b). To start, notice that any group is the directed union of its finitely generated subgroups so the proof of (b) reduces (thanks to $f$) to the case where $G$ is a finitely generated abelian group. So suppose $G$ is such a creature, say, $G$ is an abelian group generated by $g_1, \ldots, g_n$.

Let $\epsilon > 0$. We'll show that there is a probability $\mu_\epsilon \in \mathrm{ba}^+(G)$ such that for any $A \subseteq G$ and any $1 \leq k \leq n$,

$$|\mu_\epsilon(A) - \mu_\epsilon(g_k A)| < \epsilon.$$

Once this is accomplished any weak* limit point $\mu$ of $(\mu_\epsilon : \epsilon > 0)$ in

$$\mathcal{B}_{\mathrm{ba}}(G) = \mathcal{B}(G)$$

will fit the bill as a finitely additive left invariant probability on $G$.

Suppose $G$ is singly generated by $g_1$. Choose $N$ so big that $N > \frac{2}{\epsilon}$, and let

$$\mu_\epsilon(A) = \#\{i : 1 \leq i \leq N, g_1^i \in A\}/N.$$

$\mu_\epsilon \in \mathrm{ba}^+(G)$ is a probability,

$$\mu_\epsilon(A) = \frac{\#\{i : 1 \leq i \leq N, g_1^i \in A\}}{N},$$

and

$$\mu_\epsilon(g_1 A) == \frac{\#\{i : 1 \leq i \leq N, g_1^i \in g_1 A\}}{N},$$

so $\mu_\epsilon(A) - \mu_\epsilon(g_1 A)|$ is determined by the 'end points,' that is, if $g_1^{i_1}, \ldots, g_1^{i_h} \in A$ with $1 \leq i_1 < \cdots < i_k \leq N$ all distinct then $g_1^{i_1}, \ldots, g_1^{i_k} \in gA$ is tantamount to $g_1^{i_1 - 1}, \ldots, g_1^{i_k - 1} \in A$ which means the difference between the numerators of $\mu_\epsilon(A)$ and $\mu_\epsilon(g_1 A)$ is at most two and so

$$\mu_\epsilon(A) - \mu_\epsilon(gA)| < \frac{2}{N} < \epsilon.$$

In the same spirit if $G$ is generated by $g_1, \ldots, g_n$ again chose $N$ so big that $N > 2/\epsilon$ and now define

$$\mu_\epsilon(A) = \frac{\#\{(i_1, \ldots, i_n) : 1 \leq i_1, \ldots, i_n \leq N, g_1^{i_1}, \ldots, g_1^{i_n} \in A\}}{N^n},$$

$\mu_\epsilon$ is a probability in $\mathrm{ba}^+(G)$; since the generators commute, the difference between $\mu_\epsilon(gA)$ and $\mu_\epsilon(A)$, where $g$ is one of the generators of $G$ is of interest only when one of the $i_n$'s is 1 or $N$. So

$$|\mu_\epsilon(A) - \mu_\epsilon(g_k A)| \leq \frac{2N^{n-1}}{N^n} = \frac{2}{N} < 2.$$

Alas, *all solvable groups are amenable.*

After all, if $G$ is solvable it's because $G$ achieves a normal series, that is, a finite ascending sequence $H_1, \ldots, H_k$ of subgroups of $G$ so that $H_1$ is an invariant subgroup of $H_{i+1}$ and $H_{i+1}/H_i$ is abelian with $H_1 = \{e\}$ and $H_k = G$. Now one need only to look at (a) through (f) to conclude that solvable groups are indeed amenable.

## 6. ba and Banach Spaces

### 6.1. *Banach's characterization of weakly null sequences in $\mathcal{B}(Q)$*

**Theorem 6.1 (Banach).** *Let $Q$ be a (non-empty) set and $(x_n)$ be a (uniformly) bounded sequence in $\mathcal{B}(Q)$. Then $(x_n)$ is weakly null if and only if*

$$\lim_n \liminf_k |x_n(q_k)| = 0 \tag{7}$$

*for each sequence $(q_k)$ of points in $Q$.*

**Proof.** *Necessity:* Suppose to the contrary that there is a sequence of points $(q_k)$ in $Q$ such that

$$\limsup_n \liminf_k |x_n(q_k)| > \alpha > 0$$

for some $\alpha$. Then unraveling the meaning of lim sup's, we can find a strictly increasing sequence $(n_j)$ of positive integers such that

$$\liminf_k |x_{n_j}(q_k)| > \alpha > 0$$

for each $j$. Now turning to the meaning of lim inf, we find a subsequence $(q_{k_m})$ of $(q_k)$ such that

$$|\lim_m x_{n_j}(q_{k_m})| > \alpha > 0$$

for each $j$. Let $x^* \in \mathcal{B}(Q)^*$ be given by

$$x^*(x) = \text{LIM}((x(q_{k_m})_m))$$

where LIM $\in l^{\infty *}$ is a Banach limit. Then for each $j$,

$$|x^*(x_{n_j})| > \alpha$$

and so

$$\limsup_n |x^*(x_n)| \geq \alpha > 0.$$

It follows that $(x_n)$ is *not* weakly null in $\mathcal{B}(Q)$.

*Sufficiency*: By our earlier remark (Theorem 3.3), to test $(x_n)$'s weak nullity it suffices to check the action of $x^* \in \mathcal{B}(Q)^*$, for $x^*$ a positive linear functional of norm 1, on the sequence $(x_n)$. So suppose $x^*$ is such a functional, with

$$\limsup_n x^*(x_n) > \alpha > 0$$

where (7) holds:

$$\lim_n \liminf_k |x_n(q_k)| = 0$$

for each sequence $(q_k)$ of points in $Q$. Let $s_n$ be the sequence

$$s_n(q) = \begin{cases} x_n(q) & \text{if } x_n(q) \geq 0 \\ 0 & \text{otherwise} \end{cases}$$

and let $t_n = x_n - s_n$.

One of $\limsup_n x^*(s_n)$ and $\limsup_n x^*(t_n)$ must exceed $\frac{\alpha}{2}$; after all, $x_n = s_n + t_n$ so

$$x^*(x_n) = x^*(s_n) + x^*(t_n)$$

ensuring that

$$\limsup_n x^*(x_n) = \limsup_n (x^*(q_n) + x^*(t_n)) \leq \limsup_n x^*(s_n) + \limsup_n x^*(t_n).$$

If we clip off the 'bottoms' of $s_n$ by defining

$$y_n(q) = \begin{cases} s_n(q) & \text{if } s_n(q) \geq \frac{\alpha}{6} \\ 0 & \text{otherwise} \end{cases}$$

then

$$\|s_n - y_n\|_\infty \leq \frac{\alpha}{6};$$

what's more,
$$\limsup_n x^*(y_n) = \limsup_n x^*(s_n - (s_n - y_n))$$
$$= \limsup_n (x^*(s_n) - x^*(s_n - y_n))$$
so that
$$\limsup_n x^*(y_n) \geq \frac{\alpha}{2} - \frac{\alpha}{6} = \frac{\alpha}{3}.$$

Let
$$S_n = \left\{ q \in Q : |x_n(q)| \geq \frac{\alpha}{6} \right\},$$
and look at $\chi_{S_n}$. Since
$$\|y_n\|_\infty \leq \|s_n\|_\infty \leq \|x_n\|_\infty \leq M,$$
say, we see that for any $q \in Q$
$$\chi_{S_n}(q) \geq \frac{y_n(q)}{M}$$
so that
$$x^*(M \cdot \chi_{S_n}) \geq x^*(y_n).$$
From this we see that
$$\limsup_n x^*(\chi_{S_n}) \geq \frac{\alpha}{3M} =: \beta > 0.$$
For $E \subseteq Q$, let $F(E) = x^*(\chi_E)$; of course $F \in \mathcal{B}(Q)^*$ and
$$\limsup_n F(S_n) \geq \beta.$$

Let $n_1$ be the smallest positive integer such that
$$\limsup_n F(S_n \cap S_{n_1}) > 0.$$

**Such an $n_1$ exists by the way! This is the crux of the matter!** In fact, otherwise,
$$\lim_n F(S_n \cap S_k) = 0$$
for each $k$ so (because $F$ is additive)
$$\lim_n F(\cup_{j=1}^k (S_n \cap S_j)) = 0,$$
for each $k$ as well.

Let $k_1 = 1$. Pick $m_1 > k_1$ so large that
$$F(S_{m_1}) > \frac{\beta}{2}$$
and
$$F(S_{m_1} \cap S_{k_1}) < \frac{\beta}{4}.$$
Let $k_2 > m_1$. Pick $m_2$ so large that
$$F(S_{m_2}) > \frac{\beta}{2}$$
and
$$F(S_{m_2} \cap (S_1 \cup \cdots \cup S_{k_2})) < \frac{\beta}{4}.$$
Continuing in this fashion, producing $k_1 < m_1 < k_2 < m_2 < \cdots$, with
$$F(S_{m_j}) > \frac{\beta}{2}$$
and
$$F(S_{m_j} \cap (S_1 \cup \cdots \cup S_{k_j})) < \frac{\beta}{4}.$$
Now disjointivity: let $T_j$ be given by
$$T_j = S_{m_j} \backslash [S_{m_j} \cap (\cup_{i=1}^{k_j} S_i)].$$
By construction
$$F(T_j) > \frac{\beta}{4}$$
and this is a "no-no" since $F$ takes disjoint sequences to 0.

So $n_1$ does indeed exist such that
$$\limsup_n F(S_n \cap S_{n_1}) > 0.$$
Believe it or not.

Once faith has been established for $n_1$ we see that there are $n_2 < n_3 < \cdots < n_k < \cdots$ so that
$$\limsup_n F(S_n \cap S_{n_1} \cap \cdots \cap S_{n_k}) > 0$$
for each $k$. The all-important point here is that for each $k$ there is at least one point $q_k$ so
$$q_k \in S_{n_1} \cap S_{n_2} \cap \cdots \cap S_{n_k}.$$

Of course if $j \geq k$ then $q_j \in S_{n_k}$ and so by how the $S_n$'s were defined

$$|x_{n_k}(q_j)| \geq \frac{\alpha}{6}.$$

It soon follows that

$$\limsup_n \liminf_j |x_n(q_j)| \geq \frac{\alpha}{6}.$$

This contradicts (7) and thus the sufficiency is proven. □

## 6.2. Goldstine's Theorem and Pettis' proof of Milman-Pettis

**Theorem 6.2 (Goldstine-Pettis).** *Let $F \in X^{**}$. Then there is a $\mu \in \mathrm{ba}^+(B_X)$ such that $\|\mu\| = \|F\|$ and*

$$F(f) = \int_{B_X} f(x)\, d\mu(x),$$

*for each $f \in X^*$.*

$X^*$ is a closed linear subspace of $\mathcal{B}(B_X)$, the space of bounded scalar-valued functions defined on $B_X$. Using Hahn-Banach we can extend any $F \in X^{**}$ to a member of $\lambda \in \mathcal{B}(B_X)^* = \mathrm{ba}(2^X)$, without loss of norm. The result is that for any $f \in X^*$,

$$F(f) = \int_{B_X} f(x)\, d\lambda(x).$$

Write $\lambda = \lambda^+ - \lambda^-$, where $|\lambda|$ is the variation of $\lambda$, $\lambda^+ = \frac{|\lambda|+\lambda}{2}$, $\lambda^- = \frac{|\lambda|-\lambda}{2}$; of course both $\lambda^+, \lambda^- \in \mathrm{ba}^+(B_x)$.

Let $E \subseteq B_X$ and define the function $r : 2^{B_X} \to 2^{B_X}$ by rE, the reflection of $E$ through 0 by

$$\mathrm{r}E = \{x \in B_X : -x \in E\}.$$

$|\lambda^- \circ r|$ and $|\lambda^-|$ are the same. Further if $f \in X^*$ then

$$\int_{B_X} -f(x)\, d\lambda^- \circ r(x) = \int_{B_X} f(-x)\, d\lambda^- \circ r(x) = \int_{B_X} f(x)\, d\lambda^-(x),$$

equations easily established by looking first, as usual, at simple functions. It follows that if we let

$$\mu = \lambda^+ + \lambda^- \circ r$$

we get a member of $\text{ba}^+(2^{B_X})$ for which
$$||\mu|| = ||\lambda^+|| + ||\lambda^- \circ r|| = ||\lambda^+|| + ||\lambda^-|| = ||\lambda|| = ||F||$$
with
$$F(f) = \int_{B_X} f(x)\, d\mu(x)$$
for every $f \in X^*$.

**Corollary 6.1 (Goldstine's Theorem).** *For any Banach space $X$, $B_X$ is weak\* dense in $B_{X^{**}}$.*

After all, if $f \in X^*$ vanishes on $B_X$ then for each $F \in B_{X^{**}}$,
$$F(f) = \int_{B_X} f(x)\, d\mu(x) = 0,$$
since $\mu \in \text{ba}^+(B_X)$.

We turn now to Pettis' remarkable proof of the Milman-Pettis Theorem to the effect that uniformly convex Banach spaces are reflexive. This famous result was a bridge between the geometry of a Banach space and its linear isomorphic structure; justifiably famous in its day and the start of a long line of similar connections between these theories/viewpoints.

We recall that a Banach space $X$ is **uniformly convex** if given $\epsilon > 0$ there is a $\delta > 0$ such that if $||x|| = 1 = ||y||$ and $||x - y|| \geq \delta$ then $||x + y|| \leq 2(1 - \epsilon)$.

If $X$ is uniformly convex and $x^* \in X^*$ then there is $x \in S_X$ so that $x^*(x) = ||x^*||$.

In fact, pick $(x_n) \subseteq S_X$ so $1 - \frac{1}{n} < x^*(x_n) \leq 1$. Then $(x_n)$ is a Cauchy sequence. Why?

Let $\epsilon > 0$, choose $\delta > 0$ so that if $x, y \in S_X$, $||x-y|| > \epsilon$ then $||x+y|| < 2-\delta$. Then
$$x^*(x_m + x_n) = x^*(x_m) + x^*(x_n) > 2 - \frac{1}{n} - \frac{1}{m};$$
since $x^* \in S_X$, $||x_m + x_n|| \geq 2 - \delta$ (if we choose $m, n$ big enough). So $||x_m - x_n|| \leq \epsilon$ for such a choice. Then $x = \lim_n x_n$ plainly satisfies $x^*(x) = ||x^*||$ and $x \in S_X$.

**Note.** The $x$ as above is *unique* if $x^* \neq 0$.

Indeed, given another $x' \in S_X$ so that
$$x^*(x) = ||x^*|| = x^*(x'),$$
if $||x - x'|| = \rho > 0$ then
$$||x + x'|| \leq 2 + \delta_\rho < 2.$$
So (supposing $x^* \in S_{X^*}$)
$$2 = x^*(x) + x^*(x') = x^*(x + x') \leq ||x + x'|| < 2,$$
OOPS.

Here's another geometric fact of life in uniformly convex spaces.

**Lemma 6.1.** *Let $X$ be a uniformly convex Banach space. Then given $\epsilon > 0$ there is a $\delta_\epsilon > 0$ so that if $x, y \in B_X$ with $x \in S_X$ and $x^* \in X^*$ with $x^*(x) = ||x^*||$, should $||x - y|| > \epsilon$ then*
$$x^*(y) \leq (1 - \delta)||x^*||.$$

**Theorem 6.3 (Milman, Pettis).** *Every uniformly convex Banach space is reflexive.*

**Proof.** (Pettis) Let $F \in X^{**}$ and suppose $||F|| = 1$. Pick $(f_n) \subseteq S_{X^*}$ so that
$$1 = ||F|| \geq F(f_n) > 1 - \frac{1}{n}.$$
For each $n$, choose $x_n \in S_X$ so that
$$f_n(x_n) = 1 = ||f_n||.$$

<u>Claim:</u> $(x_n)$ *is Cauchy.*

By the Goldstine - Pettis theorem there is a $\mu \in ba^+(2^{B_X})$ so that $||\mu|| = 1$ and
$$F(f) = \int_{B_X} f(x) \, d\mu(x),$$
for each $f \in X^*$.

Keeping in mind the situation,
$$1 - \frac{1}{n} < F(f_n) = \int_{B_X} f_n(x) \, d\mu(x),$$

set
$$S_{n,\epsilon} = \{x \in B_X : ||x - x_n|| < \epsilon\}.$$

Then
$$\int_{B_X} f_n(x)\,d\mu(x) = \int_{S_{n,\epsilon}} f_n(x)\,d\mu(x) + \int_{B_X \setminus S_{n,\epsilon}} f_n(x)\,d\mu(x).$$

Suppose $\epsilon > 0$. Choose $\delta = \delta_\epsilon > 0$ so that if $x, y \in B_X$ with $||x|| = 1$ and $x^* \in X^*$ with $x^*(x) = ||x^*||$ then $||x - y|| \geq \epsilon$ yields $x^*(y) \leq (1 - \delta_\epsilon)||x^*||$.

Applying this to our current situation, we see

$$\text{if } x \in B_X \setminus S_{n,\epsilon}, \text{ then } f_n(x) \leq 1 - \delta_\epsilon. \tag{8}$$

Now (8) and the fact that $|f_n(x)| \leq ||f_n|| = 1$ for $x \in B_X$ tells us

$$1 - \frac{1}{n} < \int_{S_{n,\epsilon}} f_n(x)\,d\mu(x) + \int_{B_X \setminus S_{n,\epsilon}} f_n(x)\,d\mu(x)$$
$$\leq \int_{S_{n,\epsilon}} f_n(x)\,d\mu(x) + (1 - \delta_\epsilon)\mu(B_X \setminus S_{n,\epsilon})$$
$$\leq \mu(S_{n,\epsilon}) + (1 - \delta_\epsilon)\mu(B_X \setminus S_{n,\epsilon})$$
$$= \mu(B_X) - \delta_\epsilon \mu(B_X - S_{n,\epsilon}).$$

So
$$\frac{1}{n} > \delta_\epsilon \mu(B_X \setminus S_{n,\epsilon}),$$

or
$$\mu(B_X \setminus S_{n,\epsilon}) < \frac{1}{n\delta_\epsilon}.$$

Well choose $n_\epsilon$ so that if $m, n \geq n_\epsilon$ then
$$\frac{2}{\delta_\epsilon} < m, n.$$

It follows that $S_{m,\epsilon} \cap S_{n,\epsilon} \neq \emptyset$. After all,
$$\mu(S_{m,\epsilon} \cap S_{n,\epsilon}) = \mu(B_X \setminus ((B_X \setminus S_{m,\epsilon}) \cup (B_X \setminus S_{n,\epsilon})))$$
$$= \mu(B_X) - \mu(B_X \setminus S_{m,\epsilon} \cup B_X \setminus S_{n,\epsilon})$$
$$\geq 1 - \mu(B_X \setminus S_{m,\epsilon}) - \mu(B_X \setminus S_{n,\epsilon})$$
$$> 1 - \frac{1}{2} - \frac{1}{2} = 0.$$

Hence if $m, n \geq n_\epsilon$, $S_{m,\epsilon} \cap S_{n,\epsilon} \neq \emptyset$ which means
$$||x_m - x_n|| \leq 2\epsilon,$$

and $(x_n)$ is Cauchy.

Therefore $x_0 = \lim_n x_n$ exists. If we let
$$S_\epsilon = \{x \in B_X : ||x - x_0|| < \epsilon\}$$
then for $n$ large enough,
$$S_{n,\frac{\epsilon}{2}} \subseteq S_\epsilon,$$
so
$$B_X \setminus S_\epsilon \subseteq B_X \setminus S_{n,\epsilon/2}.$$
Hence
$$0 \leq \mu(B_X \setminus S_\epsilon) \leq \mu(B_X \setminus S_{n,\epsilon/2}) \leq \frac{1}{n\delta_{\epsilon/2}}.$$
This is true for each $n$ so
$$\mu(B_X \setminus S_\epsilon) = 0.$$
And this is so for each $\epsilon > 0$.

Now we're ready to see that $F(f) = f(x_0)$ for each $f \in X^*$. Indeed if $f \in X^*$ then for each $\epsilon > 0$,

$$|F(f) - f(x_0)| = \left| \int_{B_X} f(x)\, d\mu(x) - \int_{B_X} f(x_0)\, d\mu(x) \right|$$
$$\leq \int_{B_X} |f(x) - f(x_0)|\, d\mu(x)$$
$$= \int_{S_\epsilon} |f(x) - f(x_0)|\, d\mu(x) + \int_{B_X \setminus S_\epsilon} |f(x) - f(x_0)|\, d\mu(x)$$
$$= \int_{S_\epsilon} |f(x) - f(x_0)|\, d\mu(x) \leq ||f|| \int_{S_\epsilon} ||x - x_0||\, d\mu(x)$$
$$\leq ||f|| \epsilon \mu(S_\epsilon) \leq \epsilon ||f||. \qquad \square$$

## 6.3. Phillips' Lemma and some of its consequences

**Lemma 6.2 (Phillips' Lemma).** *Let* $(\mu_n) \subseteq ba(\mathbb{N})$ *and suppose*
$$\lim_{n \to \infty} \mu_n(\Delta) = 0 \qquad (9)$$
*for each* $\Delta \subseteq \mathbb{N}$. *Then*
$$\lim_{n \to \infty} \mu_n(A) = 0 \qquad (10)$$

uniformly for all finite sets $A \subseteq \mathbb{N}$ and so

$$\lim_{n \to \infty} \sum_k |\mu_n(k)| = 0. \tag{11}$$

**Proof.** Suppose (10) does *not* hold.

Then for some exceptional $\epsilon_0 > 0$ and a suitably badly behaved subsequence $(\nu_n)$ of $(\mu_n)$ we can find a sequence $(A_n)$ of finite subsets of $\mathbb{N}$ so that

$$|\nu_n(A_n)| > 2\epsilon_0$$

for each $n$.

Our first aim is to show (with our supposition that (10) fails) to select a subsequence $(\nu'_n)$ of $(\nu_n)$ and a sequence $(B_n)$ of pairwise disjoint finite subsets of $\mathbb{N}$ so that for each $n$

$$|\nu'_n(B_n)| > \epsilon_0$$

yet

$$|\nu'_n|(B_1 \cup \cdots B_n) < \frac{\epsilon_0}{4}.$$

Our selection process will make frequent use of the following consequences of (9):

$$\lim_n |\nu_n(A)| = 0$$

for each *finite* subset $A$ of $\mathbb{N}$.

To start, let $B_1 = A_1$, and $\nu'_1 = \nu_1$. There is an $N_2 > N_1 = 1$ so that

$$|\nu_{N_2}(B_1)| < \frac{\epsilon_0}{4};$$

after all, (9) is supposed. Notice

$$|\nu_{N_2}(A_{N_2} \setminus B_1)| \geq |\nu_{N_2}(A_{N_2})| - |\nu_{N_2}(B_1)| > 2\epsilon_0 - \frac{\epsilon_0}{4} > \epsilon_0.$$

Let $B_2 = A_{N_2} \setminus B_1$ and $\nu'_2 = \nu_{N_2}$.

Suppose we've chosen well and have chosen $\nu'_1, \ldots, \nu'_k$ as well as our finite sets $B_1, \ldots, B_k$ that are pairwise disjoint and so that

$$|\nu'_j(B_j)| > \epsilon_0$$

yet
$$|\nu'_j(B_1 \cup \cdots \cup B_{j-1})| < \frac{\epsilon_0}{4}$$
for $j = 1, 2, \ldots, k$. Of course $\nu'_j = \nu_{N_j}$ where $N_1 < N_2 < \cdots < N_k$.

Now we can find $N_{k+1} > N_k$ so that
$$|\nu_{N_{k+1}}(B_1 \cup \cdots \cup B_k)| < \frac{\epsilon_0}{4}.$$
Let
$$B_{k+1} = A_{N_{k+1}} \setminus (B_1 \cup \cdots \cup B_k)$$
and $\nu'_{k+1} = \nu_{N_{k+1}}$. Then
$$|\nu_{k+1}(B_{k+1})| = |\nu_{N_{k+1}}(A_{N_{k+1}} \setminus (B_1 \cup \cdots \cup B_k))|$$
$$\geq |\nu_{k+1}(A_{N_{k+1}})| - |\nu_{N_{k+1}}(B_1 \cup \cdots \cup B_k)|$$
$$\geq 2\epsilon_0 - \frac{\epsilon_0}{4} > \epsilon_0.$$

Our first step has been taken.

Next, we will extract a (further) subsequence $(\nu''_n)$ of $(\nu'_n)$ and sequences $(W_n)$ and $(C_n)$ of sets so that the sequence $(C_n)$ consists of pairwise disjoint finite subsets of $\mathbb{N}$ and $(W_n)$ is a descending sequence of infinite subsets of $\mathbb{N}$ so that

(1) $C_n \subseteq W_n \setminus W_{n+1}$
(2) $|\nu''_N(C_n)| > \epsilon_0$
(3) $\nu''_n(W_{n+1}) < \frac{\epsilon_0}{4}$ and
(4) $|\nu''_n|(C_1 \cup \cdots C_{n-1}) < \frac{\epsilon_0}{4}$.

Our building blocks will be the sequence $(B_n)$ of pairwise disjoint finite subsets of $\mathbb{N}$. Set
$$W_1 = \cup_n B_n, C_1 = B_1, \nu''_1 = \nu_1.$$
Since $\nu''_1$, has bounded variation, there is an infinite union $W_2$ of a subsequence of $B_2, B_3, \ldots$ so that
$$|\nu''_1|(W_2) < \frac{\epsilon_0}{4}.$$
Let $C_2$ be the first $B_k$ ($k > 2$) appearing as a subset of $W_2$, say $C_2 = B_{N_2}$. Let $\nu''_2 = \nu'_{N_2}$.

(a), (b), and (d) are plainly satisfied for $n = 1$. Let's see how to proceed from step $k$ to step $(k+1)$, $k > 1$. Assume $\nu_j'', C_j$ and $W_j$ have all been constructed for $j = 1, 2, \ldots, k$ so as to satisfy (a), (b), (c), and (d) at each stage, where for each $j$, $C_j$ is the first $B_k$, say $C_j = B_{N_j}$, appearing in the set $W_j$. Let $W_{k+1}$ be an infinite union of $B_k$'s that lie inside $W_k$ after $B_{N_k}$, chosen that

$$|\nu_k''|(W_{k+1}) < \frac{\epsilon_0}{4}.$$

Here's the trick we're using: $\mathbb{N}$ can be written as an infinite union of infinitely many pairwise disjoint infinite subsets of $\mathbb{N}_n$; looking at the infinitely large collection of $B_k$'s that appear in $W_k$ with $k > N_k$, we can collect these $B_k$'s into an infinite collection of pairwise disjoint infinite unions of subsequences of $(B_k : B_k \subseteq W_k, k > \mathbb{N}_k)$ according to the decomposition $\mathbb{N} = \cup_n \mathbb{N}_n$ of $\mathbb{N}$. One of these infinite unions necessarily has the property that $|\nu_k''|$ is no more than $\frac{\epsilon_0}{4}$ since the alternative - that $|\nu_k''|$ exceeds $\frac{\epsilon_0}{4}$ for each of an infinite family of pairwise disjoint sets will force $|\nu_k''(\mathbb{N})|$ to be $\infty$.

Let $C_{k+1} = B_{N_{k+1}}$ be the first $B_k$ appearing as a subset of $W_{k+1}$ and $\nu_{k+1}'' = \nu_{N_{k+1}}'$.

Each of (a), (b), (c), and (d) are easily seen: (a) is how we chose $W_{n+1}$, to start after $C_n$; (b) is the result of our very first step:

$$|\nu_n''(C_n)| = |\nu_{N_n}'(B_{N_n})| > \epsilon_0;$$

(d) follows from the realization that

$$|\nu_{n+1}''|(C_1 \cup \cdots \cup C_n) = |\nu_{N_{n+1}}'(B_1 \cup \cdots \cup B_{N_n})| < \frac{\epsilon_0}{4}$$

and (c) is part and parcel of how we chose $W_{n+1}$.

Now we look at

$$Q = \bigcup_k C_k.$$

Then

$$|\nu_n''(Q)| = |\nu_n''(C_1 \cup \cdots \cup C_{n-1}) + \nu_n''(C_n) + \nu_n''(\cup_{k=n+1}^\infty C_k)|$$
$$\geq |\nu_n''(C_n)| - |\nu_n''|(C_1 \cup \cdots \cup C_{n-1}) - |\nu_n''|(\cup_{k=n+1}^\infty C_k)$$
$$\geq |\nu_n''(C_n)| - |\nu_n''|(C_1 \cup \cdots \cup C_{n-1}) - |\nu_n''(W_{n+1})|$$
$$> \epsilon_0 - \frac{\epsilon_0}{4} - \frac{\epsilon_0}{4} = \frac{\epsilon_0}{2},$$

and this is for all $n$. This flies in the face of (9), that
$$\lim_n |\nu_n''(Q)| = \lim_n |\mu_n(Q)| = 0.$$
The second assertion (11),
$$\lim_k \sum_k |\mu_n(k)| = 0,$$
follows easily from (10) if one but realizes that
$$\sum_k |\mu_n(k)| \leq 4\sup\{|\mu_n(A)| : A \subseteq \mathbb{N} \text{ is finite}\}.$$
□

Phillips' Lemma seems to be like a bolt of lightening from an almost clear blue sky; where'd it come from? In fact, it had a curious origin.

A (bounded) subset $K$ of a Banach space $X$ is *limited* if whenever $(x_n^*)$ is a weak*–null sequence in $X^*$ (that is, $\lim_n x_n^*(x) = 0$ for each $x \in X$) we have
$$\limsup_n \sup_{x \in K} |x_n^*(x)| = 0.$$
Alternatively, $K$ is limited in $X$ whenever given a bounded linear operator $u : X \to c_0$, $u(K)$ is relatively compact.

The total boundedness of relatively norm compact sets leads to the following.

**Fact 6.4.** Relatively compact sets in $X$ are limited.

An easy consequence of the Arzela-Ascoli Theorem is the following.

**Fact 6.5.** If $X$ is a separable Banach space and $K$ is a limited subset of $X$ then $K$ is relatively compact.

The following is easily established.

**Fact 6.6.** If $Y$ is a complemented closed linear subspace of the Banach space $X$ then a (bounded) set $K \subseteq Y$ is limited in $Y$ if and only if it is limited in $X$.

Here's an unexpected dividend of Phillips' Lemma - the first example of a *natural* uncomplemented subspace of a classical Banach space.

**Corollary 6.2 (Phillips).** $c_0$ *is not a complemented subspace of* $l^\infty$.

After all, $c_0$ is separable and the set $K = \{e_n\}$ of coordinate vectors is not relatively compact, hence not limited in $c_0$. Phillip's Lemma shows that $K$ is limited in $l^\infty$.

**Remark 6.1.** It was the fact that $\{e_n\}$ is *not* relatively compact but *is* limited in $l^\infty$ that motivated Phillip's to prove his 'Lemma.' Earlier, both Gelfand and Mazur had each (erroneously) asserted that the limited subsets of *any* Banach space were precisely the relatively norm compact subsets of the space. Great minds (often) think alike! Go figure.

### References

1. S. Banach, *Théorie des Opérationes Linéaires*, Monografie Matematyczne, Warsaw (1932).
2. J. Diestel and J.J. Uhl Jr., *Vector Measures*, American Mathematical Society, New York, Mathematical Surveys and Monographs **15** (1977).
3. N. Dunford and J.T. Schwartz, *Linear Operators, Part I*, Interscience, New York and London, (1958).
4. G. Fitchenholtz and L.V. Kantorovitch, *Sur les Opérations Linéaires L'espace des Fonctions Bornées*, Studia Math., **5** (1934), 65–98.
5. H.H. Goldstine, *Weakly Complete Banach Spaces*, Duke Math J., **4** (1938), 125–141.
6. A. Grothendieck, *Espaces Vectorielles Topologies*, Inst. Mat. Pura. Appl. Univ. Sao Paulo (1954).
7. T.H. Hildebrandt, *On Bounded Functional Operations*, Trans. American Math. Soc., **36** (1934), 868–875.
8. R.E. Huff, *Vector Measures*, Penn State lecture notes (1973).
9. B.J. Pettis, *A Proof that every Uniformly Convex Space is Reflexive*, Duke Math J., **5** (1939), 249–253.
10. R.S. Phillips, *On Linear Transformations*, Transactions American Math. Soc., **48** (1940), 516–541.
11. W. Schachermayer, *On Some Classical Measure-Theoretic Theorems for non-$\sigma$-complete Boolean Algebras*, Dissertations Math **214** (1982), 1–33.
12. T. Schlumprecht, *Limited Sets in Banach Spaces*, Ph.D. dissertation, Munich, (1988).
13. K. Stromberg, *The Banach–Tarski Paradox*, American Math. Monthly, **86** (1979), 151–161.
14. S. Wagon, *The Banach–Tarski Paradox*, Cambridge University Press Encyclopedia of Mathematics (1985).

# Sampling and recovery of bandlimited functions and applications to signal processing

Th. Schlumprecht

*Department of Mathematics, Texas A&M University College Station, TX 77843, USA*
*e-mail: thomas.schlumprecht@math.tamu.edu*

Bandlimited functions, i.e square integrable functions on $\mathbb{R}^d$, $d \in \mathbb{N}$, whose Fourier transforms have bounded support, are widely used to represent signals. One problem which arises, is to find stable recovery formulae, based on evaluations of these functions at given sample points. We start with the case of equally distributed sampling points and present a method of Daubechies and DeVore to approximate bandlimited functions by quantized data.

In the case that the sampling points are not equally distributed this method will fail. We are suggesting to provide a solution to this problem in the case of scattered sample points by first approximating bandlimited functions using linear combinations of shifted Gaussians. In order to be able to do so we prove the following interpolation result.

Let $(x_j : j \in \mathbb{Z}) \subset \mathbb{R}$ be a *Rieszbasis sequence*. For $\lambda > 0$ and $f \in PW$, the space of square-integrable functions on $\mathbb{R}$, whose Fourier transforms vanish outside of $[-1, 1]$, there is a unique sequence $(a_j) \in \ell_2(\mathbb{Z})$, so that the function

$$I_\lambda(f)(x) := \sum a_j e^{-\lambda \|x-x_j\|_2^2}, \qquad x \in \mathbb{R}$$

is continuous, square integrable, and satisfies the interpolatory conditions $I_\lambda(f)(x_k) = f(x_k)$, for all $k \in \mathbb{Z}$. It is shown that $I_\lambda(f)$ converges to $f$ in $L_2(\mathbb{R}^d)$ and uniformly on $\mathbb{R}$, as $\lambda \to 0^+$.

## 1. Introduction

We first introduce some basic notations and recall some results. For $1 \le p \le \infty$ and $B \subset \mathbb{R}^d$, $m(B) > 0$, we let

$$L_p(B) = \left\{ f : B \to \mathbb{C} \text{ measurable} : \|f\|_p = \left( \int_B |f| dx \right)^{1/p} < \infty \right\},$$

if $1 \le p < \infty$, and

$$L_\infty(B) = \{ f : B \to \mathbb{C} \text{ measurable} : \|f\|_\infty = \sup\{r > 0 : m(|f| \ge r) > 0\} \},$$
$$C_b(\mathbb{R}^d) = \{ f : \mathbb{R}^d \to \mathbb{C} \text{ bounded and continuous} \},$$
$$C_0(\mathbb{R}^d) = \{ f : \mathbb{R} \to \mathbb{C} \text{ continuous and } \lim_{\|t\|_2 \to \pm\infty} f(t) = 0 \}.$$

The *Fourier transform* of $f \in L_1(\mathbb{R}^d)$ is defined point wise by

$$\hat{f}(x) = \int_{\mathbb{R}^d} f(t) e^{-i\langle x,t\rangle} dt, \qquad x \in \mathbb{R}^d.$$

Note that

$$(\hat{\cdot}) : L_1(\mathbb{R}^d) \to C_0(\mathbb{R}^d), \quad f \to \hat{f},$$

is well defined, bounded and linear with $\|(\hat{\cdot})\|_{L(L_1, C_0)} = 1$.

In general $\hat{f} \notin L_1(\mathbb{R}^d)$. But if $\hat{f} \in L_1(\mathbb{R}^d)$ then *the Inverse Transform* is given by

$$f(t) = [\hat{f}]^\vee = \frac{1}{(2\pi)^d} \int_{\mathbb{R}} \hat{f}(x) e^{i\langle x,t\rangle} dx, \quad t \in \mathbb{R}^d.$$

The linear map

$$L_2(\mathbb{R}^d) \cap L_1(\mathbb{R}^d) \to L_2(\mathbb{R}^d) \cap L_1(\mathbb{R}^d), \quad g \mapsto \hat{g}$$

is an isomorphy with respect to $\|\cdot\|_2$, and $\|\hat{g}\|_2 = (2\pi)^{d/2} \|g\|_2$. We can therefore extend this map to an operator

$$\mathcal{F} : L_2(\mathbb{R}^d) \to L_2(\mathbb{R}^d), \; g \mapsto \mathcal{F}[g], \tag{1}$$

with $\|\mathcal{F}[g]\|_2 = (2\pi)^{1/2} \|g\|_2, \; g \in L_2(\mathbb{R}^d)$.

From the inverse formula on $L_1(\mathbb{R}^d)$ we deduce that for $g \in L_2(\mathbb{R})$

$$\mathcal{F}^{-1}[g] = \|\cdot\|_2 - \lim_{N\to\infty} H_N, \text{ where } H_N(t) = \frac{1}{(2\pi)^d} \int_{[-N,N]^d} g(x) e^{ixt} dt, \; t \in \mathbb{R}.$$

**Definition 1.1.** For $B \subset \mathbb{R}^d$, bounded with positive measure, define

$$PW_B := \{\hat{g} : g \in L_2(\mathbb{R}^d) \text{ and } g = 0 \text{ almost everywhere outside } B\}.$$

Elements of $PW_B$ are called *bandlimited* or *Paley Wiener functions*.

On the one hand $PW_B$ is a Hilbert space with respect to the usual scalar product on $L_2(\mathbb{R})$, on the other hand its elements are continuous functions. Moreover

**Proposition 1.1.** *Let $B \subset \mathbb{R}^d$ be bounded and have positive measure.*

a) *$PW_B$ is a closed linear subspace of $L_2(\mathbb{R})$, and thus a Hilbert space with respect to the usual scalar product on $L_2(\mathbb{R})$.*

b) *The elements of $PW_B$ have an analytic extention onto all of $\mathbb{C}^d$. In particular Dirac measures are well defined continuous functionals on $PW_B$.*

c) $f \in PW_B$ then $f_{x_j} \in PW_B$, for $j = 1, 2 \ldots d$.

**Definition 1.2.** A sequence $(e_j)$ in a Hilbert space $H$ is called a *Riesz basis* if it is an unconditional basis of $H$. Usually we will then denote the coordinate for $(e_j)$ by $(e_j^*)$. Thus for every $x \in H$

$$x = \sum_{j=1}^{\infty} \langle e_j^*, x \rangle e_j,$$

and this convergence is unconditional in $H$.

A sequence $(x_j : j \in \mathbb{N}) \subset \mathbb{R}^d$ is called *Riesz basis sequence* for $L_2(B)$, $m(B) > 0$, if the sequence of exponentials $(e^{-i\langle x_j, (\cdot) \rangle} : j \in \mathbb{N})$ is a Riesz basis for $L_2(B)$.

From the boundedness of the coordinate functionals $(e_k^*)$ we can easily deduce that a Riesz basis sequence $(x_j) \subset \mathbb{R}^d$ is *uniformly separated*, which means that

$$q = \inf_k \|x_k - x_{k+1}\| = 2 > 0.$$

Indeed, otherwise we could choose a subsequence $k_j \subset \mathbb{Z}$ such that $\lim_{j \to \infty} \|x_{k_j} - x_{k_j+1}\|_2 = 0$, and thus, we deduce (putting $e_k = e^{i\langle x_k, (\cdot) \rangle}$, for $k \in \mathbb{Z}$)

$$\limsup_{j \to \infty} \|e_{k_j}^*\| \le \limsup_{j \to \infty} \left( e_{k_j}^* \frac{e_{k_j+1} - e_{k_j}}{\|e_{k_j+1} - e_{k_j}\|} \right) = \infty,$$

which is a contradiction.

If $d = 1$ then we can and will therefore assume that $x_j$ is indexed by $\mathbb{Z}$ and strictly increasing.

**Remark 1.1.** Note that Riesz bases sequences and *sampling bases* for Paley Wiener functions (see end of this section) are closely connected. Indeed if $(x_k)$ is a Riesz basis sequence for $L_2(B)$ and if $e_k^* \in L_2(B)$, $k \in \mathbb{N}$, are the $k$-th coordinate functional. Then it follows that $(e_k^*)$ is also a Riesz basis of $L_2(B)$ whose coordinate functionals are $(e^{-i\langle x_k, (\cdot) \rangle})$, and we can therefore write for $f \in PW_B$ that

$$\hat{f}(x) = \sum_{j \in \mathbb{N}} \langle \hat{f}, e^{-i\langle x_j, (\cdot) \rangle} \rangle_B e_j^*(x)$$

$$= \sum_{j \in \mathbb{N}} \int_{\mathbb{R}^d} \hat{f}(t) e^{i\langle x_j, t \rangle} dt \, e_j^*(x) \quad (\text{since } \operatorname{supp}(\hat{f}) \subset B)$$

$$= (2\pi)^d \sum_{j \in \mathbb{N}} f(x_j) e_j^*(x),$$

where $\langle \cdot, \cdot \rangle_B$ denotes the usual scalar product in $L_2(B)$ and where above equalities hold for almost all $x \in B$.

Thus, using again the inverse formula, we obtain

$$f(x) = (2\pi)^d \sum_{j \in \mathbb{Z}} f(x_k)[e_j^*]^\vee(x), \tag{2}$$

which implies that the Dirac measures in $x_k$, $k \in \mathbb{N}$, are the coordinate functionals for $([e_j^*]^\vee)$ which is a Riesz basis of $PW_B$ since $\mathcal{F}^{-1}$ is an isomorphism.

**Remark 1.2.** Since unconditional bases in Hilbert spaces are unique, up to equivalence (c.f. [1, Theorem 8.3.5]), every Riesz basis $(e_j)$ of $H$ must be equivalent to an orthonormal basis, and thus, there are numbers $0 < a < b$ so that for every $x$

$$a^2 \|x\|^2 \leq \sum_j |\langle x, e_j^* \rangle|^2 \leq b^2 \|x\|^2. \tag{3}$$

**Definition 1.3.** A sequence $(f_j)$ in a Hilbert space $H$ is called *frame for $H$* or *Hilbert frame for $H$* if there are $0 < a < b$ so that

$$a^2 \|x\|^2 \leq \sum_{j \in \mathbb{N}} |\langle x, f_j \rangle|^2 \leq b^2 \|x\|^2. \tag{4}$$

Reconstruction of $x \in H$ from $(\langle x, f_j \rangle)_{j \in \mathbb{N}}$:

$\Theta : H \to \ell_2,$  $x \mapsto (\langle x, f_j \rangle)_{j \in \mathbb{N}}$   *Analysis operator,*

$\Theta^* : \ell_2 \to H,$  $(\xi_j)_{j \in \mathbb{N}} \mapsto \sum_{j \in \mathbb{N}} \xi_j f_j$   adjoint

$S = \Theta^* \circ \Theta : H \to H,$  $x \mapsto \sum_{j \in \mathbb{N}} \langle x, f_j \rangle f_j$   Frame transform.

Since

$$a^2 \|x\|^2 \leq \sum_{j \in \mathbb{N}} |\langle x, f_j \rangle|^2 = \left\langle x, \sum_{j \in \mathbb{N}} \langle x, f_j \rangle f_j \right\rangle = \langle x, S(x) \rangle \leq b^2 \|x\|^2,$$

$S$ is a positive and invertible operator with

$$a \text{Id}_H \leq S \leq b \text{Id}_H.$$

Thus,

$$x = S \circ S^{-1}(x) = \sum_{j \in \mathbb{N}} \langle S^{-1}(x), f_j \rangle f_j = \sum_{j \in \mathbb{N}} \langle x, S^{-1}(f_j) \rangle f_j, \tag{5}$$

and the series converges unconditionally in $L_2$. As in the case of unconditional bases we deduce from the Uniform Boundedness Principle that there is a constant $R$ so that

$$\left\|\sum_{j\in\mathbb{N}} a_j \langle x, S^{-1}(f_j)\rangle f_j\right\| \le R\|x\| \qquad (6)$$

for all $x \in H$ and all $(a_j)$, $|a_j| \le 1$, if $j \in \mathbb{N}$.

We call (the smallest) $R$ the *unconditional constant of* $(f_j)$) and we call $S^{-1}(f_j)$ *the coordinate functionals with respect to* $(f_j)$. In the case that $a = b$ we say that $(f_i)$ is a *tight frame*. Note that in this case we obtain

$$x = \frac{1}{a} S(x) = \frac{1}{a} \sum_{j\in\mathbb{N}} \langle x, f_j\rangle f_j. \qquad (7)$$

We call the equations (5), and (7) the *expansion of $x$ with respect to the frame* $(f_j)_{j\in\mathbb{N}}$.

An important special case is $L_2(B))$, $m(B) > 0$, and we ask when for a given sequence $(x_n : n \in \mathbb{N}) \subset \mathbb{R}^d$ the sequence of exponentials

$$\mathbb{R}^d \to \mathbb{C}, \quad \xi \mapsto e^{-i\langle \xi, x_n\rangle},$$

is a frame for $L_2(B)$ with bounds $0 < a \le b$. In that case we call $(x_n : n \in \mathbb{N})$ a *Fourier Frame for* $L_2(B)$. Using the inverse formula for the Fourier transform we obtain for any $f \in PW_B$

$$f(x) = \frac{1}{(2\pi)^d} \int \hat{f}(\xi) e^{i\langle x,\xi\rangle} d\xi = \frac{1}{(2\pi)^d} \langle \hat{f}, e^{-i\langle x, \cdot\rangle}\rangle_B.$$

Thus $(x_n) \subset \mathbb{R}^d$ is a Fourier Frame for $L_2(B)$ with bounds $0 < a < b$ if and only if for all $f \in PW_B$

$$\frac{a^2}{(2\pi)^d} \le \sum_n |f(x_n)|^2 \le \frac{b^2}{(2\pi)^d}. \qquad (8)$$

**Main Goal 1.1.** Bandlimited functions are used to represent signals. These signal need to be measured, stored and reproduced. It is therefore paramount to find *good bases or frames* for the space $PW_B$. Here are some properties one seeks:

a) Bases or frames which consist of translations of the same function (*Translation bases*) and/or the corresponding functionals are evaluations at *sampling points* (*Sampling bases*). I.e. there is a strictly increasing sequence $(x_n : n \in \mathbb{Z})$ and a function $g$ on $\mathbb{R}^d$ so that

$$f(x) = \sum_{n\in\mathbb{Z}} f(x_n) g(x - x_n). \qquad (9)$$

In case that the *sampling points* $(x_n)$ are not equally distributed this is in general not achievable. So one requires the existence of a matrix $B = (b_{(m,n)} : m, n \in \mathbb{Z})$ so that

$$f(x) = \sum_{n \in \mathbb{Z}} (B \circ (f(x_j) : j \in \mathbb{Z}))_n g(x - x_n), \quad x \in \mathbb{R}^d. \qquad (10)$$

b) *Stability.* Assuming the functions values $f(x_n)$ were measured with some error one seeks to control the error of the function values of $f$, as represented by (9) or (10). More precisely assume that $\varepsilon > 0$ and assume that $(\tilde{f}_j)$ is such that $\|\tilde{f}_j - f(x_j)\| \leq \varepsilon$. Let

$$\tilde{f}(x) = \sum_{n \in \mathbb{Z}} \tilde{f}_n g(x - x_n), \text{ respectively}$$

$$\tilde{f}(x) = \sum_{n \in \mathbb{Z}} (B \circ (\tilde{f}_j) : j \in \mathbb{Z}))_n g(x - x_n).$$

For which bases/frames is it possible to estimate $\|f - \tilde{f}\|_\infty$?

(c) *Quantization.* For which frames can in (9) and (10) the $f(x_j)$'s be replaced by *quantized coefficients*, i.e. integer values of some $q > 0$, and still approximate sufficiently $f$? Are there quantization algorithms easy to implement, which are easy to implement?

## 2. Equally distributed sampling points in $\mathbb{R}$

Assume that $f \in PW_\pi$. Thus

$$\text{supp}(\hat{f}) = \{x \in \mathbb{R} : \hat{f}(x) \neq 0\} \subset [-\pi, \pi].$$

We will first derive the *Shannon Whittaker formula*. Since $(g_n)_{n \in \mathbb{Z}}$, with

$$g_n(\xi) = \frac{1}{\sqrt{2\pi}} e^{-in\xi}, \quad |\xi| \leq \pi, n \in \mathbb{Z},$$

is an orthonormal basis of $L_2([-\pi, \pi])$, we can write

$$\hat{f}(\xi) = \sum_{n \in \mathbb{Z}} c_n e^{-in\xi} \text{ for } |\xi| \leq \pi, \qquad (11)$$

with

$$c_n = \frac{1}{\sqrt{2\pi}} \langle \hat{f}, g_n \rangle = \frac{1}{2\pi} \int_{-\pi}^{\pi} \hat{f}(\xi) e^{in\xi} d\xi = \frac{1}{2\pi} \int_{-\infty}^{\infty} \hat{f}(\xi) e^{in\xi} d\xi = f(n)$$

and thus

$$\hat{f}(\xi) = \sum_{n \in \mathbb{Z}} f(n) e^{-in\xi} \chi_{[-\pi, \pi]}(\xi). \qquad (12)$$

Using the inverse formula yields the *Shannon Whittaker Formula*:

$$f(x) = \frac{1}{2\pi} \int_{-\pi}^{\pi} \sum_{n \in \mathbb{Z}} f(n) e^{-in\xi} e^{ix\xi} d\xi \qquad (13)$$

$$= \frac{1}{2\pi} \sum_{n \in \mathbb{Z}} f(n) \int_{-\pi}^{\pi} e^{-in\xi} e^{ix\xi} d\xi$$

[We can interchange $\sum$ and $\int$ because the inverse Fourier transform is a bounded operator on $L_2(\mathbb{R})$]

$$= \frac{1}{2\pi} \sum_{n \in \mathbb{Z}} f(n) \left[\frac{-i}{x-n} e^{i\xi(x-n)}\right]_{\xi=-\pi}^{\pi}$$

$$= \frac{1}{2\pi} \sum_{n \in \mathbb{Z}} f(n) \frac{2\sin(x\pi - n\pi)}{x-n} = \sum_{n \in \mathbb{Z}} f(n) \operatorname{sinc}(x\pi - n\pi).$$

**Remark 2.1.** We obtain a sampling basis for $PW_\pi$. But the representation in (12) has bad pointwise properties. It is not pointwise unconditionally converging and highly unstable under errors in measuring $f(n)$.

In order to overcome this highly unstable representation we will pass to a redundant representation and follow the description of the *Analog-Digital Conversion* in [4].

We fix $\mu_0 > 1$ and let $\mu_0 \le \mu$. We can view $L_2[-\pi, \pi]$ (in a natural way) as a subspace of $L_2[-\mu\pi, \mu\pi]$. The sequence $(g_n^{(\mu)} : n \in \mathbb{Z})$, with

$$g_n^{(\mu)}(\xi) = \frac{1}{\sqrt{2\mu\pi}} e^{-in\xi/\mu}, \qquad |\xi| \le \mu\pi, n \in \mathbb{Z},$$

is an orthonormal basis for $L_2[-\mu\pi, \mu\pi]$ and as in (12) we obtain

$$\hat{f}(\xi) = \frac{1}{\mu} \sum_{n \in \mathbb{Z}} f\left(\frac{n}{\mu}\right) e^{-in\xi/\mu} \chi_{[-\mu\pi, \mu\pi]}(\xi) \text{ a.s.} \qquad (14)$$

**Remark 2.2.** We can view the sequence $(g_n^{(\mu)}|_{[-\pi,\pi]})_{n \in \mathbb{Z}}$ as a tight frame for $L_2([-\pi, \pi])$. It is the image of an orthonormal basis of $L_2([-\mu\pi, \mu\pi])$ under the canonical projection from $L_2([-\mu\pi, \mu\pi])$ onto $L_2([-\pi, \pi])$.

Since $f$ was assumed to be in $PW_\pi$ it follows that the series (14) vanishes almost surely on the set $[-\mu\pi, -\pi] \cup [\pi, \mu\pi]$. We consider a function $g : \mathbb{R} \to \mathbb{R}$ whose Fourier transform $\hat{g}$ has the following properties:

- $\hat{g}$ is $C_\infty$,
- $\hat{g}(\zeta) = 1/2\pi$, if $|\zeta| \le \pi$,

- $\hat{g}(\zeta) = 0$, if $|\zeta| \geq \mu_0 \pi$, and
- $0 \leq \tilde{g}(\zeta) \leq 1/2\pi$ if $\pi \leq |\zeta| \leq \mu_0 \pi$.

Thus, (14) does not change if we multiply both sides by $2\pi \hat{g}$

$$\hat{f}(\xi) = \frac{1}{\mu} \sum_{n \in \mathbb{Z}} f\left(\frac{n}{\mu}\right) e^{-in\xi/\mu} 2\pi \hat{g}(\xi).$$

Using again the inverse formula for Fourier transforms we arrive to the *Redundant Whittaker Shannon Formula*

$$f(x) = \frac{1}{\mu} \sum_{n \in \mathbb{Z}} f\left(\frac{n}{\mu}\right) \int_{-\infty}^{\infty} e^{ix\xi - in\xi/\mu} 2\pi \hat{g}(\xi) d\xi \qquad (15)$$

$$= \frac{1}{\mu} \sum_{n \in \mathbb{Z}} f\left(\frac{n}{\mu}\right) g\left(x - \frac{n}{\mu}\right).$$

Note that for $z \in \mathbb{R}$, $z \neq 0$,

$$g(z) = \frac{1}{2\pi} \int_{-\infty}^{\infty} \hat{g}(\xi) e^{iz\xi} d\xi$$

$$= \frac{1}{2\pi} \int_{-\infty}^{\infty} \hat{g}'(\xi) \frac{-i}{z} e^{iz\xi} d\xi$$

$$= -\frac{1}{z^2} \frac{1}{2\pi} \int_{-\infty}^{\infty} \hat{g}''(\xi) e^{iz\xi} d\xi$$

$$= \ldots$$

We conclude that $\lim_{z \to \pm\infty} |z|^k g(z) = 0$ for all $k \in \mathbb{N}$ and that the series in (15) converges absolutely and faster than any series of the form $\sum_{n \in \mathbb{N}} n^{-k}$. Moreover, assume that $(\tilde{f}_n : n \in \mathbb{Z}) \subset \mathbb{C}$ is a sequence with

$$\varepsilon := \sup_{n \in \mathbb{Z}} \left| f\left(\frac{n}{\mu}\right) - \tilde{f}_n \right| < \infty,$$

and let

$$\tilde{f}(x) = \sum_{n \in \mathbb{Z}} \tilde{f}_n g\left(x - \frac{n}{\mu}\right) \qquad x \in \mathbb{R}.$$

Since for $x \in \mathbb{R}$ and $n \in \mathbb{Z}$ we obtain

$$\mu^{-1}\left|g\left(x - \frac{n}{\mu}\right)\right| \leq \int_{x-\frac{n}{\mu}}^{x-\frac{n-1}{\mu}} |g(\xi)| d\xi + \mu^{-1} \sup_{x-n/\mu \leq \xi \leq x-(n-1)/\mu} \left| g\left(x - \frac{n}{\mu}\right) - g(\xi) \right|$$

$$\leq \int_{x-\frac{n}{\mu}}^{x-\frac{n-1}{\mu}} |g(\xi)| d\xi + \mu^{-1} \int_{x-\frac{n}{\mu}}^{x-\frac{n-1}{\mu}} |g'(\xi)| d\xi,$$

it follows that

$$|\tilde{f}(x) - f(x)| \le \frac{\varepsilon}{\mu} \sum_{n \in \mathbb{Z}} \left| g\left(x - \frac{n}{\mu}\right) \right| \qquad (16)$$

$$\le \varepsilon \sum_{n \in \mathbb{Z}} \left[ \int_{x-\frac{n}{\mu}}^{x-\frac{n-1}{\mu}} |g(\xi)| d\xi + \mu^{-1} \int_{x-\frac{n}{\mu}}^{x-\frac{n-1}{\mu}} |g'(\xi)| d\xi \right]$$

$$\le \varepsilon (\|g\|_{L_1} + \mu^{-1} \|g'\|_{L_1}),$$

which proves stability of the representation (15).

We now turn to the problem to *quantize* the representation (15) and to replace the coefficients $f(x_n)$, $n \in \mathbb{N}$, by some integer multiples of a fixed number $q > 0$. The simplest way to do so would be to replace $f(x_n)$ by its closest integer multiple of $q$, i.e. choose $k_n \in \mathbb{Z}$ so that $|f(n/\mu) - qk_n|$ is minimal. The approximation formula (16) would then give (choosing $\tilde{f}_n = k_n q$)

$$\|\tilde{f}(x) - f(x)\| \le \sup_{n \in \mathbb{N}} |\tilde{f}_n - f(n/\mu)| (\|g\|_{L_1} + \mu^{-1} \|g'\|_{L_1}) \qquad (17)$$

$$\le q(\|g\|_{L_1} + \mu^{-1} \|g'\|_{L_1}).$$

This approximation does not use the redundancy in (15), i.e. the fact that the representation is not unique. The *First Order Sigma - Delta Procedure*, which we will describe now, takes into account previous errors and tries to cancel them. It thereby achieves a much better approximation.

In order to simplify our exposition we assume that our Paley-Wiener function $f$ is real valued and bounded by 1, i.e $\|f\|_\infty \le 1$. We will attempt to replace the sampling values $f(n/\mu)$ by either 1 or $-1$, which corresponds to an A/D conversion (analogue to digital).

**Definition 2.1.** [First Order Sigma-Delta Algorithm]

Assume $f \in PW_\pi$ is real valued and $\|f\|_\infty \le 1$. We are defining recursively sequences $(u_n : n \in \mathbb{Z}) \subset \mathbb{R}$ and $(q_n : n \in \mathbb{Z}) \subset \{\pm 1\}$:

First we let $u_0 = 0$. Assuming that for some $n \in \mathbb{N}$ $u_{n-1}$ has been determined we put:

$$q_n = \text{sign}(u_{n-1} + f(n/\mu)) \text{ and } u_n = u_{n-1} + f(n/\mu) - q_n. \qquad (18)$$

If $n \in -\mathbb{N}$ and if $u_{n+1}$ has been defined then

$$u_n = u_{n+1} - f((n+1)/\mu) - \text{sign}(u_{n+1} - f((n+1)/\mu)). \qquad (19)$$

If $n \in -\mathbb{N}$ or $n = 0$

$$q_n = \text{sign}(u_{n-1} + f(n/\mu)). \qquad (20)$$

**Lemma 2.1.** *Assume* $f \in PW_\pi$ *is real valued with* $\|f\|_\infty \leq 1$, *and that the sequences* $(u_n : n \in \mathbb{Z})$ *and* $(q_n : n \in \mathbb{Z})$ *are defined as in Definition 2.1. Then for every* $n \in \mathbb{Z}$

$$|u_n| < 1 \text{ and} \tag{21}$$
$$u_n - u_{n-1} = f(n/\mu) - q_n. \tag{22}$$

**Proof.** Assume that $n \in \mathbb{N}$ and that $|u_{n-1}| < 1$. Since $\|f\|_{L_\infty} \leq 1$ it follows that $|u_{n-1} + f(n/\mu)| < 2$ and thus

$$|u_n| = |u_{n-1} + f(n/\mu) - q_n| = |u_{n-1} + f(n/\mu) - \text{sign}(f(n/\mu) + u_{n-1})| < 1.$$

(22) follows from definition of $u_n$ in the case that $n \in \mathbb{N}$.

Assuming that $n \in -\mathbb{N}$ and that $|u_{n+1}| < 1$, it follows from the assumption on $f$ that $|u_{n+1} - f((n+1)/\mu)| < 2$ and thus

$$|u_n| = \big|u_{n+1} - f((n+1)/\mu) - \text{sign}\big(u_{n+1} - f((n+1)/\mu)\big)\big| < 1.$$

In order to show (22) we rewriting (19) into

$$u_n = u_{n-1} + f(n/\mu) + \text{sign}\big(u_n - f(n/\mu)\big).$$

Since $|u_n| < 1$ it follows that $\text{sign}\big(u_{n-1} + f(n/\mu)\big) = -\text{sign}\big(u_n - f(n/\mu)\big)$, which yields

$$q_n = \text{sign}\big(u_{n-1} + f(n/\mu)\big) = -\text{sign}\big(u_n - f(n/\mu)\big),$$

and

$$u_n - u_{n-1} = f(n/\mu) + \text{sign}\big(u_n - f(n/\mu)\big) = f(n/\mu) - q_n.$$

This equation is also true for $n = 0$, since

$$f(0) - q_0 = f(0) - \text{sign}(u_{-1} + f(0))$$
$$= f(0) + \text{sign}(u_0 - f(0)) = -u_1 = u_0 - u_1. \quad \square$$

**Proposition 2.1.** *If* $f \in PW_\pi$ *is real valued with* $\|f\|_\infty \leq 1$ *and if for* $\mu > 1$ *the sequences* $(u_n)_{n \in \mathbb{Z}}$ *and* $(q_n)_{n \in \mathbb{Z}}$ *are defined as in (18), (19) and (20) then*

$$\left|f(t) - \frac{1}{\mu}\sum_{n \in \mathbb{Z}} q_n g\left(t - \frac{n}{\mu}\right)\right| \leq \frac{1}{\mu}\|g'\|_{L_1} \text{ for all } t \in \mathbb{R}.$$

**Proof.** For $t \in \mathbb{R}$

$$\left| f(t) - \frac{1}{\mu} \sum_{n \in \mathbb{Z}} q_n g\left(t - \frac{n}{\mu}\right) \right|$$

$$= \frac{1}{\mu} \left| \sum_{n \in \mathbb{Z}} \left[ f\left(\frac{n}{\mu}\right) - q_n \right] g\left(t - \frac{n}{\mu}\right) \right| \quad \text{by (16)}$$

$$= \frac{1}{\mu} \left| \sum_{n \in \mathbb{Z}} (u_n - u_{n-1}) g\left(t - \frac{n}{\mu}\right) \right| \quad \text{by (22)}$$

$$= \frac{1}{\mu} \left| \sum_{n \in \mathbb{Z}} u_n \left[ g\left(t - \frac{n}{\mu}\right) - g\left(t - \frac{n+1}{\mu}\right) \right] \right|$$

$$\leq \frac{1}{\mu} \sum_{n \in \mathbb{Z}} \left| g\left(t - \frac{n}{\mu}\right) - g\left(t - \frac{n+1}{\mu}\right) \right| \quad \text{by (21)}$$

$$\leq \frac{1}{\mu} \sum_{n \in \mathbb{Z}} \int_{t - \frac{n+1}{\mu}}^{t - \frac{n}{\mu}} |g'(x)| dx = \frac{1}{\mu} \|g'\|_{L_1}. \qquad \square$$

## 3. Scattered sampling points

Now we assume that instead of having equally distributed sampling points (i.e $x_n = n$ or more generally $x_n = n/\mu$, for $n \in \mathbb{Z}$) we sample the function at non necessarily equally distributed points $(x_n : n \in \mathbb{Z}) \subset \mathbb{R}$. The wellknown "$\frac{1}{4}$- Theorem of Kadec's states a criterion for the sequence of exponentials $(e^{-ix_n(\cdot)} : n \in \mathbb{Z})$ to be a Riesz basis (i.e. unconditional basis) for $L_2[-1,1]$ (we change now our scaling from the interval $[-\pi, \pi]$ to $[-1, 1]$ because this produces shorter formulae).

**Theorem 3.1 ([6]).** *Assume that $(x_n : n \in \mathbb{N}) \subset \mathbb{R}$ and that*

$$L := \sup_{n \in \mathbb{N}} |x_n - n| < \frac{1}{4}. \qquad (23)$$

*Then $(e^{ix_n(\cdot)} : n \in \mathbb{Z})$ is a Riesz basis for $L_2[-1,1]$.*

**Corollary 3.1.** *Assume that $(x_n : n \in \mathbb{Z}) \subset \mathbb{R}$ and that*

$$L := \sup_{n \in \mathbb{N}} |x_n - n| < \frac{1}{4} \qquad (24)$$

*and that $\mu \geq 1$. Then $(e^{ix_n(\cdot)/\mu} : n \in \mathbb{Z})$ is a frame for $L_2[-1,1]$.*

The following observation follows from the easy fact that the tensor product of Riesz bases is a Riesz basis of the Hilbert space tensor product.

**Proposition 3.1.** *If $d \in \mathbb{N}$ and if for $j = 1, 2, \ldots d$ the sequences $(x_n^{(j)} : n \in \mathbb{Z})$ are Riesz bases or Fourier frame for $L_2[-1,1]$ then the sequence $(\overline{x}_{\overline{n}} : \overline{n} \in \mathbb{Z}^d)$, with $\overline{x}_{\overline{n}} = (x_{n_1}^{(1)}, x_{n_2}^{(2)}, \ldots x_{n_d}^{(d)})$, for $\overline{n} = (n_1, n_2, \ldots n_d) \in \mathbb{Z}^d$, is a Riesz bases sequence or Fourier frame for $L_2[-1,1]^d$.*

**Remark 3.1.** Proposition 3.1 together with Theorem 3.1 prove the existence of Riesz bases sequences and determine geometrical conditions for a sequence $(x_k) \subset \mathbb{R}^d$ to be a Riesz basis sequence for $L_2[-1,1]^d$. By scaling we can get Riesz basis sequences for $L_2(\prod_{j=1}^d [a_j, b_j])$, with $a_j < b_j$, $j = 1, 2 \ldots d$.

We do not know of any Riesz basis sequence for $L_2(B)$ for other sets $B \subset \mathbb{R}^d$, for example we do not know if there is a Riesz basis sequence for $L_2(B_2)$, where $B_2$ is the Eucledian unit ball in $\mathbb{R}^d$.

Using the following deep Theorem by Beurling, one obtains frames which are not necessarily tensor products of frames on subsets of $\mathbb{R}^d$. We first need the following notation: For a countable set $\Lambda \subset \mathbb{R}^d$ we put

$$r(\Lambda) := \frac{1}{2} \inf_{\mu, \mu' \in \Lambda, \, \mu \neq \mu'} \|\mu - \mu'\|_2, \text{ and}$$

$$R(\Lambda) := \sup_{\xi \in \mathbb{R}} \inf_{\mu \in \Lambda} \|\xi - \mu\|_2.$$

We call $\Lambda$ *uniformly discrete* if $r(\Lambda) > 0$.

**Theorem 3.2.** *[[3]] If $\Lambda$ is a uniformly discrete set with $R(\Lambda) < \frac{\pi}{2}$ then it is a Fourier frame for $L_2(B_2)$, with $B_2 = \{x \in \mathbb{R} : \|x\|_2 \leq 1\}$, and thus also a Fourier frame for $L_2(B)$ with $B \subset B_2$.*

*Moreover the frame bounds only depend on $r(\Lambda)$ and $R(\Lambda)$ and $d$.*

The following result is a generalization of the redundant version of the Whittaker Shannon formula (15) and was recently obtained by my student Aaron Bailey.

**Theorem 3.3.** *[2] Let $B \subset \mathbb{R}^d$ be symmetric, convex and bounded, with $B^\circ \neq \emptyset$. Assume that $(x_n : n \in \mathbb{N}) \subset \mathbb{R}^d$ is a Fourier frame for $L_2(B)$, and assume that $S$ is the corresponding frame operator, i.e.:*

$$S : L_2(B) \to L_2(B) \quad f \mapsto \sum_{n \in \mathbb{N}} \langle f, e^{-ix_n(\cdot)} \rangle_B e^{-i\langle x_n, \cdot \rangle}.$$

*Let $\mu \geq \mu_0 > 0$ Then there is a $g \in PW_{\mu_0 B}$ so that for all $f \in PW_{\mu B}$*

$$f(x) = \frac{1}{\mu} \sum_{n \in \mathbb{N}} \left[ G \circ \left( f\left(\frac{x_k}{\mu}\right) : k \in \mathbb{N} \right) \right]_n g\left(x - \frac{x_n}{\mu}\right), \quad x \in \mathbb{R}^d, \quad (25)$$

where $G$ is the Grammian matrix for $S$ with respect to the frame $(e^{-i\langle x_n,\cdot\rangle} : n \in \mathbb{N})$, i.e. $G = (G_{(n,k)})_{n,k \in \mathbb{N}}$ with

$$G_{(n,k)} = \langle S^{-1}(e^{-i\langle x_n,\cdot\rangle}), S^{-1}(e^{ix_k(\cdot)})\rangle_B, \quad n,k \in \mathbb{N}.$$

**Remark 3.2 ([2]).** *The representation (25) is in general not even in the one dimensional case stable under perturbations of $f(x_n)$'s because in general $G$ is not bounded as operator on $\ell_\infty$.*

We will now pass to approximative representations of Paley Wiener functions using translates of Gaussians.

For $\lambda > 0$ we define the *Gaussian function* $g_\lambda : \mathbb{R} \to \mathbb{R}$ by

$$g_\lambda(x) = e^{-\lambda x^2}, \quad \text{for all } x \in \mathbb{R},$$

and recall that

$$\hat{g}_\lambda(u) = \left(\frac{\pi}{\lambda}\right)^{1/2} e^{-u^2/(4\lambda)}, \quad \text{for all } u \in \mathbb{R}. \tag{26}$$

Let $(x_j)_{j\in\mathbb{Z}}$ be a Riesz basis sequence for $L_2[-1,1]$, $\lambda > 0$ and let $f \in PW_1$. We would like to *interpolate $f$ at the points $(x_j)$ using Gaussians shifted by $x_j$*, this means we would like to define $I_\lambda(f)$ to be a function of the form

$$I_\lambda(f) : \mathbb{R} \to \mathbb{R}, \quad \xi \mapsto \sum_{j\in\mathbb{Z}} a_n^{(\lambda)} g_\lambda(\xi - x_k),$$

which satisfies the interpolation condition:

$$I_\lambda(f)(x_j) = f(x_j), \quad j \in \mathbb{Z}.$$

To show the existence of $I_\lambda(f)$ as well as the fact that $I_\lambda(f) \in L_2(\mathbb{R}) \cap C_0(\mathbb{R})$, we need the following result. A sketch of a proof can be found in [5]. For the sake of completeness we include a proof in the appendix.

**Theorem 3.4.** *(See Appendix) Let $\lambda > 0$, and let $(x_j : j \in \mathbb{Z})$ be a sequence of real numbers satisfying the following condition: there exists a positive number $q$ such that $x_{j+1} - x_j \geq q$ for every $j \in \mathbb{Z}$. Let $A := A_\mu$ be a bi-infinite matrix whose entries are given by $A(j,k) := e^{-\mu(x_j-x_k)^2}$, $j,k \in \mathbb{Z}$. Then there exist positive constants $\beta_1$ and $\gamma_1$, depending on $\mu$ and $q$, such that $|A^{-1}(s,t)| \leq \beta_1 e^{-\gamma_1|s-t|}$, $s,t \in \mathbb{Z}$.*

Theorem 3.4 allows us to deduce:

**Proposition 3.2.** *Let $\lambda$, $(x_j : j \in \mathbb{Z})$, $q$, and $A$ be as above. The operator $A^{-1}$ acts boundedly on every $\ell_p(\mathbb{Z})$, $1 \leq p \leq \infty$.*

**Proof.** Thanks to the symmetry of $A^{-1}$ and the M. Riesz Convexity Theorem, or the Riesz-Thorin Interpolation Theorem, it is sufficient to verify that $A^{-1}$ is a bounded operator on $\ell_\infty(\mathbb{Z})$. The latter fact is a consequence of the observation that, for any fixed $s \in \mathbb{Z}$,

$$\sum_{t \in \mathbb{Z}} |A^{-1}(s,t)| \leq \beta_1 \sum_{t \in \mathbb{Z}} e^{-\gamma_1 |s-t|} \leq \frac{2}{1 - e^{-\gamma_1}}. \qquad \square$$

The interpolation operators, whose study will occupy the rest of the paper, are introduced in the following Proposition; its proof will be given in the next section.

**Proposition 3.3.** *Let $\lambda > 0$, and let $(x_j : j \in \mathbb{Z}) \subset \mathbb{R}$ be a Riesz basis sequence for $PW_1$. For any $f \in PW_1$, there exists a unique square-summable sequence $(a_j^{(\lambda)} : j \in \mathbb{Z})$ such that*

$$\sum_{j \in \mathbb{N}} a_j^{(\lambda)} g_\lambda(x_k - x_j) = f(x_k), \quad k \in \mathbb{Z}. \qquad (27)$$

*The* Gaussian Interpolation Operator $I_\lambda : PW_1 \to L_2(\mathbb{R})$, *defined by*

$$I_\lambda(f)(\cdot) = \sum_{j \in \mathbb{N}} a_j^{(\lambda)} g_\lambda((\cdot) - x_j), \qquad (28)$$

*where $(a_j^{(\lambda)} : j \in \mathbb{N})$ satisfies (27), is a well-defined, bounded linear operator from $PW_1$ to $L_2(\mathbb{R})$. Moreover, $I_\lambda(f) \in C_0(\mathbb{R})$.*

We now state the main result.

**Theorem 3.5.** *[[9]] Suppose that $(x_j : j \in \mathbb{Z})$ is a Riesz basis sequence for $PW_1$. Let $I_\lambda$, $\lambda > 0$, be the associated Gaussian Interpolation Operator. Then for every $f \in PW_1$ we have $f = \lim_{\lambda \to 0^+} I_\lambda(f)$ in $L_2(\mathbb{R})$ and uniformly on $\mathbb{R}$.*

## 4. Proof of Proposition 3.3 and Theorem 3.5

Our proof of Theorem 3.5 is different from the one given in [9], it allows to be extended to the multidimensional case (see remarks in the next section).

Since $(x_k)$ is a Riesz basis sequence, it is uniformly separated by some $q > 0$. For $f \in PW_1$ the sequence

$$\left( f(x_k) : k \in \mathbb{Z} \right) = \left( \langle \hat{f}, e^{-ix_k(\cdot)} \rangle_{[-1,1]} / 2\pi : k \in \mathbb{Z} \right)$$

is in $\ell_2(\mathbb{Z})$ and, thus it follows from Proposition 3.2 that there is a sequence $(a_k^{(\lambda)} : k \in \mathbb{Z}) \in \ell_2(\mathbb{Z})$ satisfying (27).

As shown in [8, Lemma 2.1] there is a number $\nu > 0$ which only depends on $\lambda$ and $q$ so that $\sum_j g_\lambda(x - x_j) \le \nu$. This implies that the series $\sum_j a_j^{(\lambda)} g_\lambda(x - x_j)$ is uniformly bounded and since each summand is continuous the uniform convergence implies that $I_\lambda$ is continuous. Since $\lim_{j \to \pm \infty} a_j^{(\lambda)} = 0$ it follows moreover that $I_\lambda \in C_0(\mathbb{R})$.

For $m \in \mathbb{N}$ we define a linear bounded operator $A_m$ on $L_2[-1,1]$ as follows: Let $(e_k^*) \subset L_2[-1,1]$ be the coordinate functionals as introduced in Definition 1.3, i.e., for every $h \in L_2[-1,1]$,

$$h = \sum_{k \in \mathbb{Z}} \langle h, e_k^* \rangle_{[-1,1]} e^{-ix_k(\cdot)} = \sum_{k \in \mathbb{Z}} \int_{-1}^{1} h(\xi)\overline{e_k^*(\xi)}\, d\xi\, e^{ix_k(\cdot)}. \tag{29}$$

Note that for $a \in \mathbb{R}$ we have

$$\left\| \sum_{k \in \mathbb{Z}} \langle h, e_k^* \rangle_{[-1,1]} e^{-i(\cdot)x_k} \right\|_{L_2(a+[-1,1])} \tag{30}$$

$$= \left\| \sum_{k \in \mathbb{Z}} \langle h, e_k^* \rangle_{[-1,1]} e^{-i(\cdot+a)x_k} \right\|_{L_2[-1,1]}$$

$$= \left\| \sum_{k \in \mathbb{Z}} e^{-iax_k} \langle h, e_k^* \rangle_{[-1,1]} e^{-i(\cdot)x_k} \right\|_{L_2[-1,1]}$$

$$\le R\|h\|_{L_2[-1,1]},$$

where $R$ is the unconditional constant of the basis $(e^{-i(\cdot)x_k})$. We can therefore extend $h$ to a locally integrable and almost everywhere defined function on all of $\mathbb{R}$, by simply putting

$$E(h)(x) = \sum_{k \in \mathbb{Z}} \langle h, e_k^* \rangle_{[-1,1]} e^{-ixx_k}, \quad x \in \mathbb{R}. \tag{31}$$

Let $m \in \mathbb{N}$, and define $A_m : L_2[-1,1] \to L_2[-1,1]$ by

$$A_m(h)(\xi) = E(h)(2^m(\xi))\chi_{[-1,-\frac{1}{2}]\cup[\frac{1}{2},1]}(\xi) \tag{32}$$

For $h \in L_2[-1,1]$ it follows from (30) that

$$\|A_m(h)\|_{L_2[-1,1]}^2 = \int_{[-1,-\frac{1}{2}]\cup[\frac{1}{2},1]} |E(h)(2^m u)|^2 du \tag{33}$$

$$= 2^{-m} \int_{2^m[-1,1]\setminus 2^{m-1}[-1,1]} |E(h)(v)|^2 dv$$

$$\le R^2 \|h\|_{L_2[-1,1]}^2,$$

where the last inequality follows form the fact that we need not more than $2^m$ translates of $[-1,1]$ to cover $2^m[-1,1] \setminus 2^{m-1}[-1,1]$.

**Proof of Proposition 3.3.** Let $\lambda > 0$. By (3) and Proposition 3.2, there is a positive constant $\kappa$ so that, for each $f \in PW_1$, there is a sequence $(a_j^{(\lambda)}) \in \ell_2$ satisfying (27) and the estimate

$$\|(a_j^{(\lambda)})\|_2 \leq \kappa \|f\|_2. \tag{34}$$

In order to show that $I_\lambda(f)$ is in $L_2(\mathbb{R})$ we first consider the function

$$w: \mathbb{R} \ni x \mapsto \left(\frac{\pi}{\lambda}\right)^{1/2} e^{-x^2/(4\lambda)} \sum_{k \in \mathbb{Z}} a_k^{(\lambda)} e^{-ixx_k},$$

and note that it belongs to $L_2(\mathbb{R}) \cap L_1(\mathbb{R})$. Then we notice that, applying $\mathcal{F}^{-1}$ (which is an isomorphism on $L_2$) gives us $I_\lambda(f)$. Moreover, using (30), (4), and (34), we arrive at the estimate

$$\|w\|_{L_2(\mathbb{R})} \leq C' \|f\|_{L_2(\mathbb{R})}, \tag{35}$$

where $C'$ depends only on $\lambda$ and $R$. This proves that $I_\lambda$ is a bounded operator □

**Proof of Theorem 3.5.**

Now fix $f \in PW_1$ and write $I_\lambda(f)$ as

$$I_\lambda(f)(\cdot) = \sum_{j \in \mathbb{N}} a_j^{(\lambda)} g_\lambda((\cdot) - x_j).$$

Recall from the preceding paragraph that the Fourier transform of $I_\lambda(f)$ is given by

$$\mathcal{F}[I_\lambda(f)](u) = \left(\frac{\pi}{\lambda}\right)^{1/2} e^{-u^2/(4\lambda)} \sum_{j \in \mathbb{N}} a_j^{(\lambda)} e^{-ix_j u}, \quad u \in \mathbb{R}. \tag{36}$$

The proof of Theorem 3.5 proceeds in three steps.
**Step 1.** We claim that there is a constant $D_1 < \infty$ and $\lambda_0 > 0$, only depending on $(x_j)$, so that

$$\sup_{0 < \lambda \leq \lambda_0} \|I_\lambda(f)\|_2 \leq D_1 \|f\|_2.$$

We start by defining

$$H_\lambda(u) = \left(\frac{\pi}{\lambda}\right)^{1/2} \sum_{j \in \mathbb{N}} a_j^{(\lambda)} e^{-ix_j u} = e^{u^2/(4\lambda)} \mathcal{F}[I_\lambda(f)](u), \quad u \in \mathbb{R},$$

and let $h_\lambda = H_\lambda|_{[-1,1]} \in L_2[-1,1]$. Thus, $E(h_\lambda) = H_\lambda$.

Suppose that $k \in \mathbb{Z}$. The Inverse formula for Fourier transforms implies that

$$2\pi f(x_k) = \int_{-1}^{1} \mathcal{F}[f](u) e^{ix_k u}\, du = \langle \mathcal{F}[f], e^{-ix_k(\cdot)}\rangle_{[-1,1]}. \tag{37}$$

On the other hand, equations (27) and (28) assert that

$$2\pi f(x_k) \tag{38}$$
$$= 2\pi I_\lambda(f)(x_k)$$
$$= \int_{\mathbb{R}} \mathcal{F}[I_\lambda(f)](u) e^{ix_k u}\, du \quad \text{(by (5))}$$
$$= \int_{\mathbb{R}} e^{-u^2/(4\lambda)} H_\lambda(u) e^{ix_k u}\, du$$
$$= \int_{-1}^{1} e^{-u^2/(4\lambda)} H_\lambda(u) e^{ix_k u}\, du$$
$$\quad + \sum_{m=1}^{\infty} \int_{2^m[-1,1]\setminus 2^{m-1}[-1,1]} e^{-u^2/(4\lambda)} H_\lambda(u) e^{ix_k u}\, du$$
$$= \int_{-1}^{1} e^{-u^2/(4\lambda)} H_\lambda(u) e^{ix_k u}\, du$$
$$\quad + \sum_{m=1}^{\infty} 2^m \int_{[-1,-\frac{1}{2}]\cup[\frac{1}{2},1]} e^{-2^m v^2/(4\lambda)} H_\lambda(2^m v) e^{ix_k 2^m v}\, dv$$
$$= \int_{-1}^{1} e^{-u^2/(4\lambda)} h_\lambda(u) e^{ix_k u}\, du$$
$$\quad + \sum_{m=1}^{\infty} 2^m \int_{[-1,-\frac{1}{2}]\cup[\frac{1}{2},1]} e^{-2^m v^2/(4\lambda)} A_m(h_\lambda)(v) \overline{A_m(e^{-ix_k(\cdot)})}(v)\, dv$$
$$= \langle e^{-(\cdot)^2/(4\lambda)} h_\lambda, e^{-ix_k(\cdot)}\rangle_{[-1,1]}$$
$$\quad + \sum_{m=1}^{\infty} 2^m \langle e^{-2^m(\cdot)/(4\lambda)} A_m(h_\lambda), A_m(e^{-ix_k(\cdot)})\rangle_{[-1,1]}$$
$$= \langle \mathcal{F}[I_\lambda(f)], e^{-ix_k(\cdot)}\rangle_{[-1,1]}$$
$$\quad + \sum_{m=1}^{\infty} \langle 2^m A_m^*(e^{-(2^m(\cdot))^2/4(\lambda)} A_m(h_\lambda)), e^{-i\langle x_k,(\cdot)\rangle}\rangle_{[-1,1]}$$
$$= \langle \mathcal{F}[I_\lambda(f)] + \sum_{m=1}^{\infty} 2^m A_m^*(e^{-(2^m(\cdot))^2/4(\lambda)} A_m(h_\lambda)), e^{-ix_k,(\cdot)}\rangle_{[-1,1]}.$$

As $(e^{ix_k(\cdot)} : k \in \mathbb{Z})$ is a frame for $L_2[-1,1]$ (in particular a complete system), equations (37) and (38) lead to the identity

$$\mathcal{F}[f] = \mathcal{F}[I_\lambda(f)] + \sum_{m=1}^{\infty} 2^m A_m^* \left(e^{-(2^m(\cdot))^2/4(\lambda)} A_m(h_\lambda)\right) \quad \text{a.e. on } [-1,1]. \quad (39)$$

Suppose now that $h \in L_2[-1,1]$ and $m \in \mathbb{N}$. We deduce from (33) that

$$\|2^m A_m^*\left(e^{-(2^m(\cdot))^2/(4\lambda)} A_m(h)\right)\|^2_{L_2[-1,1]}$$
$$\leq 2^{2m} R^2 \|e^{-(2^m(\cdot))^2/(4\lambda)} A_m(h)\|^2_{L_2[-1,1]}$$
$$\leq 2^{2m} R^2 \|e^{-2^{2m-2}/(4\lambda)} A_m(h)\|^2_{L_2[-1,1]}$$
$$\text{(because } \operatorname{supp} A_m(h) \subset [-1, -\tfrac{1}{2}] \cup [\tfrac{1}{2}, 1])$$
$$\leq 2^{2m} R^4 e^{-2^{2m-2}/(2\lambda)} \|h\|^2_{L_2[-1,1]},$$

whence

$$\|2^m A_m^*\left(e^{-(2^m(\cdot))^2/(4\lambda)} A_m\right)\|_{L_2[-1,1]} \leq 2^m R^2 e^{-2^{2m-2}/(4\lambda)}.$$

Therefore the linear operator

$$\tau_\lambda : L_2[-1,1] \to L_2[-1,1], \quad h \mapsto \sum_{m \in \mathbb{N}} 2^m A_m^*\left(e^{-\|2^m(\cdot)\|^2_2/(4\lambda)} A_m(h)\right)$$

is bounded. In fact, as there are numbers $\lambda_0 > 0$ and $D$, such that

$$\sum_{m \in \mathbb{N}} 2^m e^{-2^{2m-2}/(4\lambda)} \leq D e^{-1/(4\lambda)}, \quad \lambda \in (0, \lambda_0], \quad (40)$$

the operator norm of $\tau_\lambda$ obeys the following estimate:

$$\|\tau_\lambda\| \leq R^2 D e^{-1/(4\lambda)} \text{ whenever } \lambda < \lambda_0. \quad (41)$$

As the operator $\tau_\lambda$ is positive, (39) yields

$$\|\mathcal{F}[f]\|_2 \|h_\lambda\|_2 \geq \langle \mathcal{F}[f], h_\lambda \rangle_{[-1,1]}$$
$$= \langle e^{-(\cdot)^2/(4\lambda)} h_\lambda, h_\lambda \rangle_{[-1,1]} \geq e^{-1/(4\lambda)} \|h_\lambda\|^2_{L_2[-1,1]}.$$

Consequently,

$$\|h_\lambda\| \leq e^{1/(4\lambda)} \|\mathcal{F}[f]\|_2. \quad (42)$$

Thus, from (39) and (41) we get

$$\|\mathcal{F}[I_\lambda(f)]\|_{[-1,1]}\|_2 \leq \|\mathcal{F}[f]\|_2 + \|\tau_\lambda(h_\lambda)\|_2 \leq \|\mathcal{F}[f]\|_2 + DR^2 \|\mathcal{F}[f]\|_2. \quad (43)$$

Our next step is to estimate $\|\mathcal{F}[I_\lambda(f)]|_{\mathbb{R}\setminus[-1,1]}\|_2$. Equation (36) implies that

$$\|\mathcal{F}[I_\lambda(f)]|_{\mathbb{R}\setminus[-1,1]}\|_2^2 \tag{44}$$
$$= \int_{\mathbb{R}\setminus[-1,1]} e^{-u^2/(2\lambda)} |H_\lambda(u)|^2 \, du$$
$$= \sum_{m=1}^{\infty} \int_{2^m[-1,1]\setminus 2^{m-1}[-1,1]} e^{-u^2/(2\lambda)} |H_\lambda(u)|^2 \, du$$
$$= \sum_{m=1}^{\infty} 2^m \int_{[-1,-\frac{1}{2}]\cup[\frac{1}{2},1]} e^{-2^{2m}v^2/(2\lambda)} |A_m(h_\lambda)(v)|^2 \, dv$$
$$\leq \sum_{m=1}^{\infty} 2^m e^{-2^{2m}/(8\lambda)} \|A_m(h_\lambda)\|_2^2$$
$$\left(\text{as } \mathrm{supp}\, A_m(h) \subset \left[-1,-\frac{1}{2}\right] \cup \left[\frac{1}{2},1\right]\right)$$
$$\leq R^2 \|h_\lambda\|_2^2 \sum_{m=1}^{\infty} 2^m e^{-2^{2m}/(8\lambda)} \quad \text{(by (33))}$$
$$\leq e^{1/(2\lambda)} R^2 \|\mathcal{F}[f]\|_2^2 \sum_{m=1}^{\infty} e^{-2^{2m}/(8\lambda)} 2^m \quad \text{(by (42))}.$$

By changing $\lambda_0$ and $D$, if need be, one obtains, as in (40),

$$\sum_{m=1}^{\infty} e^{-2^{2m}/(8\lambda)} 2^m \leq D e^{-1/(2\lambda)}, \quad \lambda \in (0, \lambda_0]. \tag{45}$$

Combining (43) and (44) proves our claim.

**Step 2.** For all $0 < \lambda < \lambda_0$,

$$\|f - I_\lambda(f)\|_2 = (2\pi)^{-1/2} \|\mathcal{F}[f] - \mathcal{F}[I_\lambda(f)]\|_2 \tag{46}$$
$$\leq D_2 \|e^{-(1-(\cdot)^2)/(4\lambda)} \mathcal{F}[f]\|_{L_2[-1,1]},$$

for some constant $D_2$.

**Remark 4.1.** Note that (46) and the Dominated Convergence Theorem imply that

$$\lim_{\lambda \to 0^+} I_\lambda(f) = f \text{ in } L_2(\mathbb{R}).$$

Moreover, the inequality in (46) also yields the following for $0 < \beta < 1$:
$$\sup\{\|I_\lambda(f) - f\| : f \in PW_{\beta[-1,1]}, \|f\|_2 \leq 1\} \leq D_2 e^{(\beta^2-1)/(4\lambda)} \to 0, \quad (47)$$
if $\lambda \to 0^+$.

To prove (46) we define $\tilde{\tau}_\lambda = e^{1/4\lambda} \tau_\lambda$,
$$M_\lambda : L_2[-1,1] \to L_2[-1,1], \quad h \mapsto e^{-(1-(\cdot)^2)/(4\lambda)} h,$$
and
$$L_\lambda : L_2[-1,1] \to L_2[-1,1], \quad h \mapsto R \circ \mathcal{F} \circ I_\lambda \circ \mathcal{F}^{-1}(h),$$
where $R : L_2(\mathbb{R}) \to L_2[-1,1]$ is the restriction map.

**Proposition 4.1.** *The map* $\mathrm{Id} + \tilde{\tau}_\lambda \circ M_\lambda$ *is an invertible operator on* $L_2[-1,1]$, *and* $(\mathrm{Id} + \tilde{\tau}_\lambda \circ M_\lambda)^{-1} = L_\lambda$. *Furthermore, there is a constant* $\Delta > 0$, *depending only on* $(x_j)$, *so that* $\|(\mathrm{Id} + \tilde{\tau}_\lambda \circ M_\lambda)^{-1}\| \leq \Delta$, *whenever* $0 < \lambda \leq \lambda_0$.

**Proof.** Let $h \in PW_1$. From (39) we obtain
$$\mathcal{F}[h] = \mathcal{F}[I_\lambda(h)] + \tau_\lambda \left(e^{(\cdot)^2/(4\lambda)} \mathcal{F}[I_\lambda(h)]\right)$$
$$= \mathcal{F}[I_\lambda(h)] + \tilde{\tau}_\lambda \circ M_\lambda\big(\mathcal{F}[I_\lambda(h)]\big) = \big(\mathrm{Id} + \tilde{\tau}_\lambda \circ M_\lambda\big) L_\lambda(\mathcal{F}[h]).$$
This implies that $\mathrm{Id} + \tilde{\tau}_\lambda \circ M_\lambda$ is surjective and is a left inverse of the bounded operator $L_\lambda$. Next we show that $\mathrm{Id} + \tilde{\tau}_\lambda \circ M_\lambda$ is also injective. To that end, let $(\mathrm{Id} + \tilde{\tau}_\lambda \circ M_\lambda)(h) = 0$ for some $h \in L_2[-1,1]$. Then
$$0 = \langle (\mathrm{Id} + \tilde{\tau}_\lambda \circ M_\lambda)(h), M_\lambda(h) \rangle_{[-1,1]}$$
$$= \langle h, M_\lambda(h) \rangle_{[-1,1]} + \langle \tilde{\tau}_\lambda(M_\lambda(h)), M_\lambda(h) \rangle_{[-1,1]} \geq \langle h, M_\lambda(h) \rangle_{[-1,1]} \geq 0,$$
the first inequality above being a consequence of the positivity of $\tilde{\tau}_\lambda$. Hence $\langle h, M_\lambda(h) \rangle_{[-1,1]} = 0$, which implies that $h = 0$, because $M_\lambda$ is a strictly positive operator. The injectivity of $\mathrm{Id} + \tilde{\tau}_\lambda \circ M_\lambda$ follows.

Thus $\mathrm{Id} + \tilde{\tau}_\lambda \circ M_\lambda$ is invertible, and its inverse is $L_\lambda$. As the operators $L_\lambda$, $0 < \lambda \leq \lambda_0$, are uniformly bounded (Step 1), the proof is complete. □

Proposition 4.1 provides the following identity on $[-1,1]$:
$$\mathcal{F}[f] - \mathcal{F}[I_\lambda(f)] = [\mathrm{Id} - (\mathrm{Id} + \tilde{\tau}_\lambda \circ M_\lambda)^{-1}](\mathcal{F}[f]) = (\mathrm{Id} + \tilde{\tau}_\lambda \circ M_\lambda)^{-1} \circ \tilde{\tau}_\lambda \circ M_\lambda(\mathcal{F}[f]).$$
Therefore, we conclude via (41) and Proposition 4.1 that
$$\|\mathcal{F}[f] - \mathcal{F}[I_\lambda(f)]|_{[-1,1]}\|_2 \leq \|(\mathrm{Id} + \tilde{\tau}_\lambda \circ M_\lambda)^{-1}\| \|\tilde{\tau}_\lambda\| \|M_\lambda(\mathcal{F}[f])\|_2 \quad (48)$$
$$\leq \Delta R_B^2 D \|M_\lambda(\mathcal{F}[f])\|_2.$$

Now the first inequality in (44) yields

$$\|\mathcal{F}[I_\lambda(f)]|_{\mathbb{R}\setminus[-1,1]}\|_2^2 \tag{49}$$
$$\leq \sum_{m=1}^\infty 2^m e^{-2^{2m}/(8\lambda)} \|A_m(h_\lambda)\|_2^2$$
$$\leq R^2 \sum_{m=1}^\infty 2^m e^{-2^{2m}/(8\lambda)} \|e^{(\cdot)^2/(4\lambda)} \mathcal{F}[I_\lambda(f)]|_{[-1,1]}\|_2^2 \quad \text{by (33)}$$
$$= R^2 \sum_{m=1}^\infty 2^m e^{(4-2^{2m})/(8\lambda)} \|e^{-(1-(\cdot)^2)/(4\lambda)} \mathcal{F}[I_\lambda(f)]|_{[-1,1]}\|_2^2$$
$$\leq DR^2 \big[\|e^{-(1-(\cdot)^2)/(4\lambda)} \mathcal{F}[f]\|_2$$
$$\quad + \|e^{-(1-(\cdot)^2)/(4\lambda)} (\mathcal{F}[I_\lambda(f)]|_{[-1,1]} - \mathcal{F}[f]) \|_2\big]^2 \quad \text{(by (45))}$$
$$\leq DR^2 \big[\|e^{-(1-(\cdot)^2)/(4\lambda)} \mathcal{F}[f]\|_2 + \|\mathcal{F}[I_\lambda(f)]|_{[-1,1]} - \mathcal{F}[f]\|_2\big]^2$$
$$\leq 2DR_B^2 \big[\|e^{-(1-(\cdot)^2)/(4\lambda)} \mathcal{F}[f]\|_2^2 + \|\mathcal{F}[I_\lambda(f)]|_{[-1,1]} - \mathcal{F}[f]\|_2^2\big]$$
$$\leq 2DR_B^2[1 + \Delta^2 R^2 D] \|M_\lambda(\mathcal{F}(f))\|_2^2.$$

Combining this with (1) and 48 we obtain

$$\|I_\lambda(f) - f\|_2^2$$
$$= 2\pi \|\mathcal{F}[I_\lambda(f) - \mathcal{F}[f]\|_2^2$$
$$= 2\pi \big[\|\mathcal{F}[I_\lambda(f)]|_{[-1,1]} - \mathcal{F}[f]\|_2^2 + \|\mathcal{F}[I_\lambda(f)]|_{\mathbb{R}\setminus[-1,1]}\|_2^2\big]$$
$$\leq D_2 \|M_\lambda(f)\|_2^2,$$

for some constant $D_2$ only depending on $(x_j)$.

**Step 3.** There exist constants $\lambda_1 \in (0, \lambda_0]$ and $D_3$ so that

$$|I_\lambda(f)(x) - f(x)| \leq D_3 \|e^{((\cdot)^2-1)/(4\lambda)} \mathcal{F}[f]\|_2,$$

for all $x \in \mathbb{R}$. In particular $\lim_{\lambda \to 0^+} I_\lambda(f) = f$ uniformly on $\mathbb{R}$.

Let $x \in \mathbb{R}$. We use (5) to write

$$|I_\lambda(f)(x) - f(x)| \tag{50}$$
$$= \frac{1}{2\pi} \Big|\Big[\int_{-1}^1 [\mathcal{F}[I_\lambda(f)](u) - \mathcal{F}[f](u)] e^{ixu} du$$
$$\quad + \int_{\mathbb{R}\setminus[-1,1]} \mathcal{F}[I_\lambda(f)](u) e^{ixu} du\Big]\Big|$$
$$\leq \frac{1}{2\pi} \big[\|\mathcal{F}[I_\lambda(f)]|_{[-1,1]} - \mathcal{F}[f]\|_1 + \|\mathcal{F}[I_\lambda(f)]|_{\mathbb{R}\setminus[-1,1]}\|_1\big].$$

The first term converges to 0 if $\lambda$ approaches 0 because $\|\cdot\|_1 \leq m[-1,1]\|\cdot\|_2$ on $L_1[-1,1]$ and because of the Step 2. That the second term converges to 0, can be shown with arguments similar to the arguments which proved that $\left\|\mathcal{F}[I_\lambda(f)]|_{\mathbb{R}\setminus[-1,1]}\right\|_2$ converges to 0.

## 5. Multidimensional versions of Theorem 3.5

We consider now the multidimensional Gaussian interpolation operator. Let $d \in \mathbb{N}$, and let $(x_j : j \in \mathbb{N}) \subset \mathbb{R}^d$ be a Rieszbasis sequence for $L_2(B)$, where we assume that $B$ is bounded, convex, symmetric, and has a non empty open kernel.

The *d-dimensional Gaussian function with scaling parameter* $\lambda > 0$ is defined by

$$g_\lambda^{(d)}(x_1, x_2, \ldots, x_d) = e^{-\lambda \|x\|^2} = e^{-\lambda \sum_{j=1}^d x_j^2}, \quad x = (x_1, x_2, \ldots, x_d) \in \mathbb{R}^d.$$

The existence of the more dimensional interpolation operator is cannot be established in such an elementary way as in the one dimensional case but can be deduced from the following Theorem

**Theorem 5.1.** *[cf. [7, Theorem 2.3]]*
*Let $\lambda$ and $q$ be fixed positive numbers. There exists a number $\theta$, depending only on $d$, $\lambda$, and $q$, such that the following holds: if $(x_j)$ is any sequence in $\mathbb{R}^d$ with $\|x_j - x_k\|_2 \geq q$ for $j \neq k$, then $\sum_{j,k} \xi_j \overline{\xi}_k g_\lambda(\|x_j - x_k\|_2) \geq \theta \sum_j |\xi_j|^2$, for every sequence of complex numbers $(\xi_j)$.*

Using Theorem 5.1 we can deduce the existence of the Interpolation operator similarly as in the 1 dimensional case

**Theorem 5.2.** *Let $d \in \mathbb{N}$, let $\lambda$ be a fixed positive number, and let $(x_j : j \in \mathbb{N})$ be a Riesz-basis sequence for $L_2(B)$. For any $f \in PW_B$ there exists a unique square-summable sequence $(a(j, \lambda) : j \in \mathbb{N})$ such that*

$$\sum_{j \in \mathbb{N}} a(j, \lambda) g_\lambda^{(d)}(x_k - x_j) = f(x_k), \quad k \in \mathbb{N}. \tag{51}$$

*The Gaussian Interpolation Operator* $I_\lambda : PW_B \to L_2(\mathbb{R}^d)$, *defined by*

$$I_\lambda(f)(\cdot) = \sum_{j \in \mathbb{N}} a(j, \lambda) g_\lambda^{(d)}(\cdot - x_j),$$

*where $(a(j, \lambda) : j \in \mathbb{N})$ satisfies (51), is a well-defined, bounded linear operator from $PW_B$ to $L_2(\mathbb{R}^d)$. Moreover, $I_\lambda(f) \in C_0(\mathbb{R}^d)$.*

We can now extend Theorem 3.5 in the following special situations First let $B = Q = [-1,1]^d$ and assume that for $j = 1,\ldots d$ the sequence $(x_k^{(j)} : k \in \mathbb{Z}) \subset \mathbb{R}$ is a Riesz basis sequence for $L_2[-1,1]$, and put $x_k = (x_{k_1}^{(1)}, x_{k_2}(2), \ldots x_{k_d}^{(d)})$, for $k = (k_1, k_2, \ldots k_d) \in \mathbb{Z}^d$. It is then easy to see that $(x_k)_{k \in \mathbb{Z}^d}$ is a Riesz basis sequence for $Q$. We can therefore define $I_\lambda$ for that sequence.

**Theorem 5.3.** *Under the above assumptions it follows that for all $f \in PW_Q$*

$$f = \lim_{\lambda \to 0} I_\lambda(f) \text{ in } L_2(\mathbb{R}^d) \text{ and uniformly.}$$

The second case we can extend is the case that $B = B_2$, the Euclidean unit ball in $\mathbb{R}^d$.

**Theorem 5.4.** *Assume that $(x_k)$ is a Riesz basis sequence for $L_2(B_2)$ and let $I_\lambda$ be the interpolation operator defined above for $(x_k)$, then for all $f \in PW_{B_2}$ it follows that*

$$f = \lim_{\lambda \to 0} I_\lambda(f) \text{ in } L_2(\mathbb{R}^d) \text{ and uniformly.}$$

**Remark 5.1.** Theorem 5.4 can essentially be shown the same way as Theorem 3.5. Nevertheless there is one problem though. We do not know whether a there exists a Riesz basis sequence for $L_2(B_2)$.

We do not know whether Theorem 5.4 still holds if we only assume that $(x_k)$ is a uniformly separated Fourier frame, If so, Beurling's Theorem would provide easy to satisfy conditions on $(x_k)$ for which it holds.

## 6. Appendix: Proof of Theorem 3.4

Theorem 3.4 will follow from the following result on bi-infinite matrices which appears to be folkloric. A sketch of a proof can be found in [5], but for the sake of completeness we include a self-contained argument.

First let us observe that for any $\mu > 0$ Bochner's Theorem implies that $g_\mu$ is a positive definite function which means that for any finite sequence $(\xi_j)_{j=1}^\ell \subset \mathbb{R}$ the matrix

$$A = (e^{-\mu(x_k - x_\ell)^2} : 1 \le k, \ell n)$$

is positive definite.

This implies that also for any uniformly separated sequence $(x_j : j \in \mathbb{Z})$ the bi-infinite matrix

$$A = (e^{-\mu(x_k - x_\ell)^2} : k, \ell \in \mathbb{Z})$$

bounded as operator on $\ell_2(\mathbb{Z})$ (because of the fast decay of all rows and colums) and positive.

**Theorem 6.1.** *Suppose that $(A(j,k))_{j,k \in \mathbb{Z}}$ is a bi-infinite matrix which, as an operator on $\ell^2(\mathbb{Z})$, is self adjoint, positive and invertible. Assume further that there exist positive constants $\tau$ and $\gamma$ such that $|A(j,k)| \leq \tau e^{-\gamma|j-k|}$ for every pair of integers $j$ and $k$. Then there exist constants $\tilde{\tau}$ and $\tilde{\gamma}$ such that $|A^{-1}(s,t)| \leq \tilde{\tau} e^{-\tilde{\gamma}|s-t|}$ for every $s,t \in \mathbb{Z}$.*

For the proof of Theorem 6.1 we shall need the following pair of lemmata.

**Lemma 6.1.** *Let $(H, \langle \cdot, \cdot \rangle)$ be a Hilbert space, and let $A : H \to H$ be a bounded linear operator satisfying the following conditions: (i) $A = A^*$, and (ii) $\inf\{\langle x, Ax \rangle : \|x\| = 1\} > 0$. Let $R := I - \frac{A}{\|A\|}$, where $I$ denotes the identity. Then $R = R^*$, $\langle x, Rx \rangle \geq 0$ for every $x \in H$, and $\|R\| < 1$.*

**Proof.** The symmetry of $R$ is evident. If $\|x\| = 1$, then

$$\langle x, Rx \rangle = \|x\|^2 - \left\langle x, \frac{A}{\|A\|} x \right\rangle.$$

By assumption (ii) and the BCS inequality we see that the term on the right of the preceding equation is between 0 and 1. Therefore $\|R\| = \sup\{\langle x, Rx \rangle : \|x\| = 1\} \leq 1$. If $\|R\| = 1$, then there is a sequence $(x_n : n \in \mathbb{N})$ such that $\|x_n\| = 1$ for every $n$, and

$$1 = \lim_{n \to \infty} \langle x_n, Rx_n \rangle = \lim_{n \to \infty} \left( 1 - \left\langle x_n, \frac{A}{\|A\|} x_n \right\rangle \right).$$

But the second term on the right cannot converge to zero because of assumption (ii). □

**Lemma 6.2.** *Suppose that $(R(s,t))_{s,t \in \mathbb{Z}}$ is a bi-infinite matrix satisfying the following condition: there exist positive constants $C$ and $\gamma$ such that $|R(s,t)| \leq e^{-\gamma|s-t|}$ for every pair of integers $s$ and $t$. Given $0 < \gamma' < \gamma$, there is a constant $C(\gamma, \gamma')$, depending on $\gamma$ and $\gamma'$, such that $|R^n(s,t)| \leq C^n C(\gamma, \gamma')^{n-1} e^{-\gamma'|s-t|}$ for every $s,t \in \mathbb{Z}$.*

**Proof.** Suppose firstly that $s \neq t \in \mathbb{Z}$, and assume without loss that $s < t$.

Note that

$$\sum_{u=-\infty}^{\infty} e^{-\gamma|s-u|}e^{-\gamma'|t-u|} = \sum_{u=s}^{t} e^{-\gamma(u-s)}e^{-\gamma'(t-u)} \quad (52)$$

$$+ \sum_{u=-\infty}^{s-1} e^{-\gamma(s-u)}e^{-\gamma'(t-u)} + \sum_{u=t+1}^{\infty} e^{-\gamma(u-s)}e^{-\gamma'(u-t)}$$

$$=: \Sigma_1 + \Sigma_2 + \Sigma_3.$$

Now

$$\Sigma_1 = e^{-\gamma'(t-s)}e^{(\gamma-\gamma')s}\sum_{u=s}^{t} e^{-u(\gamma-\gamma')} \quad (53)$$

$$= e^{-\gamma'(t-s)}e^{(\gamma-\gamma')s}\sum_{v=0}^{t-s} e^{-s(\gamma-\gamma')-v(\gamma-\gamma')}$$

$$= e^{-\gamma'(t-s)}\sum_{v=0}^{t-s} e^{-v(\gamma-\gamma')} \leq \frac{e^{-\gamma'(t-s)}}{1-e^{-(\gamma-\gamma')}}.$$

Moreover,

$$\Sigma_2 = \sum_{v=1}^{\infty} e^{-\gamma v}e^{-\gamma'(t-s+v)} \quad (54)$$

$$= e^{-\gamma'(t-s)}\sum_{v=1}^{\infty} e^{-v(\gamma+\gamma')} \leq \frac{e^{-\gamma'(t-s)}}{1-e^{-(\gamma+\gamma')}},$$

whereas

$$\Sigma_3 = \sum_{v=1}^{\infty} e^{-\gamma' v}e^{-\gamma(v+t-s)} \quad (55)$$

$$= e^{-\gamma(t-s)}\sum_{v=1}^{\infty} e^{-(\gamma+\gamma')v} \leq \frac{e^{-\gamma'(t-s)}}{1-e^{-(\gamma+\gamma')}}.$$

If $s = t$, then

$$\sum_{u=-\infty}^{\infty} e^{-\gamma|s-u|}e^{-\gamma'|t-u|} = \sum_{u=-\infty}^{\infty} e^{-(\gamma+\gamma')|s-u|} \leq \frac{2}{1-e^{-(\gamma+\gamma')}}. \quad (56)$$

From (52-56) we conclude that

$$|R^2(s,t)| \leq C^2\left[\frac{1}{1-e^{-(\gamma-\gamma')}} + \frac{2}{1-e^{-(\gamma+\gamma')}}\right] =: C^2C(\gamma,\gamma').$$

The general result follows from this via induction. □

We are now ready to prove Theorem 6.1 and Theorem 3.4.

**Proof of Theorem 6.1.** We begin with the remark that assumptions (i) and (ii) of Lemma 6.1 ensure that $A$ is, in fact, boundedly invertible. Let $R = I - \frac{A}{\|A\|}$ be the matrix given in that lemma. As

$$R(j,k) = \frac{A(j,k)}{\|A\|} \text{ if } j \neq k, \quad \text{and} \quad R(k,k) = \frac{\|A\| - A(k,k)}{\|A\|},$$

there is some constant $C$ such that $|R(s,t)| \leq Ce^{-\gamma|s-t|}$ for every pair of integers $s$ and $t$. As $A = \|A\|(I - R)$, and $r := \|R\| < 1$ (Lemma 6.1), the standard Neumann series expansion yields the relations

$$A^{-1} = \|A\|^{-1} \sum_{n=0}^{\infty} R^n \tag{57}$$

$$= \|A\|^{-1} \sum_{n=0}^{N-1} R^n + \|A\|^{-1} R^N \sum_{n=0}^{\infty} R^n$$

$$= \|A\|^{-1} \sum_{n=0}^{N-1} R^n + R^N A^{-1},$$

for any positive integer $N$. As $R^0(s,t) = I(s,t) = 0$ if $s \neq t$, $s,t \in \mathbb{Z}$, we see from (57) that

$$A^{-1}(s,t) = \|A\|^{-1} \sum_{n=1}^{N-1} R^n(s,t) + [R^N A^{-1}](s,t), \quad s \neq t. \tag{58}$$

Choose and fix a positive number $\gamma' < \gamma$, and recall from Lemma 6.2 that there is a constant $C(\gamma, \gamma')$ such that $|R^n(s,t)| \leq C^n C(\gamma, \gamma')^{n-1} e^{-\gamma'|s-t|}$ for every positive integer $n$, and every pair of integers $s$ and $t$. So we may assume that there is some constant $D := D(\gamma, \gamma') > 1$ such that $|R^n(s,t)| \leq D^n e^{-\gamma'|s-t|}$ for every positive integer $n$, and every pair of integers $s$ and $t$. Using this bound in 58 provides the following estimate for every $s \neq t$:

$$|A^{-1}(s,t)| \leq \|A\|^{-1} e^{-\gamma'|s-t|} \sum_{n=1}^{N-1} D^n + \|A^{-1}\| r^N \tag{59}$$

$$\leq \|A\|^{-1} e^{-\gamma'|s-t|} \frac{D^N}{D-1} + \|A^{-1}\| r^N.$$

Let $m$ be a positive integer such that $e^{-\gamma'} D^{1/m} < 1$, and let $s, t \in \mathbb{Z}$ with $|s-t| \geq m$. Writing $|s-t| = Nm + k$, $0 \leq k \leq m-1$, we find that

$$e^{-\gamma'|s-t|} D^N = \left[ e^{-\gamma'} D^{\frac{1}{m+(k/N)}} \right]^{|s-t|} \leq [e^{-\gamma'} D^{1/m}]^{|s-t|}. \tag{60}$$

Further,
$$r^N = \left[r^{\frac{1}{m+(k/N)}}\right]^{|s-t|} \leq [r^{1/2m}]^{|s-t|}, \tag{61}$$
and combining (60) and (61) with (59) leads to the following bounds for every $|s-t| \geq m$:
$$|A^{-1}(s,t)| \leq \frac{\|A\|^{-1}}{D-1}[e^{-\gamma'}D^{1/m}]^{|s-t|} + \|A^{-1}\|[r^{1/2m}]^{|s-t|} = O(e^{-\tilde{\gamma}|s-t|}). \tag{62}$$
On the other hand, if $|s-t| < m$, we obtain
$$|A^{-1}(s,t)| \leq \|A^{-1}\| \leq (\|A^{-1}\|e^{m\tilde{\gamma}})e^{-\tilde{\gamma}|s-t|}, \tag{63}$$
and combining (62) with (63) finishes the proof. □

**Proof of Theorem 3.4.** It was observed that $A$ satisfies the assumptions of Lemma 6.1. Moreover, the hypothesis $x_{j+1} - x_j \geq q$, $j \in \mathbb{Z}$, implies that $|x_j - x_k| \geq |j-k|q$ for every pair of integers $j$ and $k$. So an appeal to Theorem 6.1 yields the required result. □

### References

1. F. Albiac and N. Kalton, Topics in Banach Space Theory, Graduate Texts in Mathematics, Springer Verlag (2006).
2. A. Bailey, *Sampling and recovery of multidimensional functions via frames*, J. Math. Anal. Appl. **367** (2010) 374–388.
3. A. Beurling, *Local harmonic analysis with some applications to differential operators*, Some Recent Advances in the Basic Sciences, Vol. 1 (Proc. Annual Sci. Conf., Belfer Grad. School Sci., Yeshiva Univ., New York, 1962 – 1964) (1966) 109–125.
4. I. Daubechies and R. DeVore, *Approximating a bandlimited function using very coarsely quantized data: A family of stable sigma-delta modulators of arbitrary order*, Annals of Mathematics **158** (2003), 679–710.
5. S. Jaffard, Propriétés des matrices "bien localisées" près de leur diagonale et quelques applications, *Ann. Inst. Henri Poincaré* **7** (1990), 461–476.
6. M. I. Kadec, The exact value of the Paley-Wiener constant, *Dokl. Adad. Nauk SSSR* **155** (1964), 1243–1254.
7. F. J. Narcowich and J. D. Ward, Norm estimates for the inverses of a general class of scattered-data radial-function interpolation matrices, *J. Approx. Theory* **69** (1992), 84–109.
8. F. J. Narcowich, N. Sivakumar, and J. D. Ward, On condition numbers associated with radial-function interpolation, *J. Math. Anal. Appl.* **186** (1994), 457–485.
9. Th. Schlumprecht and N. Sivakumar, *On the sampling and recovery of bandlimited functions via scattered translates of the Gaussian*, J. Approx. Theory **159** (2009) 109–153.

# PART B
# Plenary speakers

# Isometric shifts between spaces of continuous functions

Jesús Araujo*

*Departamento de Matemáticas, Estadística y Computación*
*Facultad de Ciencias, Universidad de Cantabria, Spain*
*e-mail: araujoj@unican.es*

*In memoriam of Professor Antonio Aizpuru.*

## 1. Introduction

We present an account of old and new results on isometric shifts between spaces of $\mathbb{K}$-valued continuous functions (being $\mathbb{K} = \mathbb{R}$ or $\mathbb{C}$).

The notion of isometric shift is just a natural generalization to Banach spaces of that of shift operator in $\ell^2$ and other sequence spaces, $(x_1, x_2, \ldots) \mapsto (0, x_1, x_2, \ldots)$ (see [6,14]). We say that a linear operator $T : E \to E$ is an *isometric shift* if

(1) $T$ is an isometry,
(2) The codimension of $T(E)$ in $E$ is 1,
(3) $\bigcap_{n=1}^{\infty} T^n(E) = \{0\}$.

Two basic spaces that can be seen both as spaces of sequences and spaces of continuous functions are $\ell^\infty$ and $c$, consisting of all $\mathbb{K}$-valued bounded and convergent sequences, respectively. $\ell^\infty$ and $c$ are easily seen to be isometrically isomorphic to $C(\beta \mathbb{N})$ and $C(\mathbb{N} \cup \{\infty\})$, respectively (here we consider spaces $C(X)$ endowed with the supremum norm $\|\cdot\|_\infty$). Obviously, in both cases the usual shift defined in the sequence spaces is also an isometric shift in the above sense when considering them as function spaces.

We focus on some questions that have made it possible to develop the theory of isometric shifts between spaces of continuous functions. Suppose

---

*Research partially supported by the Spanish Ministry of Science and Education (Grant number MTM2006-14786).

that $X$ is a compact Hausdorff space and there exists a shift operator $T : C(X) \to C(X)$, where $C(X)$ denotes the space of all continuous functions from $X$ to $\mathbb{K}$ endowed with the supremum norm $\|\cdot\|_\infty$.

**Question 1** Can $X$ have an infinite connected component?
**Question 2** Must $X$ be separable?
**Question 3** Can $X$ be the Cantor set **K** or a related space?

Of course there is still a very natural question that we have not included above but that lies behind all the study:

**Question 4** Can we describe $T$?

Sometimes the answer to these and other questions depend on the scalar field. We will denote by $C_\mathbb{R}(X)$ and $C_\mathbb{C}(X)$ the spaces of real-valued and complex-valued continuous functions on $X$, respectively.

Some other papers have recently studied operators related to isometric shifts (also defined on other spaces of functions). Among them, we will mention for instance [3,5,8,15,16,18–20], and [22] (see also references therein).

## 2. Question 1

The original statement of Question 1 was not exactly as given above. It was posed in the following terms by Holub in [14].

**Conjecture** If $X$ has at least one infinite connected component, there is no shift or backward shift on $C_\mathbb{R}(X)$.

We do not treat here backward shifts, and just mention that the answer to the corresponding part of the question was completely solved in [19].

As for the part on isometric shifts, in fact that author gave a first partial answer [14, Theorem 2.1].

**Theorem 2.1.** *If $X$ has a finite number of components, then $C_\mathbb{R}(X)$ admits no shift.*

Obviously, we immediately deduce that $C_\mathbb{R}[a,b]$ admits no shift. But, even if we are just dealing with the case of isometric shifts between spaces $C(X)$, it is worth to see that something else can be said for other spaces on intervals (see [14, Theorem 2.3]).

**Theorem 2.2.** *Let $E$ be a Banach space consisting of real-valued continuous functions on an interval $[a,b]$ for which*

(1) $\|\cdot\|_E = \|\cdot\|_\infty + p(\cdot)$, where $p$ is a seminorm on $E$;
(2) $1 \in E$ and $p(1) = 0$;
(3) Given any (nontrivial) interval $I \subset [a, b]$, there exists an infinite-dimensional subspace of $E$ whose members have support in $I$.

Then there is no isometric shift on $E$.

When appropiate norms are given in the spaces, an immediate corollary is the following.

**Corollary 2.1.** *There is no shift on the space $C^n[a, b]$ of all real valued functions on $[a, b]$ having $n$ continuous derivatives there ($n \geq 1$).*

It must be said that the techniques used to prove Theorem 2.1 do not carry over to the complex case. In fact, some other results were given before the next theorem, which represents a large generalization of Theorem 2.1, could be proved (see [7, Theorem 6.1]).

**Theorem 2.3.** *Let $M$ be any compact manifold with or without boundary. Then $C(M)$ does not admit an isometric shift.*

In fact, the authors prove more: they show that $C(M)$ does not admit even a codimension 1 linear isometry. Notice that in general a codimension 1 linear isometry need not be an isometric shift, as the following example shows (see [7, Example 3.2]).

**Example 2.1.** Identify as usual the spaces $C(\mathbb{N} \cup \{\infty\})$ and $c$, and define $T : C(\mathbb{N} \cup \{\infty\}) \to C(\mathbb{N} \cup \{\infty\})$ by $T(x_1, x_2, x_3, \ldots) := (x_1, 0, x_2, x_3, \ldots)$ for each $(x_n) \in c$.

Nevertheless, Question 1 was settled in the negative in [10]. There is indeed a space $X$ with an infinite connected component and such that $C(X)$ admits a shift. The example given turns out to be a compactification of integers.

We finally give a generalization of Theorem 2.1 for the infinite case and $\mathbb{K} = \mathbb{R}, \mathbb{C}$, as it appears in [10, Theorem 2.7].

**Theorem 2.4.** *Suppose that $X$ has a countably infinite number of components, all of whom are infinite. Then $C(X)$ does not admit an isometric shift.*

In view of all the above, the following question appears to be very natural, assuming that there exists a shift operator on $C(X)$:

**Question 1b** Can $X$ be infinite and connected?

## 3. Question 4 and new questions

The new results that changed dramatically the way to look at isometric shifts, allowing in particular to obtain Theorem 2.3, are those concerning their representation. Using a result of Holsztyński ([13]), Gutek, Hart, Jamison, and Rajagopalan classified isometric shifts into two types, called type I and type II (see [10]).

First, If $T : C(X) \to C(X)$ is an isometric shift, then there exist a closed subset $Y \subset X$, a continuous and surjective map $\phi : Y \to X$, and a function $a \in C(Y)$, $|a| \equiv 1$, such that $(Tf)(x) = a(x) \cdot f(\phi(x))$ for all $x \in Y$ and all $f \in C(X)$.

Then the classification is as follows:

- $T$ is said to be of type I if $Y$ can be taken to be equal to $X \setminus \{p\}$, where $p \in X$ is an isolated point.
- $T$ and is said to be of type II if $Y$ can be taken equal to $X$.

Moreover, if $T$ is of type I, then the map $\phi : X \setminus \{p\} \to X$ is indeed a homeomorphism.

In fact, that classification is not mutually exclusive, as there can be examples of isometric shifts that are of both types I and II, as the authors show.

The description given above helps the authors to provide, in the same paper, a first partial answer to Question 2.

**Theorem 3.1.** *If $C(X)$ admits an isometric shift of type II, then $X$ is separable.*

**Corollary 3.1.** *If $C(X)$ admits an isometric shift and $X$ has no isolated points, then $X$ is separable.*

Can the same be said with respect to shift operators of type I? Suppose then that $T$ is of type I, and denote $\mathbf{1} := p$, $\mathbf{2} := \phi^{-1}(\mathbf{1})$, and in general $\mathbf{n} := \phi^{-1}(\mathbf{n}-\mathbf{1}) = \phi^{1-n}(\mathbf{1})$ for each $n \geq 2$. Since $p$ is isolated, all points in the set $\mathcal{N} := \{\mathbf{1}, \mathbf{2}, \mathbf{3}, ...\}$ are also isolated, and consequently $X$ contains a copy of $\mathbb{N}$.

We write $T = T[a, \phi, \Delta]$ to describe an isometric shift $T : C(X) \to C(X)$, where $X$ is compact and contains $\mathcal{N}$. It means that $\phi : X \setminus \{\mathbf{1}\} \to X$ is a homeomorphism, satisfying in particular $\phi(\mathbf{n}+\mathbf{1}) = \mathbf{n}$ for all $n \in \mathbb{N}$. It also means that $a \in C(X \setminus \{\mathbf{1}\})$, $|a| \equiv 1$, and that $\Delta$ is a continuous linear functional on $C(X)$ with $\|\Delta\| \leq 1$. The description of $T$ we have is $(Tf)(x) = a(x)f(\phi(x))$, when $x \neq 1$, and $(Tf)(1) = \Delta(f)$, for every $f \in C(X)$.

Taking into account the above and that Question 2 remains unsolved only for type I isometric shifts, we can ask a new question, very much related but much simpler.

**Question 2a** If $C(X)$ admits an isometric shift of type I, must $\mathcal{N}$ be dense in $X$?

Examples of isometric shifts for which $\text{cl}_X(\mathcal{N}) = X$ are called *primitive*. In fact the answer to Question 2a is negative and was given first by Farid and Varadarajan in [7]. They proved the following.

**Theorem 3.2.** *Let $n \in \mathbb{N}$. For $X := \mathbb{N} \cup \{\infty\}$, there exists an isometric shift $T : C(X) \to C(X)$ such that $X \setminus \text{cl}_X(\mathcal{N})$ is a set of $n$ isolated points.*

We compare it with the next two results, given later in [11], were the sharp distinction between the real and complex cases is made clear.

**Theorem 3.3.** *For $X := \mathbb{N} \cup \{\infty\}$, if $T : C_\mathbb{R}(X) \to C_\mathbb{R}(X)$ is a type I isometric shift, then $X \setminus \mathcal{N}$ is finite.*

**Theorem 3.4.** *For $X := \mathbb{N} \cup \{\infty\}$, there exists an isometric shift $T : C_\mathbb{C}(X) \to C_\mathbb{C}(X)$ such that $X \setminus \mathcal{N}$ is infinite.*

All examples given so far of spaces $X$ for which $C(X)$ admits type I isometric shifts are compactifications of integers, that is, $\mathbb{N} \cup \{\infty\}$, $\beta\mathbb{N}$, and the counterexample to Question 1 are compactifications of integers. It is interesting to see that in some cases the denseness of isolated points allows defining isometric shifts (see [11, Theorem 2.1]).

**Theorem 3.5.** *Let $X$ be an infinite metric space and let $D$ be a dense set of isolated points. If $X \setminus D$ is connected, then $C(X)$ admits a primitive isometric shift.*

**Question 2b** If $C(X)$ admits an isometric shift of type I, must $X$ have a dense set of (countably many) isolated points?

The answer to this question is again "no", although up to 2001 all known examples were compactifications of integers. Two very different spaces were given in [4] and [21]. The example in [21] involves the Cantor set but still is totally disconnected. As for that in [4], it satisfies $X \setminus \text{cl}_X(\mathcal{N}) = \mathbb{T}$ (the unit circle in $\mathbb{C}$), thus giving a new negative answer to Question 1 as well.

Looking at the above examples, assuming that $C(X)$ admits an isometric shift of type I, we can ask, as Gutek and Norden did in [11],

**Question 2c** How big can $X \setminus \mathrm{cl}_X(\mathcal{N})$ be?

This will be studied by considering how complex $\phi$ can be. Related to this, we can ask also

**Question 2d** How complex can $X$ be if $a$ is simple?

## 4. Questions 2c and 1 again

How to measure the complexity of $T$, $f \mapsto a \cdot f \circ \phi$? We can do it in two ways, namely:

- Forcing the map $\phi$ to be complex.
- Avoiding freedom in the choice of the map $a \in C(X \setminus \{\mathbf{1}\})$, in particular, setting $a \equiv 1$.

Concerning the first way, we give the following definition.

**Definition 4.1.** Let $X$ be compact, and suppose that $T = T[a, \phi, \Delta] : C(X) \to C(X)$ is a *non-primitive* isometric shift of type I. For $n \in \mathbb{N}$, we say that $T$ is *$n$-generated* if $n$ is the least number with the following property: There exist $n$ points $x_1, \ldots, x_n \in X \setminus \mathrm{cl}_X(\mathcal{N})$ such that the set

$$\{\phi^k(x_i) : k \in \mathbb{Z}, i \in \{1, \ldots, n\}\}$$

is dense in $X \setminus \mathrm{cl}_X(\mathcal{N})$.

We say that $T$ is $\infty$-generated if it is not $n$-generated for any $n \in \mathbb{N}$.

Notice that the above definition does not make sense when $T$ is an isometric shift which is not of type I. On the other hand, it is proved in [10, Theorem 2.5] that for such isometric shifts, there exists a point $x \in X$ such that $\{\phi^k(x) : k \in \mathbb{Z}\}$ is dense in $X$. In particular, if $T$ is a non-primitive isometric shift of both types I and II, then it is 1-generated.

Obviously, $\infty$-generated examples are necessarily very complex. The first example of such a shift was given in [11, Theorem 3.5].

**Theorem 4.1.** *Let $X := \beta \mathbb{N} + \mathcal{N} \cup \{\infty\}$. There exists an isometric shift on $C(X)$ which is $\infty$-generated.*

Our goal here is to give spaces with many infinite connected components admitting isometries which are $n$-generated for different $n$. This cannot be done in general, and we must put some restrictions on the number of components. For this, we need some definitions.

We say that $\mathbb{P} \subset \mathbb{N}$ is an *initial subset* if either $\mathbb{P} = \mathbb{N}$ or $\mathbb{P} = \{1, \ldots, N\}$ for some $N \in \mathbb{N}$.

**Definition 4.2.** Given an initial subset $\mathbb{P}$, a sequence $(p_n)_{n \in \mathbb{P}}$ of natural numbers is said to be $\mathbb{P}$-compatible if

- if $\mathbb{P} = \{1\}$, then $p_1 > 1$, and
- if $\mathbb{P} \neq \{1\}$, then $p_{n+1}$ is an even multiple of $p_n$ for every $n$ (allowing the possibility $p_1 = 1$).

The following result gives us a picture of how complicated $X$ and $T$ can be. In it the symbols $+$ and $\sum$ denote the topological sum of spaces. Of course, when there are many connected components we have to take a compactification so as to ensure that $X$ is compact.

Even if the space $X$ given in Theorem 4.1 is a compactification of integers, some of the ideas in it were taken to prove part of Theorem 4.2. Theorem 4.2 and Corollaries 4.1, 4.2 and 4.3 were proved in [2].

Below $\mathbb{T}^0$ denotes an isolated point.

**Theorem 4.2.** *Let $\mathbb{P}$ be an initial subset of $\mathbb{N}$ and let $(p_n)_{n \in \mathbb{P}}$ be a $\mathbb{P}$-compatible sequence. Suppose that $(\kappa_n)_{n \in \mathbb{P}}$ and $(Z_n)_{n \in \mathbb{P}}$ are a sequence of cardinals and a sequence of (nonempty) compact spaces satisfying at least one of the following two conditions:*

- **Condition 1.** *For every $n \in \mathbb{P}$, $0 \leq \kappa_n \leq \mathfrak{c}$ and $Z_n := \mathbb{T}^{\kappa_n}$;*
- **Condition 2.** *For every $n \in \mathbb{P}$, $\aleph_0 \leq \kappa_n \leq \mathfrak{c}$ and $Z_n := K_n^{\kappa_n}$, where $K_n$ is separable.*

*Then, for $N = \operatorname{card} \mathbb{P}$, there exists a compactification $\omega X_0$ of*

$$X_0 := \sum_{n \in \mathbb{P}} \underbrace{Z_n + \cdots + Z_n}_{p_n}$$

*such that $\omega X_0 \setminus X_0$ is either countable or empty, and such that $C(X)$ admits an $N$-generated isometric shift of type I (where $X := \omega X_0 + \mathcal{N} \cup \{\infty\}$).*

The next corollary shows that things are very different with respect to the case where we do not allow $X$ to contain isolated points, and should be compared with Theorem 2.4.

**Corollary 4.1.** *Every infinite-dimensional normed space contains a compact subset $X$ with infinitely many pairwise nonhomeomorphic components such that $C(X)$ admits an $\infty$-generated isometric shift of type I.*

**Corollary 4.2.** *With the same notation as in Theorem 4.2, if $s := \sup\{\kappa_n : n \in \mathbb{P}\} = \aleph_0$ (and further each $K_n$ is metrizable if we are under Condition 2), then $X$ may be taken to be contained in $\ell^2$, endowed with the norm topology. Moreover, if we are under Condition 1 and $s < \aleph_0$, then $X$ may be taken to be contained in $\mathbb{C}^s$.*

We easily deduce the following corollary, using the case when $s = 1$.

**Corollary 4.3.** *In $\mathbb{R}^2$, we can find a compact set $X$ having a countably infinite number of components (each of them being infinite), and such that $C(X)$ admits an isometric shift which is $\infty$-generated.*

The above corollary should be compared with the following very closed result given previously in [11]:

**Theorem 4.3.** *There exists a compact subset $X$ of $\mathbb{R}^2$ and a type I isometric shift on $C(X)$ such that $X \setminus \mathcal{N}$ is infinite.*

**Remark 4.1.** Notice that each copy of a power of $\mathbb{T}$ given in Theorem 4.2 is a connected component of $X \setminus \mathrm{cl}_X(\mathcal{N})$, and they are indeed all the infinite connected components of $X$ (if we assume $\kappa_n \neq 0$ for every $n$). The same comment applies to copies of powers of $K_n$ when every $K_n$ is connected. This implies that we can construct examples where we may decide at will on the number of (different) infinite connected components of $X \setminus \mathrm{cl}_X(\mathcal{N})$.

**Remark 4.2.** In Theorem 4.2, we are not assuming that spaces $K_n$ or cardinals $\kappa_n$ are necessarily pairwise different.

## 5. Question 2d for $a = 1$

Theorem 4.2 and its corollaries in Section 4 involve spaces $X = \omega X_0 + \mathcal{N} \cup \{\infty\}$ for a suitable compact space $\omega X_0$, that is, $\mathcal{N}$ appears with its one-point compactification as a topological summand. We next see that in such a way $a$ is always different from the constant map 1. Results in this section can be found in [2].

**Proposition 5.1.** *Let $X$ be compact, and suppose that $X = X_1 + X_2$, with $X_1 \neq \emptyset$ and $\mathcal{N}$ being a (countable) dense subset of $X_2$. Suppose also that $\phi : X \setminus \{\mathbf{1}\} \to X$ is a homeomorphism (with $\phi(\mathbf{n+1}) = \mathbf{n}$ for $\mathbf{n} \in \mathcal{N}$). Let*

$T : C(X) \to C(X)$ be a codimension 1 linear isometry such that $(Tf)(x) = f(\phi(x))$ for every $f \in C(X)$ and $x \in X \setminus \{1\}$. Then $T$ is not an isometric shift.

Consequently, when we try to obtain $a = 1$, we need to embed $\mathcal{N}$ in $X$ in a different way. This can be done and leads to the following result.

**Theorem 5.1.** $C(X)$ admits an isometric shift of type I for which $a \equiv 1$ in the following cases, being $X = W \cup \mathcal{N}$:

- if $W$ is a separable infinite power of a compact space with at least two points,
- if $W$ is a compact $n$-manifold (with or without boundary), for $n \geq 2$,
- if $W$ is the Sierpiński curve.

Also in the case when the weight $a$ is equal to 1, it is possible to find examples where the number of infinite connected components is $n$, for every $n \in \mathbb{N}$.

**Corollary 5.1.** Let $n \in \mathbb{N}$. Let $Y_0$ be a connected and compact space with more than one point, and suppose that $\phi : Y_0 \to Y_0$ is a homeomorphism having a periodic point, and such that $\phi^n$ is transitive. Then there exist a compact space $X$ and an isometric shift of type I on $C(X)$ with $a \equiv 1$, such that $X \setminus \mathcal{N}$ consists exactly of $n$ connected components with more than one point, each homeomorphic to $Y_0$.

## 6. Questions 3 and 1b

The question on whether there is or not an isometric shift on $C(\mathbf{K})$ was raised in [10], where some hints were also given about the existence of such a map. According to the representation we have given, we first see that, if there is one, then it cannot be of type I. The answer to the question (for the real case) appears in [12].

**Theorem 6.1.** There is an isometric shift on $C_{\mathbb{R}}(\mathbf{K})$.

Once again, when a sequence is adjoined to the Cantor set, we can find many different isometric shifts, all of them of type I (see [2]).

**Theorem 6.2.** Suppose that $(x_n)$ is an eventually nonconstant sequence in $\mathbb{R} \setminus \mathbf{K}$ which converges to a point $L \in \mathbb{R}$. Let $X := \mathbf{K} \cup \{x_n : n \in \mathbb{N}\} \cup \{L\}$. We have

- If $L \notin \mathbf{K}$, then for each $n \in \mathbb{N} \cup \{\infty\}$, there exists an isometric shift of type I on $C(X)$ which is n-generated.
- If $L \in \mathbf{K}$, then there exists an isometric shift of type I on $C(X)$ which also satisfies the additional condition that $a \equiv 1$.

**Remark 6.1.** Recall that, as mentioned by the authors, by [11, Theorem 1.9] we can conclude that if $X$ consists of a convergent sequence adjoined to a non-separable Cantor cube, then $C(X)$ does not admit an isometric shift. This is not true in the separable case, as shown in [21, Example 20] for an isometric shift of both types I and II. Theorem 5.1 and Theorem 6.2 say also the contrary in the separable case for isometric shifts which are not of type II. Theorem 6.2 provides indeed completely different families of isometric shifts of type I.

Some other special examples were also studied in [12], where the two following results appear (in particular, an answer to Question 1b).

**Theorem 6.3.** *There is a Peano continuum $X$ such that $C_\mathbb{C}(X)$ admits an isometric shift.*

**Theorem 6.4.** *There is a one-dimensional, connected, compact, metric space $X$ such that $C_\mathbb{C}(X)$ admits an isometric shift.*

## 7. Question 2

Question 2 was first raised in [10], and not much was known about its final answer until very recently. As we mentioned above, the question was known to be positive for type II isometric shifts, so that if $X$ has no isolated points and $C(X)$ admits a shift, then $X$ must be separable (see Corollary 3.1). Something similar can be said if we know that $C(X)$ admits an isometric shift $T$ that is also disjointness preserving, that is, $(Tf)(Tg) \equiv 0$ whenever $fg \equiv 0$, as it was shown in [10, Theorem 5.1]; in this case $X$ must be separable. Another positive answer was given in [4], and can be stated as follows.

**Theorem 7.1.** *Let $M$ be a complete metric space, and suppose that $C(\beta M)$ admits an isometric shift. Then $M$ is separable.*

Almost at the same time, another result giving a clue on a general property spaces must satisfy so as to admit isometric shifts, was given in [11, Theorem 1.4].

**Theorem 7.2.** *If $C(X)$ admits an isometric shift of type I, then $X$ has the countable chain condition.*

Related to this, we also have that if $C(X)$ admits an isometric shift of type I, then $C_0(X \setminus \text{cl}_X(\mathcal{N}))$ (the space of $\mathbb{K}$-valued continuous functions vanishing at infinity) must have cardinality at most equal to $\mathfrak{c}$ (see [11, Theorem 1.9]).

Question 2 was recently solved in the negative in [1]. The following theorems provide examples of $X$ not separable for which $C(X)$ admits an isometric shift (necessarily of type I). All the examples given in [1] are based on $\mathfrak{M}$, the maximal ideal space of the algebra $L^\infty(\mathbb{T})$ of all Lebesgue-measurable essentially bounded *complex*-valued functions on $\mathbb{T}$. It is not hard to see that $\mathfrak{M}$ is not separable. Consequently, since it has no isolated points, $C(\mathfrak{M})$ admits no isometric shift (Corollary 3.1). A different conclusion is obtained when we adjoin to $\mathfrak{M}$ a convergent sequence and its limit.

A first example is the following (see [1, Theorem 3.1]).

**Theorem 7.3.** $C(\mathfrak{M} + \mathcal{N} \cup \{\infty\})$ *admits an isometric shift.*

$\mathfrak{M}$ is homeomorphic to an infinite closed subset of $\beta \mathbb{N} \setminus \mathbb{N}$, and consequently its cardinal must be $2^\mathfrak{c}$ (see [17] and [9, Corollary 9.2]). Now we can obtain more examples, in particular with many infinite connected components, as is the case of the following result ([1, Theorem 3.2]), which has $2^\mathfrak{c}$ infinite connected components (see again Question 1).

**Theorem 7.4.** *Let $\kappa$ be any cardinal such that $1 \leq \kappa \leq \mathfrak{c}$. Then $C(\mathfrak{M} \times \mathbb{T}^\kappa + \mathcal{N} \cup \{\infty\})$ admits an isometric shift.*

We can also give examples with just one infinite component (see [1, Theorem 3.3]).

**Theorem 7.5.** *Let $\kappa$ be any cardinal such that $1 \leq \kappa \leq \mathfrak{c}$. Then $C(\mathfrak{M} + \mathbb{T}^\kappa + \mathcal{N} \cup \{\infty\})$ admits an isometric shift.*

All the above results are valid both for $\mathbb{K} = \mathbb{R}$ and $\mathbb{K} = \mathbb{C}$. We next see some depending on the scalar field (see [1, Theorems 5.1, 5.2, and Example 5.3]).

**Theorem 7.6.** *Let $\mathbb{K} = \mathbb{C}$. Suppose that $n \in \mathbb{N}$, and that $(\kappa_j)_{j=1}^n$ is a finite sequence of cardinals satisfying $0 \leq \kappa_j \leq \mathfrak{c}$ for every $j$. Let $X_n := \mathbb{T}^{\kappa_1} + \ldots + \mathbb{T}^{\kappa_n} + \mathcal{N} \cup \{\infty\}$. Then*

- $C_{\mathbb{C}}(X_n)$ admits an $n$-generated isometric shift.
- $C_{\mathbb{C}}(\mathfrak{M} + X_n)$ admits an isometric shift.

The above is in general not true in the real case.

**Example 7.1.** If $\mathbb{K} = \mathbb{R}$, then neither $C_{\mathbb{R}}\left(\mathbb{T} + \mathbb{T}^2 + \mathbb{T}^3 + \mathcal{N} \cup \{\infty\}\right)$ nor $C_{\mathbb{R}}\left(\mathfrak{M} + \mathbb{T} + \mathbb{T}^2 + \mathbb{T}^3 + \mathcal{N} \cup \{\infty\}\right)$ admit an isometric shift.

## 8. A final question

We end with the following question, which was posed in [12] and does not seem to have been answered yet:

**Question** Is there a *finite-dimensional* Peano continuum $X$ such that $C_{\mathbb{C}}(X)$ admits an isometric shift?

## References

1. J. Araujo, *On the separability problem for isometric shifts on $C(X)$*. J. Funct. Anal. **256** (2009) 1106-1117.
2. J. Araujo, *Examples of isometric shifts on $C(X)$*. Preprint.
3. J. Araujo and J.J. Font, *Codimension 1 linear isometries on function algebras*. Proc. Amer. Math. Soc. **127** (1999), 2273-2281.
4. J. Araujo and J.J. Font, *Isometric shifts and metric spaces*. Monatshefte Math. **134** (2001), 1-8.
5. L.-S. Chen, J.-S. Jeang, and N.-C. Wong, *Disjointness preserving shifts on $C_0(X)$*. J. Math. Anal. Appl. **325** (2007), 400-421.
6. R. M. Crownover, *Commutants of shifts on Banach spaces*. Michigan Math. J. **19** (1972), 233-247.
7. F.O. Farid and K. Varadajaran, *Isometric shift operators on $C(X)$*. Can. J. Math. **46** (1994), 532-542.
8. J.J. Font, *Isometries on function algebras with finite codimensional range*. Manuscripta Math. **100** (1999), 13-21.
9. L. Gillman and M. Jerison, *Rings of continuous functions*. Springer Verlag, New York, 1976.
10. A. Gutek, D. Hart, J. Jamison and M. Rajagopalan, *Shift operators on Banach spaces*. J. Funct. Anal. **101** (1991), 97-119.
11. A. Gutek and J. Norden, *Type 1 shifts on $C(X)$*. Topology Appl. **114** (2001), 73-89.
12. R. Haydon, *Isometric shifts on $C(K)$*. J. Funct. Anal. **135** (1996), 157-162.
13. H. Holsztyński, *Continuous mappings induced by isometries of spaces of continuous functions*. Studia Math. **26** (1966), 133-136.
14. J.R. Holub, *On shift operators*. Canad. Math. Bull. **31** (1988), 85-94.
15. K. Izuchi, *Douglas algebras which admit codimension 1 linear isometries*. Proc. Amer. Math. Soc. **129** (2001), 2069-2074.

16. J.-S. Jeang and N.-C. Wong, *Isometric shifts on $C_0(X)$*. J. Math. Anal. Appl. **274** (2002), 772-787.
17. S. Negrepontis, *On a theorem by Hoffman and Ramsay*. Pacific J. Math. **20** (1967), 281-282.
18. M. Rajagopalan, T. M. Rassias, and K. Sundaresan, *Generalized backward shifts on Banach spaces $C(X,E)$*. Bull. Sci. Math. **124** (2000), 685-693.
19. M. Rajagopalan and K. Sundaresan, *Backward shifts on Banach spaces $C(X)$*. J. Math. Anal. Appl. **202** (1996), 485-491.
20. M. Rajagopalan and K. Sundaresan, *An account of shift operators*. J. Anal. **8** (2000), 1-18.
21. M. Rajagopalan and K. Sundaresan, *Generalized shifts on Banach spaces of continuous functions*. J. Anal. **10** (2002), 5-15.
22. T. M. Rassias and K. Sundaresan, *Generalized backward shifts on Banach spaces*. J. Math. Anal. Appl. **260** (2001), 36-45.

# Uniform algebras of symmetric holomorphic functions

Richard M. Aron

*Department of Mathematics, Kent State University, Kent, OH 44242, USA*
*e-mail: aron@math.kent.edu*

Pablo Galindo*

*Departamento de Análisis Matemático, Universidad de Valencia, 46100 Burjasot Valencia, Spain*
*e-mail: Pablo.Galindo@uv.es*

Dedicated to the memory of Antonio Aizpuru.

We give a survey of results dealing with the Banach algebra structure of the set of uniformly continuous and symmetric holomorphic functions on the ball of some classical Banach spaces.

*Keywords*: Symmetric holomorphic functions, maximal ideal space, Banach algebras of holomorphic functions on the ball.

## Introduction

In this note, we first review the work of R. Alencar, A. Zagorodnyuk, and the authors on the algebra of uniformly continuous holomorphic functions on the open unit ball $B$ of some common infinite dimensional Banach spaces [1]. In addition, we give a complete characterization of the set of homomorphisms on the set of such holomorphic functions when $B$ is the open unit ball of $\mathbb{C}^n$, endowed with a symmetric norm. In this section, we provide some necessary background on the topic of uniformly continuous, symmetric, holomorphic mappings. In section 2, we discuss the *spectrum* (also known as the maximal ideal space) when the underlying Banach space is infinite dimensional, while in section 3, we discuss the same problem for $\mathbb{C}^n$.

---

*The first author was partially supported by Ministerio de Ciencia e Innovación MTM 2008-03211. The second author was partially supported by Project MTM 2007-064521, MEC-FEDER (Spain).

## 1. Symmetric holomorphic functions

We begin with a short summary of standard definitions. Let $X$ be a complex Banach space with open unit ball $B$, and let $n \in \mathbb{N}$. A function $P : X \to \mathbb{C}$ is said to be a (continuous) $n$–*homogeneous polynomial* if there is a (necessarily unique) symmetric continuous $n$–linear form $A : X \times \cdots \times X \to \mathbb{C}$ such that $P(x) = A(x,...,x)$ for all $x \in X$. A function $f : B \to \mathbb{C}$ is called *holomorphic* if there is a sequence of $n$–homogeneous polynomials $P_n : X \to \mathbb{C}$ such that

$$f(x) = \sum_{n=0}^{\infty} P_n(x), \quad \text{for all } x \in B.$$

It suffices that the convergence of the above series be pointwise, and standard arguments show that it is then automatically locally uniformly convergent.

We consider Banach sequence spaces $X$ with a *symmetric* norm, that is, for all permutations $\sigma : \mathbb{N} \to \mathbb{N}$, and $x = (x_n) \in B$, also $(x_{\sigma(1)},...,x_{\sigma(n)},...) \in B$.

A holomorphic function $f : B \to \mathbb{C}$ is called *symmetric* if for all $x \in B$ and all permutations $\sigma : \mathbb{N} \to \mathbb{N}$, the following holds:

$$f(x_1,...,x_n,...) = f(x_{\sigma(1)},...,x_{\sigma(n)},...).$$

Note that this notion of symmetric is different from that of symmetric multilinear form, in the previous paragraph.

Our interest throughout this paper will be in the set

$\mathcal{A}_{us}(B) =$
$\{f : B \to \mathbb{C} \mid f \text{ is holomorphic, uniformly continuous, and symmetric on } B\}$.

The following result is straightforward.

**Proposition 1.1.** $\mathcal{A}_{us}(B)$ *is a unital commutative Banach algebra under the supremum norm. Each function* $f \in \mathcal{A}_{us}(B)$ *admits a unique (automatically symmetric) extension to* $\overline{B}$.

Let us give some examples of $\mathcal{A}_{us}(B)$ when $B$ is the open unit ball of some classical Banach spaces $X$.

**Example 1.1.** $X = c_0$. We will use the following result, which we believe is originally due to W. Bogdanowicz.

**Theorem 1.1 ([2]).** *Let $P : c_0 \to \mathbb{C}$ be an $n$–homogeneous polynomial and $\varepsilon > 0$. Then there is $N \in \mathbb{N}$ and an $n$–homogeneous polynomial $Q : \mathbb{C}^N \to \mathbb{C}$ such that for all $x = (x_1, ..., x_N, x_{N+1}, ...) \in B$, $|P(x) - Q(x_1, ..., x_N)| < \varepsilon$.*

One consequence of the preceding result is the following Corollary.

**Corollary 1.1.** *For all $n \in \mathbb{N}, n \geq 1$, the only $n$–homogeneous symmetric polynomial $P : c_0 \to \mathbb{C}$ is $P = 0$.*

**Proof.** Given $\varepsilon > 0$, let $N$ and $Q$ be as above. For any $x = (x_1, ..., x_N, ...) \in c_0$, first choose $M \in \mathbb{N}$ so that $|P(x) - P(x_1, ..., x_M, 0, 0, ...)| < \varepsilon$. Since $P$ is symmetric,

$$|P(x) - P(\underbrace{0, ..., 0}_{N}, x_1, ..., x_M, 0, 0, ...)| < \varepsilon,$$

and so $|P(x) - Q(0, ..., 0)| < 2\varepsilon$. However, since $Q$ is homogeneous, $Q(0, ..., 0) = 0$, and so $|P(x)| < 2\varepsilon$. Since $\varepsilon$ and $x$ are arbitrary, the result follows. □

Since any function $f \in \mathcal{A}_{us}(B)$ can be uniformly approximated on $B$ by finite sums of symmetric homogeneous polynomials, it follows that $\mathcal{A}_{us}(B)$ consists of just the constant functions when $B$ is the open unit ball of $c_0$.

**Example 1.2.** $X = \ell_p$ for some $p, 1 \leq p < \infty$.
*The linear $(n = 1)$ case.* Let $\varphi \in \ell_p^*$ be a symmetric 1–homogeneous polynomial on $\ell_p$; that is, $\varphi$ is a symmetric continuous linear form. Since $\varphi$ can be regarded as a point $(y_1, ..., y_m, ...) \in \ell_{p^*}$ and since $y_j = \varphi(e_j) = \varphi(e_1)$ for all $j$, we see that $y_1 = \cdots y_m = \cdots$. Therefore, the set of symmetric linear forms $\varphi$ on $\ell_1$ consists of the 1–dimensional space $\{b(1, ..., 1, ...) \mid b \in \mathbb{C}\}$. For $p > 1$, the above shows that there are no non-trivial linear symmetric forms on $\ell_p$.
*The quadratic $(n = 2)$ case.* Let $P : \ell_p \to \mathbb{C}$ be a symmetric 2–homogeneous polynomial, and let $A : \ell_p \times \ell_p \to \mathbb{C}$ be the unique symmetric bilinear form associated to $P$, such that $P(x) = A(x, x)$ for all $x \in \ell_p$. Now, $P(e_1) = P(e_j)$ for all $j \in \mathbb{N}$. Moreover,

$$P(e_1 + e_2) = A(e_1 + e_2, e_1 + e_2) = A(e_1, e_1) + 2A(e_1, e_2) + A(e_2, e_2)$$

$$= P(e_1) + 2A(e_1, e_2) + P(e_2)$$

and likewise
$$P(e_j + e_k) = P(e_j) + 2A(e_j, e_k) + P(e_k),$$
for all $j$ and $k$ in $\mathbb{N}$. Therefore $A(e_j, e_k) = A(e_1, e_2)$.

So, for all $N$,
$$P(x_1, ..., x_N, 0, 0, ...) = a \sum_{j=1}^{N} x_j^2 + b \sum_{j \neq k} x_j x_k,$$
where $a = P(e_1)$ and $b = A(e_j, e_k)$.

From this, we can conclude that for $X = \ell_1$, the space of symmetric 2–homogeneous polynomials on $\ell_1$, $\mathcal{P}_s(^2\ell_1)$, is 2–dimensional with basis $\{\sum_j x_j^2, \sum_{j \neq k} x_j x_k\}$. On the other hand, the corresponding space $\mathcal{P}_s(^2\ell_2)$ of symmetric 2–homogeneous polynomials on $\ell_2$, is 1–dimensional with basis $\{\sum_j x_j^2\}$. For $1 < p < 2, \mathcal{P}_s(^2\ell_p)$ is also the one-dimensional space generated by $\sum_j x_j^2$, while $\mathcal{P}_s(^2\ell_p) = \{0\}$ for $p > 2$.

This argument can be extended to all $n$ and all $p$, and we can conclude that for all $n, p$, the space of symmetric $n$–homogeneous polynomials on $\ell_p$, $\mathcal{P}_s(^n\ell_p)$, is finite dimensional. Consequently, since for all $f \in \mathcal{A}_{us}(B)$, $f$ is a uniform limit of symmetric $n$–homogeneous polynomials, we have reasonably good knowledge about the functions in $\mathcal{A}_{us}(B)$. Speaking heuristically, $\mathcal{A}_{us}(B)$, for $B$ the open unit ball of an $\ell_p$ space, is a "small" algebra.

For a thorough study of symmetric polynomials we refer to [3] where detailed proofs of the mentioned examples can be found.

## 2. The spectrum of $\mathcal{A}_{us}(B)$

Recall that the *spectrum* (or *maximal ideal space*) of a Banach algebra $\mathcal{A}$ with identity $e$ is the set $\mathcal{M}(\mathcal{A}) = \{\varphi : \mathcal{A} \to \mathbb{C} \mid \varphi \text{ is a homomorphism and } \varphi(e) = 1\}$. We recall that if $\varphi \in \mathcal{M}(\mathcal{A})$, then $\varphi$ is automatically continuous with $\|\varphi\| = 1$. Moreover, when we consider it as a subset of $\mathcal{A}^*$ with the weak-star topology, $\mathcal{M}(\mathcal{A})$ is compact.

We will examine $\mathcal{M}(\mathcal{A}_{us}(B))$ when $B = B_{\ell_p}$. The most obvious element in $\mathcal{M}(\mathcal{A}_{us}(B))$ is the evaluation homomorphism $\delta_x$ at a point $x$ of $\overline{B}$ (recalling that since the functions in $\mathcal{A}_{us}(B)$ are uniformly continuous, they have unique continuous extensions to $\overline{B}$). Of course, if $x, y \in B$ are such that $y$ can be obtained from $x$ by a permutation of its coordinates, then

$\delta_x = \delta_y$. It is natural to wonder whether $\mathcal{M}(\mathcal{A}_{us}(B))$ consists of only the set of equivalence classes $\{\delta_{\tilde{x}} \mid x \in \overline{B}\}$, where $x \sim y$ means that $x$ and $y$ differ by a permutation.

**Example 2.1.** For each $n \in \mathbb{N}$, define $F_n : B \to \mathbb{C}$ by $F_n(x) = \sum_{j=1}^{\infty} x_j^n$. To simplify, we take $B = B_{\ell_2}$ (so that $F_n$ will be defined only for $n \geq 2$). It is known that the algebra generated by $\{F_n \mid n \geq 2\}$ is dense in $\mathcal{A}_{us}(B)$. For each $k \in \mathbb{N}$, let

$$v_k = \frac{1}{\sqrt{k}}(e_1 + \cdots + e_k).$$

It is routine that each $v_k$ has norm 1, that $\delta_{v_k}(F_2) = 1$ for all $k \in \mathbb{N}$, and that for all $n \geq 3$,

$$\delta_{v_k}(F_n) = F_n(v_k) = \frac{1}{(\sqrt{k})^n} k \to 0 \text{ as } k \to \infty.$$

Since $\mathcal{M}(\mathcal{A}_{us}(B))$ is compact, the set $\{\delta_{v_k} \mid k \in \mathbb{N}\}$ has an accumulation point $\varphi \in \mathcal{M}(\mathcal{A}_{us}(B))$. It is clear that $\varphi(F_2) = 1$ and that $\varphi(F_n) = 0$ for all $n \geq 3$. It is not difficult to verify that $\varphi \neq \delta_x$ for every $x \in \overline{B}$. This construction could be altered slightly, by letting $v_k = \frac{1}{\sqrt{k}}(\alpha_1 e_1 + \cdots + \alpha_k e_k)$, where each $|\alpha_j| \leq 1$. Thus, with this method we give a small number of additional homomorphisms in $\mathcal{M}(\mathcal{A}_{us}(B))$ that do not correspond to point evaluations.

It should be mentioned that we do not know whether $\mathcal{M}(\mathcal{A}_{us}(B_{\ell_p}))$ contains other points. However, we are able to give a different characterization of $\mathcal{M}(\mathcal{A}_{us}(B_{\ell_p}))$ (which may be no more enlightening). In order to do this, we first simplify our notation by considering only $B_{\ell_1}$. For each $n \in \mathbb{N}$, define $\mathcal{F}^n : B_{\ell_1} \to \mathbb{C}^n$ as follows:

$$\mathcal{F}^n(x) = (F_1(x), ..., F_n(x)) = (\sum_j x_j, ..., \sum_j x_j^n).$$

Let $D_n = \mathcal{F}^n(B_{\ell_1})$, and let $[D_n]$ be the polynomially convex hull of $D_n$ (see, e.g., [4]). Let

$$\Sigma_1 = \{(b_i)_{i=1}^{\infty} \in \ell_{\infty} : (b_i)_{i=1}^{n} \in [D^n], \text{ for all } n \in \mathbb{N}\}.$$

In other words, $\Sigma_1$ is the inverse limit of the sets $[D_n]$, endowed with the natural inverse limit topology.

**Theorem 2.1.** $\Sigma_1$ *is homeomorphic to* $\mathcal{M}(\mathcal{A}_{us}(B_{\ell_1}))$.

The analogous results, and the analogous definitions, are valid for $\Sigma_p$ and $\mathcal{M}(\mathcal{A}_{us}(B_{\ell_p}))$.

The basic steps in the proof of Theorem 2.1 are as follows: First, since the algebra generated by $\{F_n \mid n \geq 1\}$ is dense in $\mathcal{A}_{us}(B_{\ell_1})$, each homomorphism $\varphi \in \mathcal{M}(\mathcal{A}_{us}(B_{\ell_1}))$ is determined by its behavior on $\{F_n\}$. Next, every symmetric polynomial $P$ on $\ell_1$ can be written as $P = Q \circ \mathcal{F}^n$ for some $n \in \mathbb{N}$ and some polynomial $Q : \mathbb{C}^n \to \mathbb{C}$. Finally, to each $(b_i) \in \Sigma_1$, one associates $\varphi = \varphi_{(b_i)} : \mathcal{A}_{us}(B_{\ell_1}) \to \mathbb{C}$ by $\varphi(P) = Q(b_1, ..., b_n)$. This turns out to be a well-defined homomorphism, and the mapping $(b_i) \in \Sigma_1 \rightsquigarrow \varphi_{(b_i)} \in \mathcal{M}(\mathcal{A}_{us}(B_{\ell_1}))$ is a homeomorphism.

## 3. The spectrum of $\mathcal{A}_{us}(B)$ in the finite dimensional case

We now turn to $\mathcal{A}_{us}(B)$, where $B$ is the open unit ball of $\mathbb{C}^n$, endowed with a *symmetric* norm. Because of finite dimensionality, $\mathcal{A}_{us}(B) = \mathcal{A}_s(B)$, where $\mathcal{A}_s(B)$ is the Banach algebra of symmetric holomorphic functions on $B$ that are continuous on $\overline{B}$.

Unlike the infinite dimensional case, the following result holds.

**Theorem 3.1.** *Every homomorphism* $\varphi : \mathcal{A}_s(B) \to \mathbb{C}$ *is an evaluation at some point of* $\overline{B}$.

We describe below the main ideas in the proof of this result.

**Proposition 3.1.** *Let* $C \subset \mathbb{C}^n$ *be a compact set. Then $C$ is symmetric and polynomially convex if and only if $C$ is polynomially convex with respect to only the* symmetric *polynomials.*

In other words, $C$ is symmetric and polynomially convex if and only if

$$C = \{z_0 \in \mathbb{C}^n \mid |P(z_0)| \leq \sup_{z \in C} |P(z)|, \quad \text{for all symmetric polynomials } P\}.$$

For $i \in \mathbb{N}$, let

$$R_i(x) = \sum_{1 \leq k_1 < \cdots k_i \leq n} x_{k_1} \cdots x_{k_j}.$$

**Proposition 3.2.** *Let $B$ be the open unit ball of a symmetric norm on $\mathbb{C}^n$. Then the algebra generated by the symmetric polynomials $R_1, ..., R_n$ is dense in* $\mathcal{A}_s(B)$.

**Lemma 3.1.** *(Nullstellensatz for symmetric polynomials). Let $P_1, ..., P_m$ be symmetric polynomials on $\mathbb{C}^n$ such that*

$$\ker P_1 \cap \cdots \cap \ker P_m = \emptyset.$$

*Then there are symmetric polynomials $Q_1, ..., Q_m$ on $\mathbb{C}^n$ such that $\sum_{j=1}^{m} P_j Q_j \equiv 1$.*

We are now ready to prove Theorem 3.1. Consider the symmetric polynomials $P_1 = R_1 - \varphi(R_1), ..., P_m = R_m - \varphi(R_m)$. If $\ker P_1 \cap \cdots \cap \ker P_m = \emptyset$, then Lemma 3.1 implies that there are symmetric polynomials $Q_1, ..., Q_m$ on $\mathbb{C}^n$ such that $\sum_{j=1}^{m} P_j Q_j \equiv 1$. This is impossible, since $\varphi(P_j Q_j) = 0$. Therefore, there exists some $x \in \mathbb{C}^n$ such that $P_j(x) = 0$ for all $j$, which means $\varphi(R_j) = R_j(x)$ for all $j$. By Proposition 3.2, $\varphi(P) = P(x)$, for all symmetric polynomials $P : \mathbb{C}^n \to \mathbb{C}$.

So, for all such $P$, $|\varphi(P)| = |P(x)| \leq \|P\|$. This means that $x$ belongs to the symmetrical polynomial convex hull of $\overline{B}$. Since $\overline{B}$ is symmetric and convex, it is symmetrically polynomially convex (by Proposition 3.1). Thus $x \in \overline{B}$.

### References

1. R. Alencar, R. M. Aron, P. Galindo, and A. Zagorodnyuk, *Algebras of symmetric holomorphic functions on $\ell_p$*, Bull. Lond. Math. Soc. **35** (2003), 55-64.
2. W. M. Bogdanowicz, "Weak continuity of polynomial functionals on the space of sequences convergent to zero" (in Russian with English translation), Bulletin de l'Academie Polonaise des Sciences **5** (1957):243-246.
3. M. González, R. Gonzalo and J. Jaramillo, *Symmetric polynomials on rearrangement invariant function spaces*, J. London Math. Soc. **(2) 59**, (1999) 681-697.
4. L. Hörmander, *An introduction to complex analysis in several variables*, North Holland (1990).

# Some results on the local theory of normed spaces since 2002 (1997)

F. J. García-Pacheco*

*Department of Mathematical Sciences, Texas A&M University, College Station Texas, 77843-3368, USA*
*e-mail: fgarcia@math.tamu.edu*

*This work is dedicated to the beloved memory of Prof. Antonio Aizpuru.*

We will review a series of results of Prof. Antonio Aizpuru on the local theory of normed spaces since 2002. However, we will begin by introducing 1997 Aizpuru's E-Property.

## 1. Aizpuru's first contribution to the geometry of Banach spaces

Aizpuru always felt a profound admiration for the work of Bade. In his 1971 lecture notes (see [7]), Bade proved an interesting theorem on the extremal structure of the unit ball of spaces of continuous functions.

**Theorem 1.1 (Bade, 1971).** *Let $K$ be a compact Hausdorff space. The following conditions are equivalent:*

*(1)* $\mathsf{B}_{\mathcal{C}(K)} = \overline{\mathrm{co}}\left(\mathrm{ext}\left(\mathsf{B}_{\mathcal{C}(K)}\right)\right)$.
*(2)* $K$ *is 0-dimensional.*

This result of Bade motivated him for the following definition (see [7]).

**Definition 1.1 (Bade, 1971).** *Let $X$ be a Banach space. We say that $X$ has the Bade Property exactly when*
$$\mathsf{B}_X = \overline{\mathrm{co}}\left(\mathrm{ext}\left(\mathsf{B}_X\right)\right).$$

---

*The author would like to express his gratefulness to the organizing committee of the IV CIDAMA, in particular, to Prof. Rambla-Barreno, Prof. Pérez-Fernández, Prof. León-Saavedra, and Prof. Benítez-Trujillo for the nice hospitality and environment.

This definition of Bade motivated Aron and Lohman to introduce a stronger version of the Bade Property (see [6]).

**Definition 1.2 (Aron and Lohman, 1987).** *Let $X$ be a Banach space. We say that $X$ has the $\lambda$-Property exactly when every $x \in \mathsf{B}_X$ admits an amenable triple $(e, y, \lambda) \in \text{ext}(\mathsf{B}_X) \times \mathsf{B}_X \times (0,1]$, in other words, $x = \lambda e + (1-\lambda)y$.*

Indeed, Aron and Lohman showed that their $\lambda$-Property is strictly stronger than the Bade Property.

**Remark 1.1 (Aron and Lohman, 1987).** *If $X$ is a Banach space with the $\lambda$-Property, then $X$ has the Bade Property. The converse does not hold. Indeed, $\mathcal{C}(\mathsf{B}_\mathbb{C}, \mathbb{C})$ has the Bade Property but lacks the $\lambda$-Property.*

However, Aron and Lohman's $\lambda$-Property coincides with the Bade Property in the class of spaces of continuous functions (see [14]).

**Theorem 1.2 (Oates, 1990).** *Let $K$ be a compact Hausdorff space. The following conditions are equivalent:*

*(1) $\mathcal{C}(K)$ has the Bade Property.*
*(2) $\mathcal{C}(K)$ has the $\lambda$-Property.*
*(3) $K$ is 0-dimensional.*

The $\lambda$-Property also made a deep impact in Aizpuru, so he proposed himself to separate these two properties. The answer did not take long to come: The E-Property (see [1]). In one of our long and entertaining conversations, Aizpuru mentioned to me that the letter E in the name of his property refers to "extreme".

**Definition 1.3 (Aizpuru, 1997).** *Let $X$ be a Banach space. We say that $X$ has the E-Property exactly when every norm-attaining functional on $X$ attains its norm at an extreme point of $\mathsf{B}_X$.*

Aizpuru noted first that his E-Property lies between the $\lambda$-Property and the Bade Property, leading to the fact that these three properties are equivalent in the class of spaces of continuous functions in virtue of Oates' Theorem.

**Remark 1.2 (Aizpuru, 1997).** *Let $X$ be a Banach space. Then:*

*(1) If $X$ has the $\lambda$-Property, then $X$ has the E-Property.*
*(2) If $X$ has the E-Property, then $X$ has the Bade Property.*

**Corollary 1.1 (Aizpuru, 1997).** *Let $K$ be a compact Hausdorff space. The following conditions are equivalent:*

*(1) $\mathcal{C}(K)$ has the Bade Property.*
*(2) $\mathcal{C}(K)$ has the E-Property.*
*(3) $\mathcal{C}(K)$ has the $\lambda$-Property.*
*(4) $K$ is $0$-dimensional.*

The challenge then is to show that these three properties are different from each other, in other words, Aizpuru's E-Property strictly separates the $\lambda$-Property from the Bade Property.

**Remark 1.3 (Aizpuru, 1997).**

*(1) $\mathcal{C}(\mathsf{B}_\mathbb{C}, \mathbb{C})$ has the Bade Property but lacks the E-Property.*
*(2) $(\bigoplus \ell_1^n)_p$, $1 < p < \infty$, has the E-Property but lacks the $\lambda$-Property.*

Special attention should be paid to the proof of the first part of the previous remark, where Aizpuru shows that $\mathcal{C}(\mathsf{B}_\mathbb{C}, \mathbb{C})$ lacks the E-Property by reaching the contradiction that there exists a retract from the unit disc $\mathbb{D} := \{z \in \mathbb{C} : |z| \leq 1\}$ onto its frontier, $\mathsf{S}^1$.

## 2. Aizpuru's interest in L-summands

In the early 2000's we were given the Ph.D. dissertation of Becerra-Guerrero (see [9]). In his dissertation, Becerra-Guerrero characterizes the real Hilbert spaces in terms of $\mathsf{L}^2$-summand vectors and isometric reflection vectors.

**Theorem 2.1 (Becerra-Guerrero and Rodríguez-Palacios, 1999).**
*Let $X$ be a real Banach space. The following conditions are equivalent:*

*(1) $X$ is a Hilbert spaces.*
*(2) The set of $\mathsf{L}^2$-summand vectors of $X$ has non-empty interior.*
*(3) The set of isometric reflection vectors of $X$ has non-empty interior.*

Before introducing Aizpuru's results we will define precisely the concepts of "isometric reflection vector" and "$\mathsf{L}^p$-summand vector" (see [10] and [15]).

**Definition 2.1 (Behrends, 1977; Skorik and Zaidenberg, 1991).**
*Let $X$ be a real Banach space. Let $x \in X$. Then:*

*(1) We say that $x$ is an $\mathsf{L}^p$-summand vector of $X$, $1 \leq p \leq \infty$, if there exists a closed maximal subspace $M$ of $X$ such that $X = \mathbb{R}x \oplus_p M$.*

(2) We say that $x$ is an *isometric reflection vector* of $X$ if there exists a closed maximal subspace $M$ of $X$ such that $X = \mathbb{R}x \oplus M$ and $\|\lambda x + m\| = \|\lambda x - m\|$ for all $\lambda \in \mathbb{R}$ and all $m \in M$.

In [3] we generalized the "$\mathsf{L}^2$-summand" part of Becerra-Guerrero and Rodríguez-Palacios' Theorem as follows.

**Theorem 2.2 (Aizpuru and García-Pacheco, 2006).** *Let $X$ be a real Banach space. The set of $\mathsf{L}^2$-summand vectors of $X$ is a closed vector subspace of $X$ which is $\mathsf{L}^2$-complemented in $X$.*

And in [5] we gave a shorter and more simple proof of the "isometric reflection" part of Becerra-Guerrero and Rodríguez-Palacios' Theorem.

**Theorem 2.3 (Aizpuru, García-Pacheco, and Rambla, 2004).**
*Let $X$ be a real Banach space. Then $X$ is a Hilbert space if and only if the set of isometric reflection vectors of $X$ has non-empty interior.*

The next step that Aizpuru wanted to take is investigate the extremal properties of isometric reflection vectors. One of the first things that we immediately noticed is the following.

**Remark 2.1 (Aizpuru and García-Pacheco, 2004).** *Let $X$ be a real Banach space. Let $x \in \mathsf{S}_X$ be an isometric reflection vector of $X$. If $Y$ is a 2-dimensional subspace of $X$ containing $x$, then $x$ is an isometric reflection vector of $Y$.*

The previous remark arises consequently the following question.

**Question 2.1 (Aizpuru and García-Pacheco, 2004).** *Does the converse to the previous remark remain true? In other words, is "being an isometric reflection vector" a 2-dimensional property?*

We first realized that the previous question has an obvious positive answer in the class of smooth spaces. Afterwards we solved this question in the negative (see [4]).

**Theorem 2.4 (Aizpuru and García-Pacheco, 2008; García-Pacheco, 2009).** *Let $L$ be an uncountably infinite discrete topological space. Denote by $\widehat{L}$ the one-point compactification of $L$ and by $\mathbf{1} \in \mathcal{C}\left(\widehat{L}\right)$ the constant function on $\widehat{L}$ equal to 1. Then:*

*(1) If $Y$ is a 2-dimensional subspace of $\mathcal{C}\left(\widehat{L}\right)$ containing $\mathbf{1}$, then $\mathbf{1}$ is an $\mathsf{L}^1$-summand vector of $Y$, and thus it is an strongly exposed point of $\mathsf{B}_Y$.*

*(2) $\mathbf{1}$ is not an exposed point of $\mathsf{B}_{\mathcal{C}(\widehat{L})}$.*

*(3) $\mathbf{1}$ is not an isometric reflection vector of $\mathcal{C}\left(\widehat{L}\right)$.*

## 3. Aizpuru's new adventure in lineability

In 2004 Aron and Gurariy raised an interesting question on the linear structure of non-linear subsets of a Banach space.

**Question 3.1 (Aron and Gurariy, 2004).** *Given an infinite dimensional Banach space $X$, does $\mathsf{NA}(X)$ contain an infinite dimensional vector subspace? Or at least, can $X$ be equivalently renormed to make $\mathsf{NA}(X)$ contain an infinite dimensional vector subspace?*

One of the first striking results related to the previous problem shows that the previous question has a negative answer if we replace "infinite dimensional vector subspace" by "infinite dimensional closed vector subspace" (see [8]).

**Theorem 3.1 (Bandyopadhyay and Godefroy, 2005).** *Let $X$ be an Asplund Banach space with the Dunford-Pettis property. Then the closed vector subspaces of $\mathsf{NA}(X)$ are finite dimensional. In particular, $X$ cannot be equivalently renormed to make $\mathsf{NA}(X)$ contain a closed infinite dimensional vector subspace.*

Our contribution to Aron and Gurariy's Question began with the following positive partial solution (see [2]).

**Theorem 3.2 (Acosta, Aizpuru, Aron, and García-Pacheco, 2007).** *Let $X$ be a Banach space. If $X$ admits a monotone Schauder basis, then $\mathsf{NA}(X)$ contains an infinite dimensional vector subspace. If, in addition, the monotone Schauder basis is also shrinking, then $\mathsf{NA}(X)$ contains an infinite dimensional dense vector subspace.*

The previous theorem led to the following result.

**Corollary 3.1 (García-Pacheco, 2009).** *Let $X$ be a Banach space. If $X$ admits an infinite dimensional separable quotient, then $X$ can be equivalently renormed so that $\mathsf{NA}(X)$ contains an infinite dimensional vector subspace.*

Later on, the non-separable version of the previous two results appeared.

**Theorem 3.3 (García-Pacheco, 2009).** *Let $X$ be a non-separable Banach space. If $X$ admits a monotonic projection basis $(x_i, x_i^*)_{i \in I} \subset X \times X^*$, then $\mathsf{NA}\,(X)$ contains a vector subspace of dimension $\mathrm{card}\,(I)$.*

**Corollary 3.2 (García-Pacheco, 2009).** *Let $X$ be a non-separable Banach space that admits a fundamental and biorthogonal system. Then $X$ can be equivalently renormed so that $\mathsf{NA}\,(X)$ contains a vector subspace of dimension $\mathrm{dens}\,(X)$.*

Aron and Gurariy also wondered a similar question regarding the linear structure of non-norm-attaining functionals.

**Question 3.2 (Aron and Gurariy, 2004).** *Given a non-reflexive Banach space $X$, does $X^* \setminus \mathsf{NA}\,(X)$ contain an infinite dimensional vector subspace? Or at least, can $X$ be equivalently renormed to make $X^* \setminus \mathsf{NA}\,(X)$ contain an infinite dimensional vector subspace?*

The previous question has a negative answer (see [11]).

**Theorem 3.4 (García-Pacheco, 2008).** *There exists a non-reflexive dual Banach space $X^*$ such that no equivalent dual norm on $X^*$ makes $X^{**} \setminus \mathsf{NA}\,(X^*)$ contain a 2-dimensional vector subspace.*

Following a similar line to Aron and Gurariy's, Enflo proposed the following question (now involving the topological structure of the set of norm-attaining functionals).

**Question 3.3 (Enflo, 2005).** *Let $X$ be a Banach space. Is it possible to renorm $X$ equivalently so that $X^* \setminus \mathsf{NA}\,(X)$ is nowhere dense?*

The first result on this problem answers Enflo's Problem affirmatively in the class of the real Banach spaces with separable dual (see [12]).

**Theorem 3.5 (García-Pacheco, 2009).** *Let $X$ be a real Banach space with separable dual. Then $X$ can be equivalently renormed so that $X^* \setminus \mathsf{NA}\,(X)$ is nowhere dense.*

The previous theorem is based upon the fact that every real Banach space with separable dual can be equivalently renormed so that the union of the smooth faces of the unit ball of the dual is dense in the unit sphere of the dual. This proof uses an induction argument. By using transfinite induction, one can show that every weakly compactly generated real Banach space can be equivalently renormed so that the union of the smooth faces of the unit ball of the dual is dense in the unit sphere of the dual, leading then to the following.

**Corollary 3.3 (García-Pacheco, 2009).** *Let $X$ be a weakly compactly generated real Banach space. Then $X$ can be equivalently renormed so that $X^* \setminus \mathsf{NA}(X)$ is nowhere dense.*

## 4. A geometrical question that Aizpuru and I always wondered

At the very beginning of my Ph.D. dissertation period at the University of Cádiz, Aizpuru and I noticed that the completion of a uniformly convex normed space must also be uniformly convex. We wondered whether there is a similar result for strictly convex normed spaces.

**Question 4.1 (Aizpuru and García-Pacheco, 2002).** *Does there exist an strictly convex normed space whose completion is not strictly convex?*

Our intuition always suggested us a positive answer to the previous question. However, we were never able to provide an example. Finally in 2009 we could provide such example (see [13]).

**Theorem 4.1 (García-Pacheco and Zheng, 2009).** *Every infinite dimensional strictly convex Banach space can be equivalently renormed to be non-strictly convex and to have a dense and maximal strictly convex subspace.*

After this, the natural question to wonder is whether there is a similar result for smooth normed spaces. Surprisingly, the smooth case is much easier to prove than the strictly convex case (see [13]).

**Theorem 4.2 (García-Pacheco and Zheng, 2009).** *Every infinite dimensional Gateaux-differentiable Banach space can be equivalently renormed to be non-Gateaux differentiable and to have a dense and maximal Gateaux-differentiable subspace.*

The proof of the previous result relies on the fact that every Gateaux-differentiable Banach space can be equivalently renormed so that every element in the new unit sphere is a smooth point of the new unit ball except for only one point and its opposite. In addition, we also realized that the same proof works to show that every Fréchet-differentiable Banach space can be equivalently renormed so that every element in the new unit sphere is a strongly smooth point of the new unit ball except for only one point and its opposite, leading to the following theorem.

**Theorem 4.3 (García-Pacheco and Zheng, 2009).** *Every infinite dimensional Fréchet-differentiable Banach space can be equivalently renormed to be non-Gateaux differentiable and to have a dense and maximal Fréchet-differentiable subspace.*

Obviously, the next step to take is prove a similar result to the previous one involving locally uniform rotundity instead of Fréchet differentiability.

**Theorem 4.4 (García-Pacheco and Zheng, 2009).** *Let $X$ be an infinite dimensional (real) Banach space whose unit ball contains exactly two maximal segments and the rest of the points in the unit sphere are locally uniformly rotund points. Then $X$ contains a dense and maximal locally uniformly convex subspace.*

At this time the point is to show that the previous theorem is not empty, in other words, that there exists an infinite dimensional (real) Banach space whose unit ball contains exactly two maximal segments and the rest of the points in the unit sphere are locally uniformly rotund points of the unit ball.

**Theorem 4.5 (García-Pacheco and Zheng, 2009).** *If there exists a 3-dimensional (real) Banach space whose unit ball contains exactly two maximal segments, then there exists an infinite dimensional (real) Banach space whose unit ball contains exactly two maximal segments and the rest of the points in the unit sphere are locally uniformly rotund points.*

So we were able to reduce the task of showing that Theorem 4.4 is not empty to showing that there exists a 3-dimensional (real) Banach space whose unit ball contains exactly two maximal segments. We were also able to conjecture that there exists such a 3-dimensional Banach space (whose unit ball is described in the next theorem) but unfortunately we were not able to verify our conjecture.

**Conjecture 4.1 (García-Pacheco and Zheng, 2009).** *The 3-dimensional unit ball*

$$\left\{ (x, y, z) \in \mathbb{R}^3 : \frac{x^2}{4 - z^2} + \frac{y^2}{1 - z^2} \leq 1, -1 \leq z \leq 1 \right\}$$

*contains only two maximal segments.*

We think that we can possibly develop another technique to prove the truthfulness of the previous conjecture.

## References

1. Aizpuru, A.: "On operators which attain their norm at extreme points". *Arch. Math.* **69** (1997), 333–337.
2. Acosta, M. D., Aizpuru, A., Aron, R. M., and García-Pacheco, F. J.: "Functionals that do not attain their norm". *Bull. Belg. Math. Soc. Simon Stevin* **14** 3 (2007), 407–418.
3. Aizpuru, A., and García-Pacheco, F. J.: "$L^2$-summand vectors in Banach spaces". *Proc. Amer. Math. Soc.* **134** 7 (2006), 2109–2115.
4. Aizpuru, A., and García-Pacheco, F. J.: "A short note about exposed points in real Banach spaces". *Acta Math. Sci. Ser. B Engl. Ed.* **28** 4 (2008), 797–800.
5. Aizpuru, A., García-Pacheco, F. J., and Rambla, F: "Isometric reflection vectors in Banach spaces". *J. Math. Anal. Appl.* **299** 1 (2004), 40–48.
6. Aron, R. M., and Lohman, R. H.: "A geometric function determined by extreme points of the unit ball of a normed space". *Pacific J. Math.* **2** (1987), 209–231.
7. Bade, W. G.: "The Banach space $\mathcal{C}(S)$". *Lecture Notes Ser.* **26**, 1971.
8. Bandyopadhyay, P., and Godefroy, G.: "Linear structures in the set of norm-attaining functionals on a Banach space". *J. Convex Anal.* **13** 3-4 (2006), 489–497.
9. Becerra-Guerrero, J., and Rodríguez-Palacios, A.: "Transitivity of the norm on Banach spaces". *Extracta Math.* **17** 1 (2002), 1–58.
10. Behrends, E., et al.: "$L_p$-structure in real Banach spaces". *Lecture Notes in Mathematics* **613**, Berlin-Heidelberg-New York, Springer-Verlag, 1977.
11. García-Pacheco, F. J.: "Vector subspaces of the set of non-norm-attaining functionals". *Bull. Aust. Math. Soc.* **77** 3 (2008), 425–432.
12. García-Pacheco, F. J.: "Nowhere density of the set of non-norm-attaining functionals". *Oper. Theory Adv. Appl.* **195** 1 (2009), 167–172.
13. García-Pacheco, F. J., and Zheng, B.: "Convexity and smoothness properties on non-complete spaces". *Preprint* (2009).
14. Oates, D.: "A sequentially convex hull". *Bull. London Math Soc.* **22** (1990), 467–468.
15. Skorik, A., and Zaidenberg, M.: "On isometric reflections in Banach spaces". *Math. Phys. Anal. Geom.* **4** (1997), 212–247.

# A survey on linear (additive) preserver problems

Mostafa Mbekhta*

Université de Lille I
UFR de Mathématiques
59655 Villeneuve d'Ascq Cedex
France
e-mail: mostafa.mbekhta@univ-lille1.fr

This survey is articulated around two major axis. The first one concerns the Kaplansky problem; the history of the problem and several results are presented. The second one concerns some new preserver problems (concerning the generalized inverse, Fredholm or semi-Fredholm operators). The common point of these results is that they are interesting only in the infinite dimensional situation. Several open questions are mentioned over all the paper.

*Keywords*: Generalized inverses, Jordan homomorphism, linear preserver problems.

## 1. Introduction

Linear preserver problems represent one of the most active areas of research in all of matrix (or operator) theory and the theory of Banach algebras. These problems are related to the characterization of all linear transformations on matrix algebras, linear bounded operator algebras or, more generally Banach algebras that leave invariant certain functions, subsets or relations.

Linear preserver problems go back a long time, probably to Frobenius in 1897 who characterized complex linear maps that preserve the determinant (see Theorem 1.1). Since then similar characterizations of linear maps preserving other properties have appeared in the works of many eminent mathematicians like Banach, Kadison, Hua, Kaplansky, and more. Detailed and informative surveys articles on the subject can be found in [5,12,30,35,46,47,53,55,58,66,79] and the references therein. New problems

---

*This work is partially supported by I+D MEC project no. MTM 2007-65959 and by DGI (Spain) Proyecto MTM 2007-67994.

have however emerged when trying to solve linear preserving problems that have significance in infinite dimension but are trivially true in finite dimensions. The focus of this survey is to deal with this aspect of the theory, reviewing in particular some work of the author and collaborators.

We begin this section by listing the various types of linear preserver problems.

**Problem I.** Let $F$ be a scalar-valued, vector-valued or set-valued function on a Banach algebra $\mathcal{A}$. Characterize those linear transformations $\phi$ on $\mathcal{A}$ which satisfy

$$F(\phi(x)) = F(x) \quad \text{for all } x \in \mathcal{A}.$$

**Examples:**
**(i) Scalar-valued case:**
(a) For $\mathcal{A} = \mathcal{M}_n(\mathbb{C})$ and $F(x) = \det(x)$, we have the following classical result of Frobenius [27].

**Theorem 1.1 (G. Frobenius 1897).** *A linear map $\phi$ from $\mathcal{M}_n(\mathbb{C})$ into $\mathcal{M}_n(\mathbb{C})$ preserves the determinant (i.e. $\det(\phi(x)) = \det(x)$) if and only if it takes one of the following forms*

$$\phi(x) = axb \quad \text{or} \quad \phi(x) = ax^{tr}b, \quad \text{for all } x \in \mathcal{M}_n(\mathbb{C})$$

*where $a, b \in \mathcal{M}_n(\mathbb{C})$ are non-singular matrices such that $\det(ab) = 1$, and $x^{tr}$ stands for the transpose of $x$.*

(b) Let $K$ be a compact metric space and $\mathcal{C}(K)$ the Banach space of continuous real valued functions defined on $K$ equipped with the supremun norm. The linear transformations of $\mathcal{C}(K)$ leaving invariant the function $F(x) = \|x\|$ are determined in the next theorem established by Banach [7].

**Theorem 1.2 (S. Banach 1932).** *Let $\phi : \mathcal{C}(K) \to \mathcal{C}(K)$ be a surjective linear map. Then $\phi$ is an isometry, i.e. $\|\phi(f)\|_\infty = \|f\|_\infty$, if and only if*

$$\phi(f(t)) = h(t)f(\varphi(t)) \quad \text{for all } t \in K,$$

*where $|h(t)| = 1$ and $\varphi$ is a homeomorphism of $K$.*

This result was extended by Kadison [41] to the more general case of $C^*$-algebras.

**Theorem 1.3 (R. Kadison 1951).** *Let $\mathcal{A}$ and $\mathcal{B}$ be $C^*$-algebras and $\phi : \mathcal{A} \to \mathcal{B}$ be a surjective linear map. Then $\phi$ is an isometry if and only if there is a unitary element $u \in \mathcal{B}$ and a $C^*$-isomorphism $\varphi : \mathcal{A} \to \mathcal{B}$ such that $\phi = u\varphi$.*

In the special case of the algebra $\mathcal{B}(H)$ of all bounded linear operators acting on a Hilbert space $H$, the previous theorem can be reformulated as follows.

**Theorem 1.4.** *A surjective linear map* $\phi : \mathcal{B}(H) \to \mathcal{B}(H)$ *is an isometry if and only if $\phi$ takes one of the following forms*

$$\phi(T) = UTV \quad \text{or} \quad \phi(T) = UT^{tr}V \quad \text{for all } T \in \mathcal{B}(H),$$

*where $U$, $V$ are an unitary operators of $\mathcal{B}(H)$.*

**(ii) Vector-valued case :** Let $\mathcal{A}$ be a $C^*$-algebra. The surjective linear transformations of $\mathcal{A}$ that preserve the absolute value function $F(x) = |x| = (x^*x)^{1/2}$ are described in the following theorem.

**Theorem 1.5.** *Let $\mathcal{A}$ be a $C^*$-algebra and $\phi : \mathcal{A} \to \mathcal{A}$ be a surjective linear map. Then $|\phi(x)| = |x|$ for all $x \in \mathcal{A}$ if and only if there exists an unitary element $u \in \mathcal{A}$ such that $\phi(x) = ux$ for all $x \in \mathcal{A}$.*

**Proof.** The "if" part is obvious. We prove the "only if" part. Suppose that $|\phi(x)| = |x|$ for all $x \in \mathcal{A}$, then it is easy to see that $\phi(x)^*\phi(x) = x^*x$, and so $\|\phi(x)\| = \|x\|$, for all $x \in \mathcal{A}$. Since $\phi$ is surjective, Theorem 1.3 ensures the existence of a unitary $u \in \mathcal{A}$ and a $C^*$-automorphism $\varphi : \mathcal{A} \to \mathcal{A}$ such that $\phi = u\varphi$.

Next, we will show that $\varphi(x) = x$, for any $x \in \mathcal{A}$. By our assumption on $\phi$ we infer that $\varphi(x^*x) = \varphi(x)^*\varphi(x) = x^*x$. Hence, $\varphi(x) = x$, for all positive element $x \in \mathcal{A}$. Now, if $x$ self-adjoint, i.e. $x = x^*$, by representing the $C^*$-subalgebra generated by $x$ as a algebra of continuous functions, $x$ can be written as follows, $x = x_1 - x_2$ with $x_1, x_2 \geq 0$. It follows that for each self-adjoint element $x \in \mathcal{A}$,

$$\varphi(x) = \varphi(x_1 - x_2) = \varphi(x_1) - \varphi(x_2) = x_1 - x_2 = x.$$

Thus $\varphi(x) = x$, for every self-adjoint element $x \in \mathcal{A}$. Since every element $x \in \mathcal{A}$ is a linear combination of two self-adjoint elements, we obtain $\varphi(x) = x$. Therefore $\phi(x) = ux$ for all $x \in \mathcal{A}$. This completes the proof. □

**(iii) Subspace-valued case :** Let $\mathcal{A} = \mathcal{B}(X)$ be the Banach algebra of bounded operators on a Banach space $X$ and denote by $F(x)$ either the kernel $\ker(x)$ of $x$, or the range $R(x)$ of $x$. The surjective linear maps on $\mathcal{A}$ that preserve such functions are described in the following theorem.

**Theorem 1.6 ([63]).** *Let $\phi : \mathcal{B}(X) \to \mathcal{B}(X)$ be a surjective additive map. Then the following assertions are equivalent :*

(1) $\ker(\phi(T)) = \ker(T)$ *(resp. $R(\phi(T)) = R(T)$) for all $T \in \mathcal{B}(X)$;*
(2) *there is an invertible operator $A : X \to X$ such that $\phi(T) = AT$ (resp. $\phi(T) = TA$) for all $T \in \mathcal{B}(X)$.*

**(iv) Set-valued case :** For a Banach algebra $\mathcal{A}$ and $F(x) = \sigma(x)$, the spectrum of $x$, we find the famous Kaplansky problem that will be discussed in detail later on.

**Problem II.** Let $S$ be a given subset of $\mathcal{A}$. Characterize those linear transformations $\phi$ on $\mathcal{A}$ which satisfy :

$$\text{for every } x \in \mathcal{A},\ x \in S \Longrightarrow \phi(x) \in S,$$

or

$$\text{for every } x \in \mathcal{A},\ x \in S \Longleftrightarrow \phi(x) \in S.$$

**Examples :** Several important subsets have been studied in connection with Problem II. For instance, we mention :

(1) The sets of idempotent, nilpotent, Fredholm, semi-Fredholm operators, see [6,52,53,74,75];
(2) The set $S = \mathcal{A}^{-1} = \{x \in \mathcal{A};\ x \text{ invertible}\}$;
(3) The set $S = \{x \in \mathcal{A};\ \exists y \in \mathcal{A};\ xyx = x \text{ and } yxy = y\}$ of all the elements of $\mathcal{A}$ having a generalized inverse, see [50,52].

The second example is relevant to the Kaplansky problem which is discussed further below and in section 4. The third example will be discussed in sections 2 and 3.

**Problem III.** Let $\mathcal{R}$ be a relation on $\mathcal{A}$. Characterize those linear transformations $\phi$ on $\mathcal{A}$ which satisfy :

$$\text{for } x, y \in \mathcal{A}, x\mathcal{R}y \Longrightarrow \phi(x)\mathcal{R}\phi(y),$$

or

$$\text{for } x, y \in \mathcal{A}, x\mathcal{R}y \Longleftrightarrow \phi(x)\mathcal{R}\phi(y).$$

**Examples :** Among the relations studied in the literature in connection with Problem III, we mention

(1) the commutativity relation, see [21,62,77];
(2) the similarity relation, see [35,76];
(3) the relation $\mathcal{R}$ defined by $x\mathcal{R}y$ if and only $xy = 0$, see [34].

**Problem IV.** Given a function $F : \mathcal{D}_F \subseteq \mathcal{A} \to \mathcal{A}$. Characterize those linear transformations $\phi$ on $\mathcal{A}$ which satisfy

$$F(\phi(x)) = \phi(F(x)) \quad \text{for all } x \in \mathcal{D}_F.$$

**Examples :** The problem IV was considered by several mathematicians in different situations.

(1) In the special case when $\mathcal{A}$ is a Banach algebra and $F(x) = x^{-1}$, the inverse of $x$, Problem IV is motivated by a result due to Hua [37].

**Theorem 1.7 (Hua, 1949).** *Every unital additive map $\phi$ between two fields such that $\phi(x^{-1}) = \phi(x)^{-1}$ is an isomorphism or an anti-isomorphism.*

This result has been later extended to matrix algebras over certain fields (see [25]) and to Banach algebra (see [9,54]).

(1) When $\mathcal{A}$ is a $C^*$-algebra and $F(x) = x^\dagger$, where $x^\dagger$ denotes the Moore-Penrose inverse of $x$, the problem is studied in [54,81].
(2) In [19], the authors discuss Problem IV for the function $F(x) = x^k$, where $k$ is a fixed integer not less than 2, on a matrix algebra $\mathcal{A}$. Note that when $k = 2$, $\phi$ is a Jordan homomorphism (see next).

Let $A$ and $B$ be two unital complex Banach algebras. A linear map $\phi : \mathcal{A} \to \mathcal{B}$ is called a *Jordan homomorphism* if

$$\phi(ab + ba) = \phi(a)\phi(b) + \phi(b)\phi(a) \quad \text{for all } a, b \text{ in } \mathcal{A},$$

or equivalently,

$$\phi(a^2) = \phi(a)^2 \quad \text{for all } a \text{ in } \mathcal{A},$$

see [33,39,64].

We say that $\phi$ is *unital* if $\phi(1) = 1$, where 1 is the unit for both $\mathcal{A}$ and $\mathcal{B}$.

**Remarks:**

(1) Every homomorphism is a Jordan homomorphism.
(2) Every anti-homomorphism $\phi$, i.e. $\phi(ab) = \phi(b)\phi(a)$, is a Jordan homomorphism.
(3) The converse is true in some particular cases. For instance, if $\phi : \mathcal{A} \to \mathcal{B}$ is onto and $\mathcal{B}$ is prime, that is $a = 0$ or $b = 0$ whenever $a\mathcal{B}b = \{0\}$, see [33].

(4) It is well-known that every unital Jordan homomorphism preserves invertibility, i.e. $\phi(x) \in \mathcal{B}^{-1}$ whenever $x \in \mathcal{A}^{-1}$.

Obviously one can ask if the reverse implication in the previous remark $(iv)$ holds true. A negative answer to this question, in this general form, is given in [80]. However, for semi-simple Banach algebras, we have one of the most famous linear preserver problems posed by Kaplansky [45] :

**Kaplansky's problem** : *Let $\phi$ be a unital surjective linear map between two semi-simple Banach algebras $\mathcal{A}$ and $\mathcal{B}$ which preserves invertibility. Is it true that $\phi$ is a Jordan homomorphism ?*

This problem has been solved in the finite dimensional case, in 1897, by Frobenius (Theorem 1.1).

**Theorem 1.8 (J. Dieudonné, 1943, [23]).** *Let $\phi : \mathcal{M}_n(\mathbb{C}) \to \mathcal{M}_n(\mathbb{C})$ be a surjective, unital linear map. If $\sigma(\phi(x)) \subseteq \sigma(x)$ for all $x \in \mathcal{M}_n(\mathbb{C})$, then there is an invertible element $a \in \mathcal{M}_n(\mathbb{C})$ such that $\phi$ takes one of the following forms:*

$$\phi(x) = axa^{-1} \quad or \quad \phi(x) = ax^{tr}a^{-1}.$$

In the commutative and infinite-dimensional case, the well-known Gleason-Kahane-Żelazko theorem provides an affirmative answer, see, [28,42,79].

**Theorem 1.9 (Gleason-Kahane-Zelazko, 1968).** *Let $\mathcal{A}$ be a Banach algebra and $\mathcal{B}$ a semi-simple commutative Banach algebra. If $\phi : \mathcal{A} \to \mathcal{B}$ is a linear map such that $\sigma(\phi(x)) \subseteq \sigma(x)$ for all $x \in \mathcal{A}$, then $\phi$ is an homomorphism.*

In the non-commutative case, the best known results so far are due to B. Aupetit [4] and A.R. Sourour [80].

**Theorem 1.10 (A.R. Sourour, 1996, [80]).** *Let $X, Y$ be two Banach spaces and $\phi : \mathcal{B}(X) \to \mathcal{B}(Y)$ be a linear, bijective and unital map. Then the following assertions are equivalent :*

*(1) $\sigma(\phi(T)) \subseteq \sigma(T)$ for all $T \in \mathcal{B}(X)$;*
*(2) $\phi$ is an isomorphism or an anti-isomorphism;*
*(3) either there is a bounded invertible operator $A : Y \to X$ such that $\phi(T) = A^{-1}TA$ for all $T \in \mathcal{B}(X)$; or, there is a bounded invertible operator $B : Y \to X^*$ such that $\phi(T) = B^{-1}T^*B$ for all $T \in \mathcal{B}(X)$, and in this case $X$ and $Y$ are reflexive.*

We will say that a $C^*$-algebra $\mathcal{A}$ is *of real rank zero* if the set formed by all the real linear combinations of (orthogonal) projections is dense in the set of self-adjoint elements of $\mathcal{A}$, [15,22].

It is well known that this property is satisfied by every von Neumann algebra, and in particular by the $C^*$-algebra $\mathcal{B}(H)$ of all bounded linear operators on a Hilbert space $H$, and also by the Calkin algebra $\mathcal{C}(H) = \mathcal{B}(H)/\mathcal{K}(H)$ where $\mathcal{K}(H)$ denotes the closed ideal of compact operators on $H$.

**Theorem 1.11 (B. Aupetit (2000), [4]).** *Let $\mathcal{A}$ be a $C^*$-algebra of real rank zero and $\mathcal{B}$ be a semi-simple Banach algebra. If $\phi : \mathcal{A} \to \mathcal{B}$ is a surjective linear map such that $\sigma(\phi(x)) = \sigma(x)$ for all $x \in \mathcal{A}$, then $\phi$ is a Jordan isomorphism.*

**Remark:** The Kaplansky problem is still open even if we suppose that $\mathcal{A}$ and $\mathcal{B}$ are $C^*$-algebras.

## 2. Generalized inverse preserving maps

**Definition.** An element $b \in \mathcal{A}$ is called a *generalized inverse* of $a \in \mathcal{A}$ if $b$ satisfies the following two identities

$$aba = a \quad \text{and} \quad bab = b.$$

Let $\mathcal{A}^\wedge$ denote the set of all the elements of $\mathcal{A}$ having a generalized inverse. For more details on the generalized inverse we refer the reader to [31,32,49,60] and the references therein.

**Theorem 2.1 (Kaplansky (1948) [43]).** *Let $\mathcal{A}$ be a Banach algebra:*

*(1) If $\mathcal{A} = \mathcal{A}^\wedge$ then $\mathcal{A}$ is finite-dimensional.*
*(2) Furthermore, if $\mathcal{A}$ is semi-simple, then $\mathcal{A} = \mathcal{A}^\wedge$ if and only if $\mathcal{A}$ is finite-dimensional.*

**Definition.** We say that a linear map $\phi : \mathcal{A} \to \mathcal{B}$ *preserves generalized invertibility in both directions* if $x \in \mathcal{A}^\wedge$ if and only if $\phi(x) \in \mathcal{B}^\wedge$.

**Remark:** Observe that every $n \times n$ complex matrix has a generalized inverse, and therefore, every map on a matrix algebra preserves generalized invertibility in both directions. So, we have here an example of a linear preserver problem which makes sense only in the infinite-dimensional case.

**Theorem 2.2 (M. Mbekhta, L. Rodman and P. Šemrl [50]).** *Let $H$ be an infinite-dimensional separable Hilbert space and let $\phi : \mathcal{B}(H) \to \mathcal{B}(H)$ be a bijective continuous unital linear map preserving generalized invertibility in both directions. Then $\phi(\mathcal{K}(H)) = \mathcal{K}(H)$, and the induced map $\varphi : \mathcal{C}(H) \to \mathcal{C}(H)$, i.e. $\varphi \circ \pi = \pi \circ \phi$ where $\pi : \mathcal{B}(H) \to \mathcal{C}(H)$ is the quotient map, is either an automorphism or an anti-automorphism.*

## 3. Semi-Fredholm and generalized inverse preserving maps

Let $\mathcal{F}(H)$ included in $\mathcal{B}(H)$ be the ideal of all finite rank operators. It is well-known, and easy to check, that if $A \in \mathcal{B}(H)^\wedge$, then $A + F \in \mathcal{B}(H)^\wedge$ for every $F \in \mathcal{F}(H)$. Hence, if $\psi : \mathcal{B}(H) \to \mathcal{F}(H)$ is an arbitrary linear map, then $\phi : \mathcal{B}(H) \to \mathcal{B}(H)$ preserves generalized invertibility in both directions if and only if the map $A \mapsto \phi(A) + \psi(A)$, $A \in \mathcal{B}(H)$, preserves generalized invertibility in both directions.

The proof of Theorem 2.2 above is rather long and complicated (see [50]). To explain the idea we recall that an operator $A \in \mathcal{B}(H)$ is said to be *Fredholm* if its image is closed and both its kernel and cokernel are finite-dimensional and is *semi-Fredholm* if its image is closed and its kernel or its cokernel is finite-dimensional. We denote by $\mathcal{SF}(H) \subset \mathcal{B}(H)$ the subset of all semi-Fredholm operators (see [18,59]). If $\phi : \mathcal{B}(H) \to \mathcal{B}(H)$ is a bijective continuous unital linear map preserving generalized invertibility in both directions, then it is easy to see that $\phi(\mathcal{F}(H)) = \mathcal{F}(H)$ and then, by continuity, $\phi(\mathcal{K}(H)) = \mathcal{K}(H)$. The next step was to show that $\phi$ preserves the set of operators that are generalized invertible and are not semi-Fredholm (they have the infinite-dimensional kernel and the infinite-dimensional cokernel). A certain natural partial order was introduced on the set of such operators. Some rather complicated properties of this order together with the assumption that $\phi$ is unital were used to show that the induced linear map $\varphi : \mathcal{C}(H) \to \mathcal{C}(H)$ preserves idempotents. And then one can prove using standard techniques that this induced map is either an automorphism, or an anti-automorphism. It was conjectured that the same conclusion holds true without the continuity assumption and that an analogous result can be obtained without assuming that $\phi$ is unital. Moreover, as the assumption of preserving generalized invertibility is not affected by linear perturbations mapping into $\mathcal{F}(H)$ it is natural to replace the bijectivity assumption by the surjectivity up to finite rank operators. We say that $\phi : \mathcal{B}(H) \to \mathcal{B}(H)$ is *surjective up to finite rank operators* if for every $A \in \mathcal{B}(H)$ there exists $B \in \mathcal{B}(H)$ such that $A - \phi(B) \in \mathcal{F}(H)$. Can we relax the assumptions of the main theorem from [50] as suggested above? One of the two main results of this section, Theorem 3.1, answers this question in the affirmative.

**Theorem 3.1 (M.Mbekhta and P. Šemrl [52]).** *Let $H$ be a separable Hilbert space of infinite dimension and $\phi : \mathcal{B}(H) \to \mathcal{B}(H)$ be a linear map preserving generalized invertibility in both directions. Assume that $\phi$ is surjective up to finite rank operators. Then $\phi(\mathcal{K}(H)) \subseteq \mathcal{K}(H)$ and there exist an invertible element $a \in \mathcal{C}(H)$ and either an automorphism $\tau : \mathcal{C}(H) \to \mathcal{C}(H)$ or an anti-automorphism $\tau : \mathcal{C}(H) \to \mathcal{C}(H)$ such that the induced map $\varphi : \mathcal{C}(H) \to \mathcal{C}(H)$, i.e. $\varphi \circ \pi = \pi \circ \phi$, is of the form*

$$\varphi(x) = a\tau(x), \quad x \in \mathcal{C}(H).$$

The proof of Theorem 3.1 is a direct consequence of Theorem 3.2 below on the characterization of linear maps preserving semi-Fredholm operators, see Corollary 3.1 below. Using Theorem 3.2, we will not only extend the result of Theorem 2.2 but also provide a considerably shorter proof.

We say that a map $\phi : \mathcal{B}(H) \to \mathcal{B}(H)$ preserves semi-Fredholm (Fredholm) operators in both directions if for every $A \in \mathcal{B}(H)$, the operator $\phi(A)$ is semi-Fredholm (Fredholm) if and only if $A$ is.

**Theorem 3.2 ([52]).** *Let $H$ be an infinite-dimensional separable Hilbert space and $\phi : \mathcal{B}(H) \to \mathcal{B}(H)$ be a linear map preserving semi-Fredholm operators in both directions. Assume that $\phi$ is surjective up to compact operators. Then $\phi(\mathcal{K}(H)) \subseteq \mathcal{K}(H)$ and the induced map $\varphi : \mathcal{C}(H) \to \mathcal{C}(H)$ is either an automorphism, or an anti-automorphism, multiplied by an invertible element $a \in \mathcal{C}(H)$.*

The following Lemma provides a characterization of semi-Fredholm operators in terms of generalized inverses operators

**Lemma 3.1.** *Let $A \in \mathcal{B}(H)$. Then the following are equivalent:*

*(1) $A$ is semi-Fredholm,*
*(2) for every $B \in \mathcal{B}(H)$, there exists $\delta > 0$ such that $A + \lambda B \in \mathcal{B}(H)^\wedge$ for every complex $\lambda$ with $|\lambda| < \delta$.*

As straightforward consequence of Lemma 3.1 we obtain the following corollary.

**Corollary 3.1.** *Let $H$ be an infinite-dimensional separable Hilbert space and $\phi : \mathcal{B}(H) \to \mathcal{B}(H)$ be a linear map preserving generalized invertibility in both directions. Assume that $\phi$ is surjective up to finite rank operators. Then $\phi$ preserves semi-Fredholm operators in both directions.*

For the proof of Theorem 3.2 we will need two lemmas. The first one gives a characterization of compact operators in terms of semi-Fredholm operators.

**Lemma 3.2.** *Let $K \in \mathcal{B}(H)$. Then the following assertions are equivalent:*
*(i) $K$ is compact,*
*(ii) for every $B \in S\mathcal{F}(H)$ we have $B + K \in S\mathcal{F}(H)$.*

**Corollary 3.2.** *Let $\phi : \mathcal{B}(H) \to \mathcal{B}(H)$ be a linear map preserving semi-Fredholm operators in both directions. If $\phi$ is surjective up to compact operators, then $\phi(\mathcal{K}(H)) \subseteq \mathcal{K}(H)$.*

Recall that an operator $A \in \mathcal{B}(H)$ is called *upper semi-Fredholm* if its range is closed and its kernel is finite-dimensional. It is called *lower semi-Fredholm* if its range is closed and of finite codimension.

It is well-known that $A$ is upper semi-Fredholm if and only if $A^*$ is lower semi-Fredholm.

**Lemma 3.3.** *Let $A$ be lower (resp. upper) semi-Fredholm. If $A$ is not Fredholm, then we can find a lower (resp. upper) semi-Fredholm operator $B$ such that every non-trivial linear combination $\lambda A + \mu B$, $\lambda \neq 0$ or $\mu \neq 0$, is lower (resp. upper) semi-Fredholm.*

**Corollary 3.3.** *Let $H$ be an infinite-dimensional separable Hilbert space and let $\phi : \mathcal{B}(H) \to \mathcal{B}(H)$ be a linear map preserving semi-Fredholm operators in both directions. Assume that $\phi$ is surjective up to compact operators. Then $\phi(I)$ is Fredholm (i.e. $\pi(\phi(I))$ is invertible in $\mathcal{C}(H)$).*

In the following theorem, we characterize linear maps preserving lower (resp. upper) semi-Fredholm operators in both directions.

**Theorem 3.3 ([52]).** *Let $H$ be an infinite-dimensional separable Hilbert space and let $\phi : \mathcal{B}(H) \to \mathcal{B}(H)$ be a linear map surjective up to compact operators. Then the following are equivalent:*

*(1) $\phi$ preserves upper semi-Fredholm operators in both directions;*
*(2) $\phi$ preserves lower semi-Fredholm operators in both directions;*
*(3) $\phi(\mathcal{K}(H)) \subseteq \mathcal{K}(H)$ and the induced map $\varphi : \mathcal{C}(H) \to \mathcal{C}(H)$ is an automorphism multiplied by an invertible element $a \in \mathcal{C}(H)$.*

For $T \in \mathcal{B}(H)$, the *essential spectrum*, $\sigma_e(T)$, of $T$, is defined as the spectrum of $\pi(T)$ in the Calkin algebra $\mathcal{C}(H)$, i.e. $\sigma_e(T) = \sigma(\pi(T))$. Obviously,

$$\sigma_e(T) = \{\lambda \in \mathbb{C}; \quad T - \lambda I \text{ is not Fredholm}\}.$$

We say that a linear map $\phi : \mathcal{B}(H) \to \mathcal{B}(H)$ preserves the essential spectrum if $\sigma_e(\phi(T)) = \sigma_e(T)$ for all $T \in \mathcal{B}(H)$.

For a Fredholm operator $T \in \mathcal{B}(H)$, the *index* of $T$ is given by $\operatorname{ind}(T) = \dim \operatorname{Ker} T - \operatorname{codim} \operatorname{Im} T \in \mathbb{Z}$.

**Theorem 3.4 ([53]).** *Let $H$ be an infinite-dimensional Hilbert space and let $\phi : \mathcal{B}(H) \to \mathcal{B}(H)$ be a linear map. Assume that $\phi$ is surjective up to compact operators. Then the following are equivalent:*

*(1) $\phi$ preserves the essential spectrum;*
*(2) $\phi$ preserves the set of Fredholm operators in both directions and $\phi(I) = I - K$ where $K \in \mathcal{K}(H)$;*
*(3) $\phi(\mathcal{K}(H)) \subseteq \mathcal{K}(H)$ and the induced map $\varphi : \mathcal{C}(H) \to \mathcal{C}(H)$, $\varphi \circ \pi = \pi \circ \phi$, is either an automorphism, or an anti-automorphism.*
*Furthermore, in this case, the following statements hold :*

*(a) $\phi$ preserves Fredholm operators in both directions;*
*(b) there is an $n_0 \in \mathbb{Z}$ such that either*

$$\operatorname{ind}(\phi(T)) = n_0 + \operatorname{ind}(T) \quad \text{or} \quad \operatorname{ind}(\phi(T)) = n_0 - \operatorname{ind}(T),$$

*for every Fredholm operator $T$.*

**Conjecture 3.1 ([53]).** *Let $H$ be an infinite-dimensional Hilbert space and let $\phi : \mathcal{B}(H) \to \mathcal{B}(H)$ be a linear map. Assume that $\phi$ is surjective up to compact operators. Then the following conditions are equivalent:*

*(1) $\phi$ preserves the essential spectrum;*
*(2) there exists $\psi : \mathcal{B}(H) \to \mathcal{B}(H)$ either an automorphism or an anti-automorphism and there exists a linear map $\chi : \mathcal{B}(H) \to \mathcal{K}(H)$ such that $\phi(T) = \psi(T) + \chi(T)$ for every $T \in \mathcal{B}(H)$;*
*(3) either*

  *(a) $\phi(T) = ATA^{-1} + \chi(T)$ for every $T \in \mathcal{B}(H)$ where $A$ is an invertible operator in $\mathcal{B}(H)$ and $\chi : \mathcal{B}(H) \to \mathcal{K}(H)$ is a linear map, or*
  *(b) $\phi(T) = BT^{tr}B^{-1} + \chi(T)$ for every $T \in \mathcal{B}(H)$ where $B$ is an invertible operator in $\mathcal{B}(H)$ and $\chi : \mathcal{B}(H) \to \mathcal{K}(H)$ is a linear map.*

**Remark 3.1.** Notice that the implications (2) $\iff$ (3) $\implies$ (1) hold. Indeed, the implications (3) $\implies$ (2) $\implies$ (1) are clear and (2) $\implies$ (3) follows from the fundamental isomorphism theorem [68, Theorem 2.5.19], see also [20]. Therefore, it remains to prove that (1) $\implies$ (2) or (1) $\implies$ (3).

We end this section by presenting some open questions from [52].

**Question 3.1.** It would be interesting to know if Theorem 3.1 and Theorem 3.2 hold true in the context of the Banach algebra of bounded linear operators on a complex Banach space.

Let $T \in \mathcal{B}(H)$ be generalized invertible and let $A \in \mathcal{B}(H)$ be a Fredholm operator. Then both $AT$ and $TA$ belong to $\mathcal{B}(H)^\wedge$, [52]. Thus, for a fixed Fredholm operators $A, B \in \mathcal{B}(H)$ and an arbitrary linear map $\chi : \mathcal{B}(H) \to \mathcal{F}(H)$, both linear maps

$$T \mapsto ATB + \chi(T), \quad T \in \mathcal{B}(H), \tag{1}$$

and

$$T \mapsto AT^{tr}B + \chi(T), \quad T \in \mathcal{B}(H), \tag{2}$$

preserve generalized invertibility in both directions. Here, $A^{tr}$ denotes the transpose of $A$ with respect to some arbitrary, but fixed orthonormal basis.

**Question 3.2.** Does a map $\phi$ satisfying the hypothesis of Theorem 3.1 have one of the forms (3.1) and (3.2)?

For fixed Fredholm operators $A, B \in \mathcal{B}(H)$ and an arbitrary linear map $\chi : \mathcal{B}(H) \to \mathcal{K}(H)$, both linear maps

$$T \mapsto ATB + \chi(T), \quad T \in \mathcal{B}(H), \tag{3}$$

and

$$T \mapsto AT^{tr}B + \chi(T), \quad T \in \mathcal{B}(H), \tag{4}$$

preserve Fredholm operators in both directions.

**Question 3.3.** Does a map $\phi$ satisfying the hypothesis of Theorem 3.2 have one of the forms (3.3) and (3.4)?

Note that the maps of the form (3.3) preserve lower (resp. upper) semi-Fredholm operators in both directions. This suggests the following question.

**Question 3.4.** Does a map $\phi$ satisfying the hypothesis of Theorem 3.3 have the above described form (3.3)?

## 4. Additive maps strongly preserving generalized inverses

We denote by $\mathcal{A}^{-1}$ the set of invertible elements of $\mathcal{A}$. Following [54], we shall say that an additive map $\phi : \mathcal{A} \to \mathcal{B}$ *strongly preserves invertibility* if $\phi(x^{-1}) = \phi(x)^{-1}$ for every $x \in \mathcal{A}^{-1}$. Similarly, we shall say that $\phi$ *strongly preserves generalized invertibility* if $\phi(y)$ is a generalized inverse of $\phi(x)$ whenever $y$ is a generalized inverse of $x$.

One easily checks that a Jordan homomorphism strongly preserves invertibility (resp. generalized inverses). In the present section we deal with additive maps strongly preserving invertibility (resp. generalized inverses). The motivation for this problem is Hua's theorem which states that every unital additive map $\phi$ between two fields such that $\phi(x^{-1}) = \phi(x)^{-1}$ is an isomorphism or an anti-isomorphism (see [2,37]). This result has been later extended to the algebra of matrices over some fields (see [17,25]) and recently to Banach algebras (see [54]). We mention that the following result has been obtenned by an algebraic approach in contrast with the more analytic approach in [54] where only unital continuous maps are considered.

**Theorem 4.1 (N.Boudi and M.Mbekhta, [9]).** *Let $\mathcal{A}$ and $\mathcal{B}$ be unital Banach algebras and let $\phi : \mathcal{A} \to \mathcal{B}$ be an additive map. Then $\phi$ strongly preserves invertibility if and only if $\phi(1)\phi$ is a unital Jordan homomorphism and $\phi(1)$ commutes with the range of $\phi$.*

For the special case of the complex matrix algebra $\mathcal{A} = \mathcal{M}_n(\mathbb{C})$, we derive the following corollary that provides a more explicit form of linear maps strongly preserving invertibility.

**Corollary 4.1 ([9] and [54]).** *Let $\phi : \mathcal{M}_n(\mathbb{C}) \to \mathcal{M}_n(\mathbb{C})$, be a linear map. Then the following conditions are equivalent:*
*(1) $\phi$ preserves invertibility ;*
*(2) there is a $\lambda \in \{-1, 1\}$ such that $\phi$ takes one of the following forms:*

$$\phi(x) = \lambda a x a^{-1} \quad or \quad \phi(x) = \lambda a x^{tr} a^{-1},$$

*for some invertible element $a \in \mathcal{M}_n(\mathbb{C})$.*

**Remark 4.1.** (I) The assertion (3) of Corollary 4.1 has been obtained respectively by J. Dieudonné [23], Marcus and Purves [48] under the weaker assumption that $\phi$ preserves invertibility. Note that this corollary leads to different conditions equivalent to their results. In particular, if $\phi$ is as in Corollary 4.1, then the apparently weaker condition "$\phi$ preserves invertibility" is actually equivalent to the strongler condition "$\phi$ strongly preserves invertibility".

(II) In the context of the Kaplansky conjecture, that is $\mathcal{A}, \mathcal{B}$ are semi-simple and $\phi$ is surjective, Theorem 4.1 shows that the conjecture of Kaplansky is reduced to the question if his assumption of preserving invertibility actually implies the strong preserving invertibility (as it is in the matrix case mentioned above in (I)).

**Theorem 4.2 ([9]).** *Let $\mathcal{A}$ and $\mathcal{B}$ be unital complex Banach algebras and let $\phi : \mathcal{A} \to \mathcal{B}$ be an additive map such that $1 \in Im(\phi)$ or $\phi(1)$ is invertible. Then the following conditions are equivalent:*

*(1) $\phi$ strongly preserves generalized invertibility;*
*(2) $\phi(1)\phi$ is a unital Jordan homomorphism and $\phi(1)$ commutes with the range of $\phi$.*

For unital additive maps, the following theorem follows immediately from Theorem 4.2.

**Theorem 4.3 ([9]).** *Let $\mathcal{A}$ and $\mathcal{B}$ be unital complex Banach algebras and let $\phi : \mathcal{A} \to \mathcal{B}$ be a unital additive map. Then $\phi$ strongly preserves generalized inverses if and only if $\phi$ is a Jordan homomorphism.*

For the special case of linear maps over the complex matrix algebra $\mathcal{A} = \mathcal{M}_n(\mathbb{C})$, we deduce the following corollary that gives a more explicit form of the linear maps strongly preserving generalized invertibility.

**Corollary 4.2 ([9]).** *Let $\phi : \mathcal{M}_n(\mathbb{C}) \to \mathcal{M}_n(\mathbb{C})$, be a linear map. Then $\phi$ strongly preserves generalized inverses if and only if either $\phi = 0$ or there is $\lambda \in \{-1, 1\}$ such that $\phi$ takes one of the following forms:*

$$\phi(x) = \lambda a x a^{-1} \quad or \quad \phi(x) = \lambda a x^{tr} a^{-1},$$

*for some invertible element $a \in \mathcal{M}_n(\mathbb{C})$.*

We conclude this section by stating a conjecture which arises in a natural way from Theorems 4.1 and 4.2 (see [9]).

**Conjecture 4.1.** *Let $\mathcal{A}$ and $\mathcal{B}$ be unital complex Banach algebras and let $\phi : \mathcal{A} \to \mathcal{B}$ be a linear map. If $\phi$ strongly preserves generalized invertibility, then $\phi(1)\phi$ is a Jordan homomorphism and $\phi(1)$ commutes with the range of $\phi$.*

## 5. Additive maps preserving the Moore-Penrose inverses

In the context of $C^*$-algebras, it is well known that every generalized invertible element $a$ has a unique generalized inverse $b$ for which $ab$ and $ba$ are projections. Such an element $b$ is called the *Moore-Penrose inverse* of

$a$ and denoted by $a^\dagger$. In other words, $a^\dagger$ is the unique element of $\mathcal{A}$ that satisfies:

$$aa^\dagger a = a,\ a^\dagger a a^\dagger = a^\dagger,\ (aa^\dagger)^* = aa^\dagger,\ (a^\dagger a)^* = a^\dagger a.$$

Let $\mathcal{A}^\dagger$ denote the set of all elements of $\mathcal{A}$ having a Moore-Penrose inverse. For more details on the Moore-Penrose inverse we refer the reader to [31, 32,49,60,67] and references therein.

We will say that a linear map $\phi : \mathcal{A} \to \mathcal{B}$ *strongly preserves Moore-Penrose invertibility* if $\phi(x^\dagger) = \phi(x)^\dagger$ for all $x \in \mathcal{A}^\dagger$.

Let $\mathcal{A}$ and $\mathcal{B}$ be a $C^*$-algebras. We say that a linear map $\phi : \mathcal{A} \to \mathcal{B}$ is a $C^*$-*Jordan homomorphism* if it is a Jordan homomorphism which preserves the adjoint operation, i.e. $\phi(x^*) = \phi(x)^*$ for all $x$ in $\mathcal{A}$. The $C^*$-homomorphism and $C^*$-anti-homomorphism are analogously defined.

Hua's identity (see, [2,37]) ensures that if $a$, $b$ and $a - b^{-1}$ are invertible, then $a^{-1} - (a - b^{-1})^{-1}$ is also invertible and

$$[a^{-1} - (a - b^{-1})^{-1}]^{-1} = a - aba. \tag{5}$$

The next proposition and theorem generalize the main result of [54, Section 3] established for linear and continuous maps acting on Hilbert spaces to additive maps and the continuity assumption of $\phi$ is redundant.

**Proposition 5.1.** *Let $\mathcal{A}$ and $\mathcal{B}$ be $C^*$-algebras and $\phi : \mathcal{A} \to \mathcal{B}$ an additive unital map that preserves strongly the Moore-Penrose invertibility. The following assertions hold :*

*(i) $\phi$ is a Jordan homomorphism;*
*(ii) $\phi$ preserves the set of projections (i.e., $\phi$ maps projections in $\mathcal{A}$ into projections in $\mathcal{B}$).*

**Proof.** Note that if $x$ and $\phi(x)$ are invertible then $x^{-1} = x^\dagger$ and $\phi(x)^{-1} = \phi(x)^\dagger$. Thus $\phi(x^{-1}) = \phi(x^\dagger) = \phi(x)^\dagger = \phi(x)^{-1}$. Let $u \in \mathcal{A}^{-1}$ such that $\phi(u)$ is invertible. Take $a = u + \lambda$ and $b = \lambda^{-1}u$, where $\lambda \in \mathbb{Q}$ and $\lambda \neq 0$. Then $a - b^{-1} = u + \lambda(1 - u^{-1})$, $\phi(a) = \phi(u) + \lambda$, $\phi(b) = \lambda^{-1}\phi(u)$ and $\phi(a) - \phi(b)^{-1} = \phi(u) + \lambda(1 - \phi(u)^{-1})$. For $|\lambda|$ sufficiently small, $a$, $b$, $a - b^{-1}$, $\phi(a)$, $\phi(b)$ and $\phi(a) - \phi(b)^{-1}$ are invertible. Therefore, by (5.1), $\phi(a - aba) = \phi(a) - \phi(a)\phi(b)\phi(a)$. It follows that

$$\phi(u^2) = \phi(u)^2. \tag{6}$$

Now, for any $x \in \mathcal{A}$ and $\lambda \in \mathbb{Q}$ such that $x+\lambda$ and $\phi(x+\lambda) = \phi(x)+\lambda$ are invertible, it follows from (5.2) that $\phi((x+\lambda)^2) = \phi(x+\lambda)^2$. Consequently

$\phi(x^2) = (\phi(x))^2$ for every $x \in \mathcal{A}$ i.e $\phi$ is a Jordan homomorphism, and (i) is proved.

(ii) Let $e$ be an idempotent such that $e^* = e$, in particular we have $e = e^\dagger$ and so $\phi(e) = \phi(e^\dagger) = \phi(e)^\dagger$. Moreover, because $\phi$ is a Jordan homomorphism, $\phi(e)$ is an idempotent, and since $\phi(e)\phi(e)^\dagger$ is a projection, we obtain

$$\phi(e) = \phi(e)\phi(e)^\dagger = (\phi(e)\phi(e)^\dagger)^* = \phi(e)^*.$$

Thus $\phi(e)$ is a projection in $\mathcal{B}$, as desired. □

The next theorem establishes the same equivalence as in Theorem 3.2 of [54], but with weaker conditions. The continuity of the map $\phi$ in the theorem becomes a consequence rather than a hypothesis as was unnecessarily required in [54].

**Theorem 5.1.** *Let $\mathcal{A}$ be a $C^*$-algebra of real rank zero and $\mathcal{B}$ be a prime $C^*$-algebra. Let $\phi : \mathcal{A} \to \mathcal{B}$ be a surjective, unital additive map. Then the following conditions are equivalent:*

*(1) $\phi(x^\dagger) = \phi(x)^\dagger$ for all $x \in \mathcal{A}^\dagger$;*
*(2) $\phi$ is either a $C^*$-homomorphism or a $C^*$-anti-homomorphism.*

*In this case, $\phi$ is continuous.*

**Proof.** (i) $\Longrightarrow$ (ii). It follows from the above proposition that $\phi$ is a Jordan homomorphism and preserves the set of projections. Since, $\mathcal{B}$ is semi-simple and $\phi$ is a surjective Jordan homomorphism, $\phi$ is continuous (see [64, Theorem 6.3.10]). Now, we complete the proof by using the same arguments as in the proof of [54, Theorem 3.2]. □

As an application of the above theorem in the context of the $C^*$-algebra $\mathcal{B}(H)$ of bounded linear operator on complex separable Hilbert space, we derive the following result which characterizes the surjective unital additive maps from $\mathcal{B}(H)$ onto itself that strongly preserves Moore-Penrose inverses.

Denote by $\mathcal{B}^\dagger(H)$ the set of the operators on $H$ that possess a Moore-Penrose inverse.

**Theorem 5.2.** *Let $\phi : \mathcal{B}(H) \to \mathcal{B}(H)$ be a surjective unital additive map. Then the following conditions are equivalent:*

*(1) $\phi(T^\dagger) = \phi(T)^\dagger$ for all $T \in \mathcal{B}^\dagger(H)$;*

(2) there is a unitary operator $U$ in $\mathcal{B}(H)$ such that $\phi$ takes one of the following forms

$$\phi(T) = UTU^* \quad \text{or} \quad \phi(T) = UT^{tr}U^* \quad \text{for all } T,$$

where $T^{tr}$ is the transpose of $T$ with respect to an arbitrary but fixed orthonormal basis of $H$.

**Remark 5.1.** In [81], the authors provide a characterization of the linear maps on the algebra of matrices over some fields that preserve Moore-Penrose inverse.

In connection with Theorem 5.1, we conclude this section with the following conjecture.

**Conjecture 5.1.** *Let $\mathcal{A}$ and $\mathcal{B}$ be $C^*$-algebras. Let $\phi : \mathcal{A} \to \mathcal{B}$ be a surjective unital additive map. Then the following conditions are equivalent:*

(1) $\phi(x^\dagger) = \phi(x)^\dagger$ for all $x \in \mathcal{A}^\dagger$;
(2) $\phi$ is a $C^*$-Jordan homomorphism.

## 6. Local spectrum preserver maps

Let $T \in \mathcal{B}(X)$ and let $x \in X$. A complex number $\lambda \in \mathbb{C}$ belongs to the *local resolvent set of $T$ at $x$*, denoted $\lambda \in \rho_T(x)$, if there exists an open neighborhood $U$ of $\lambda$ and an analytic function $\hat{x}_T : U \to X$ such that $(T - \mu)\hat{x}_T(\mu) = x$ for all $\mu \in U$. The *local spectrum of $T$ at $x$* is $\sigma_T(x) := \mathbb{C} \setminus \rho_T(x)$.

**Proposition 6.1.** *Let $T \in \mathcal{B}(X)$. The following properties are satisfied*

(1) $\sigma_T(\{0\}) = \emptyset$;
(2) $\sigma_T(x) \subset \sigma_{sur}(T) := \{\lambda \in \mathbb{C} : R(T - \lambda) \neq X\}$, for all $x \in X$;
(3) *there exists $x \in X$ such that $\sigma_T(x) = \sigma_{sur}(T)$;*
(4) *if $S \in \mathcal{B}(X)$ commutes with $T$, then $\sigma_T(Sx) \subset \sigma_T(x)$, for all $x \in X$;*
(5) *if $x = y + z$, then $\sigma_T(x) \subset \sigma_T(y) \cup \sigma_T(z)$.*

Recently, A. Bourhim and T. Ransford [10] studied additive maps on $\mathcal{B}(X)$, the bounded operators acting on a complex Banach space $X$, preserving the local spectrum $\sigma_T(x)$ of $T$ at $x \in X$. They showed that if $\phi : \mathcal{B}(X) \to \mathcal{B}(X)$ is an additive map such that $\sigma_{\phi(T)}(x) = \sigma_T(x)$ for all $T$ and all $x$, then $\phi(T) = T$ for all $T$.

Fixing the local spectrum for all vectors and all operators seems to be a very strong condition. So it is natural to ask the following question (see [29]):

**Problem, [29].** Let $x_0 \in X$ be a fixed nonzero vector. Can one characterize those linear maps $\phi$ on $\mathcal{B}(X)$ which preserve the local spectrum at $x_0$ (i.e $\sigma_{\phi(T)}(x_0) = \sigma_T(x_0)$ for all $T$).

**Theorem 6.1 (M. González and M. Mbekhta, [29]).** *Let $X$ be a finite dimensional complex vector space. Then a linear map $\phi : \mathcal{B}(X) \to \mathcal{B}(X)$ preserves the local spectrum at $x_0$ if and only if there exists an invertible $A \in \mathcal{B}(X)$ such that $A(x_0) = x_0$ and $\phi(T) = ATA^{-1}$ for every $T \in \mathcal{B}(X)$.*

Recently, J. Bračič and V. Müller [11], gave an affirmative answer in the infinite-dimensional case, where only continuous maps are considered.

**Theorem 6.2.** *Let $X$ be a Banach space and $x_0 \in X$ non-zero vector. Let $\phi : \mathcal{B}(X) \to \mathcal{B}(X)$ be a surjective continuous linear map. Then $\phi$ preserves the local spectrum at $x_0$ if and only if there exists an invertible $A \in \mathcal{B}(X)$ such that $A(x_0) = x_0$ and $\phi(T) = ATA^{-1}$ for every $T \in \mathcal{B}(X)$.*

**Question 6.1.** It would be interesting to know if Theorem 6.2 holds without the continuity assumption.

### References

1. B.H. Arnold, *Rings of operators on vector spaces*, Ann. of Math., **45** (1944), 24-49.
2. E. Artin, *Geometric Algebra*, Interscience, New York, 1957.
3. B. Aupetit, *A primer on spectral theory*, Springer-Verlag, New York, 1991.
4. B. Aupetit, *Spectrum-preserving linear mappings between Banach algebras or Jordan-Banach algebras*, J. London Math. Soc. **62** (2000), 917-924.
5. B. Aupetit, *Sur les transformations qui conservent le spectre*, Banach Algebra'97 (Walter de Gruyter, Berlin), (1998), 55-78.
6. Z.F. Bai and J.C. Hou, *Additive Maps Preserving Nilpotent Operators or Spectral Radius*, Acta Mathematica Sinica, English Series **21** (2005), 1167-1182.
7. S. Banach, *Théorie des opérations linéares*, Chelsea, Warsaw, 1932
8. N. Boudi and Y. Hadder, *On linear maps preserving generalized invertibility and related properties*, J. Math. Anal. Appl., **345** (2008), 20-25.
9. N. Boudi and M.Mbekhta *Additive maps preserving strongly generalized inverses*, J. Operator Theory (to appear).
10. A. Bourhim and T. Ransford. *Additive maps preserving local spectrum*. Integr. Equ. Oper. Theory **55** (2006), 377-385.

11. J. Bračič and V. Müller, *Local spectrum and local spectral radius of an operator at a fixed vector*, Studia Math., **194** (2009) 155-162.
12. M. Brešar and P. Šemrl, *Linear preservers on* $\mathcal{B}(X)$, Banach Center Publications **38** (1997), 49-58.
13. M. Brešar and P. Šemrl, *Mappings which preserve idempotents, local automorphisms, and local derivations*, Canad. J. Math. **45** (1993), 483-496.
14. M. Brešar and P. Šemrl, *Linear maps preserving the spectral radius*, J. Funct. Anal. **142** (1996), 360-368.
15. L.G. Brown and G.K. Pedersen, $C^*$-*algebras of real rank zero*, J. Funct. Anal. **99** (1991), 131-149.
16. J.W. Calkin, *Two-sided ideals and congruences in the ring of bounded operators in Hilbert space*, Ann. of Math., **42** (1941), 839-873.
17. C. Cao, X. Zhang, *Linear preservers between matrix modules over connected commutative rings*, Linear Algebra and its Applications **397** (2005), 355-366.
18. S.R. Caradus, W.E. Pfaffenberger, and B. Yood, *Calkin Algebras and Algebras of Operators on Banach Spaces*, Marcel Dekker, New York, 1974.
19. G.H. Chan and M.H. Lim, *Linear preservers on powers of matrices*, Linear Algebra Appl. **162-164** (1992), 615-626.
20. P.R. Chernoff, *Representations, automorphisms, and derivations of some operator algebras*, J. Funct. Anal. **12** (1973), 257-289.
21. Choi, M.D., Jafarian, A.A., Radjavi, H. *Linear maps preserving commutativity*, Linear Algebra and Its Applications, **87**, (1987), 227-241.
22. K.R. Davidson, $C^*$-*Algebras by Example*, American Mathematical Society, 1996.
23. J. Dieudonné *Sur une généralisation du groupe orthogonal à quatre variables*, Arch. Math (Basel) **1** (1949), 282-287.
24. R.G. Douglas, *Banach Algebra Techniques in Operator Theory*, Academic Press 1972.
25. H. Essannouni, A. Kaidi *Le théorème de Hua pour les algèbres artiniennes simples*, Linear Algebra Appl. **297** (1999), 9-22.
26. C. Feldman *The Wedderburn principal theorem in Banach algebras*, Proc. Amer. Math. Soc. **2** (1951), 771-777.
27. G. Frobenius, *Über die Darstellung der endlichen Gruppen durch lineare Substitutionen*, Sitzungsber. Königl. Preuss. Akad. Wiss. Berlin (1897), 994-1015.
28. A. Gleason, *A characterization of maximal ideals*, J. Anal. Math. **19** (1967) 171-172.
29. M. Gonzalez and M. Mbekhta *Linear maps on* $\mathcal{M}_n(\mathbb{C})$ *preserving the local spectrum*, Linear Algebra and Application **427** (2007), 176-182.
30. A. Guterman, C.-K. Li, and P. Šemrl, *Some general techniques on linear preserver problems*, *Linear Algebra Appl.* **315** (2000), 61-81.
31. R. Harte, M. Mbekhta *On generalized inverses in* $C^*$-*algebras*, Studia Math. **103** (1992), 71-77.
32. R. Harte, M. Mbekhta *Generalized inverses in* $C^*$-*algebras II*, Studia Math. **106** (1993), 129-138.
33. I.N. Herstein, *Jordan homomorphisms*, Trans. Amer. Math. Soc. **81** (1956), 331-341.

34. IJ. Hou and J. Cui, *Additive maps on standard operator algebras preserving invertibilities or zero divisors*, Linear Algebra Appl. **359** (2003), 219-233.
35. J.C. Hou and X. L. Zhang, *Additive maps preserving similarity of operators on Banach spaces*, Acta. Math. Sinica. **22** (2006), 179-186.
36. J. Hou and J. Cui, *Linear maps preserving essential spectral functions and closeness of operator ranges*, Bull. London Math. Soc. **39** (2007), 575-582.
37. L. K. Hua, *On the Automorphisms of a Sfield*, Proc. Nat. Acad. Sci. Amer. Vol.**35** (1949), 386-389.
38. T. Hungerford, *Algebra*, Springer-Verlag, New York, 1974.
39. N. Jacobson, C.E. Rickart, *Jordan homomorphism of rings*, Trans. Amer. Math. Soc. **69** (1950), 497-502.
40. R. C. James, *A non-reflexive Banach space isometric with its second conjugate space*, Nat. Acad. Sci. U.S.A. **37**, (1951), 174-177.
41. R. Kadison, *Isometries of operator algebras*, Ann. of Math. **54** (1951), 325-338.
42. J.P. Kahane and W. Zelazko, *A characterization of maximal ideals in commutative Banach algebras*, Studia Math. **29** (1968) 339-343.
43. I. Kaplansky, *Regular Banach algebras*, J. Indian Math. Soc. **12** (1948), 57-62.
44. I. Kaplansky, *Ring isomorphisms of Banach algebras*, Canad. J. Math. **6** (1954), 374-381.
45. I. Kaplansky, *Algebraic and analytic aspects of operator algebras*, Amer. Math. Soc., Providence, 1970.
46. C.K. Li and N.K. Tsing, *Linear preserver problems: A brief introduction and some special techniques*, Linear Algebra Appl. **162-164** (1992), 217-235.
47. C.K. Li and S. Pierce, *Linear preserver problems*, Amer. Math. Monthly **108** (2001), 591-605.
48. M. Marcus and R. Purves, *Linear transformations on algebras of matrices: The invariance of the elementary symmetric functions*, Canad. J. Math, **11** (1959), 383-396.
49. M. Mbekhta, *Conorme et inverse généralisé dans les $C^*$-algèbres*, Canad. Math. Bull. Vol. **35** (4) (1992) 515-522.
50. M. Mbekhta, L. Rodman, and P. Šemrl, *Linear maps preserving generalized invertibility*, Int. Equ. Op. Th. **55** (2006), 93-109.
51. M.Mbekhta *Linear maps preserving the generalized spectrum*, Extracta Mathematicae **22** (2007), 45-54.
52. M. Mbekhta and P. Šemrl, *Linear maps preserving semi-Fredholm operators and generalized invertibility*, Linear and Multilinear Algebra, **57** (2009), 55-65.
53. M. Mbekhta, *Linear maps preserving a set of Fredholm operators*, Proc. Amer. Math. Soc. **135** (2007), 3613-3619.
54. M. Mbekhta, *A Hua type theorem and linear preserver problems*, Proc. Roy. Irish Acad. **109A** (2), (2009), 109-121.
55. M. Mbekhta and J. Zemànek, *A spectral approach to the Kaplansky problem*, Banach Center Publications **75** (2007), 319-320.
56. L. Molnàr, *Two characterizations of additive *-automorphisms of $\mathcal{B}(H)$*, Bull. Austral. Math. Soc. **53** (1996), 391-400.

57. L. Molnàr, *The automorphism and isometry groups of $l_\infty(\mathbf{N}, \mathcal{B}(\mathbf{H}))$ are topologically reflexive*, Acta Sci. Math. (Szeged) **64** (1998), 671-680.
58. L. Molnàr, *Selected Preserver Problems on Algebraic Structures of Linear Operators and on Function Spaces*, Lecture Notes in Mathematics 1895, Springer-Verlag 2007.
59. V. Müller, *Spectral Theory of Linear Operators and Spectral Systems in Banach Algebras*, Birkhäuser Verlag, Basel, 2003.
60. M. Z. Nashed (ed.), *Generalized inverses and applications*, Academic Press, New York-London, 1976.
61. M. Omladič, *On operators preserving commutativity*, J. Funct. Anal., **66** (1986), 105-122.
62. M. Omladič, H. Radjavi and P. Šemrl, *Preserving commutativity*, J. Pure Appl. Algebra **156** (2001), 309-328.
63. M. Oudghiri, *Linear mapping preserving the kernel or the range of operators*, Extracta Math, 2009 (to appear).
64. T.W. Palmer, *Banach Algebras and the General Theory of *-Algebras*, Vol I, CUP, 1994.
65. C. Pearcy and D. M. Topping, *Sums of small numbers of idempotents*, Michigan Math. J. **14** (1967), 453-465.
66. S. Pierce et.al., *A survey of linear preserver problems*, Linear and Multilinear Algebra **33** (1992), 1-192.
67. R. Penrose, *A generalized inverse for matrices*, Proc. Cambridge Philos. Soc. **51** (1955), 406-413.
68. C.E. Rickart, *General Theory of Banach Algebras*, Van Nostrand, Princeton, 1960.
69. P. Šemrl, *Isomorphisms of standard operator algebras*, Proc. Amer. Math. Soc. **123** (1995), 1851-1855.
70. P. Šemrl, *Local automorphisms and derivations on $\mathcal{B}(H)$*, Proc. Amer. Math. Soc. **125** (1997), 2677-2680.
71. P. Šemrl, *Order-preserving maps on the poset of idempotent matrices*, Acta Sci. Math. (Szeged) **69** (2003), 481-490.
72. P. Šemrl, *Applying projective geometry to transformations on rank one idempotents*, J. Funct. Anal. **210** (2004), 248-257.
73. P. Šemrl, *Orthogonality preserving transformations on the set of n-dimensional subspaces of a Hilbert space*, Illinois J. Math. **48** (2004), 567-573.
74. P. Šemrl, *Maps on idempotents*, Studia Math. **169** (2005), 21-44.
75. P. Šemrl, *Maps on idempotent operators.*, Perspectives in operator theory, Banach Center Publ. **75**, Polish Acad. Sci., Warsaw, (2007), 289-301
76. P. Šemrl, *Similarity preserving linear maps*, J. Operator Theory **60** (2008), 71-83.
77. P. Šemrl, *Commutativity preserving maps*, Linear Algebra and its Applications **429** (2008), 1051-1070.
78. A.M. Sinclair, *Jordan automorphism on a semisimple Banach algebras*, Proc. Amer. Math. Soc. **24** (1970), 526-528.
79. A.R. Sourour, *The Gleason-Kahane-Żelazko theorem and its generalizations*, Banach Center Publications **30** (1994), 327-331.

80. A.R. Sourour, *Invertibility preserving linear maps on $\mathcal{L}(X)$*, Trans. Amer. Math. Soc. **348** (1996), 13-30.
81. X. Zhang, C. Cao, and C. Bu, *Additive maps preserving M-P inverses of matrices over fields*, Linear and Multilinear Algebra **46** (1999), 199-211.

# Bounded approximation properties via Banach operator ideals

Eve Oja*

Faculty of Mathematics and Computer Science, Tartu University
J. Liivi 2, EE-50409 Tartu, Estonia
e-mail: eve.oja@ut.ee

We survey some connections of the bounded approximation property with Banach operator ideals, both historical and recent ones. The exposition is self-contained and includes an introduction to Banach operator ideals.

*Keywords*: Banach spaces, bounded approximation properties, Banach operator ideals.

## 1. Introduction

Let $X$ and $Y$ be Banach spaces, both real or both complex. We denote by $\mathcal{L}(X,Y)$ the Banach space of all bounded linear operators acting from $X$ to $Y$ and by $\mathcal{F}(X,Y)$ its subspace of finite-rank operators. We write $\mathcal{L}(X)$ for $\mathcal{L}(X,X)$ and, similarly, $\mathcal{F}(X)$ for $\mathcal{F}(X,X)$. Let $I_X$ denote the identity operator on $X$.

Recall that $X$ has the *approximation property* (AP) if there exists a net $(S_\alpha) \subset \mathcal{F}(X)$ such that $S_\alpha \to I_X$ uniformly on compact subsets of $X$. If $(S_\alpha)$ can be chosen with $\sup_\alpha \|S_\alpha\| \leq \lambda$ for some $\lambda \geq 1$, then $X$ is said to have the *$\lambda$-bounded approximation property* ($\lambda$-BAP). If $\lambda = 1$, then $X$ has the *metric approximation property* (MAP). If $X$ has the $\lambda$-BAP for some $\lambda \geq 1$, then $X$ is said to have the *bounded approximation peoperty* (BAP).

The notions of the AP and the MAP were introduced and deeply studied by A. Grothendieck in 1955 in his famous Memoir [10] as "la condition d'approximation" [10, Chapter I, p. 167] and "la condition d'approximation

---

*The research was partially supported by Estonian Science Foundation Grant 7308 and Estonian Targeted Financing Project SF0180039s08.
This paper is based on a lecture delivered at the IV International Course of Mathematical Analysis in Andalucía. International Congress of Mathematical Analysis in memory of professor Antonio Aizpuru Tomás, September 8–12, 2009. The kind invitation and support for the meeting by the Organizers are gratefully acknowledged.

métrique" [ibidem, p. 178]. Grothendieck also occasionally introduced the BAP as "la variante affaiblie de la propriété d'approximation métrique" [ibidem, p. 182]. But the first source where the BAP was essentially considered seems to be Banach's book "Théorie des opérations linéaires" from 1932 (see [1, p. 237] where, in fact, the bounded compact AP was considered).

From the definitions, it is clear that the MAP implies the BAP, and the BAP implies the AP. P. Enflo [7] was the first who succeeded, in 1972, to construct a (separable reflexive) Banach space without the AP. This was the negative solution to the *approximation problem* – do all Banach spaces have the AP? S. Mazur has promised, in 1936, an exceptional prize for solving it – a live goose (see, e.g., [19, p. 231]). A white goose was indeed handed to Enflo by Mazur (see [12] for a photo of this remarkable event in Warsaw, 1972). Relying on Enflo's example, the first counterexamples showing that the AP, BAP, and MAP are, in general, different notions were constructed by T. Figiel and W. B. Johnson [8] in 1973.

The field of approximation properties continues to attract researchers because it contains interesting problems which have been open for a long time, and still are. For instance, it is not known whether the notions of the AP, BAP, and MAP are different for *dual* spaces. This is a famous open problem that goes back to Grothendieck's Memoir [10]. Sometimes it is called the "AP-implies-MAP problem", and it is as follows.

**Problem 1.1 (see, e.g., [3, Problem 3.8]).** *Does the AP of the dual space $X^*$ of a Banach space $X$ imply the MAP of $X^*$?*

Actually, there are two related subproblems, both open: it is not known whether the AP in dual spaces implies the BAP, neither is known whether the BAP in dual spaces implies the MAP. For an overview around the AP-implies-MAP problem, see [25].

Let $X$ be a Banach space with a (Schauder) basis $(e_k)$. This means that for every $x \in X$ one can find a unique sequence $(a_k)$ of scalars such that $x = \sum_{k=1}^{\infty} a_k e_k$. Let $P_n$ denote the natural projections associated with the basis, i.e. $P_n(\sum_{k=1}^{\infty} a_k e_k) = \sum_{k=1}^{n} a_k e_k$. Then $P_n \to I_X$ uniformly on the compact subsets of $X$. Hence, $X$ has the $\lambda$-BAP with $\lambda = \sup \|P_n\|$. If the basis $(e_k)$ happens to be monotone, i.e. $\lambda = 1$, then $X$ has the MAP. Thus "the BAP" and "the MAP" are "parallel" notions for "the basis" and "the monotone basis", respectively. However, it is not known how "parallel" these notions really are. For example, it is well known that a Banach space with a basis can be equivalently renormed to have a monotone basis. But the following long-standing problem is open.

**Problem 1.2 (see, e.g., [3, Problem 3.12]).** *Could a Banach space with the BAP be equivalently renormed to have the MAP?*

We shall see some new open questions in Sections 3 and 4 of the present notes. One may conclude that the essence of the BAP is still far from being clearly understood. These notes aim to survey some connections of the BAP with Banach operator ideals, both historical and very recent ones. We shall not concern the well-known type of bounded approximation properties which involves operator ideals without using the ideal norm. They are defined similarly to the BAP, only instead of the finite-rank operators $\mathcal{F}(X)$, a more general class of operators on $X$, defined by an operator ideal, is allowed. For instance, if compact operators are allowed instead of $\mathcal{F}(X)$, then $X$ has the (*bounded*) *compact* AP. This type of (bounded) approximation properties has been studied since the early 1980s (see, e.g., [18] for references).

We refer the reader to an excellent survey [3] by P. Casazza for a state-of-the-art on approximation properties, as it was ten years ago. There is a recent survey [25] on bounded approximation properties which complements the present notes in several aspects and provides an extensive list of related references. In turn, the present notes also complement [25].

Our notation is standard. A Banach space $X$ will be regarded as a subspace of its bidual $X^{**}$ under the canonical isometric embedding $j_X : X \to X^{**}$. The closed unit ball of $X$ is denoted by $B_X$. The closure of a set $A \subset X$ is denoted by $\overline{A}$. The next Section contains a mild introduction to Banach operator ideals where the relevant notation can be found.

The present survey notes are self-contained. The only prerequisite for reading them is some basic knowledge of Banach space theory. In particular, no knowledge of tensor products is assumed.

## 2. Banach operator ideals

Speaking about bounded linear operators between arbitrary Banach spaces, we can freely speak (i.e. without mentioning concrete Banach spaces), for instance, about the class of finite-rank operators, or compact operators, or weakly compact operators. Let us denote these classes respectively by $\mathcal{F}, \mathcal{K}, \mathcal{W}$. They all are subclasses of the class $\mathcal{L}$ of all bounded linear operators. An operator ideal is just the right notion to formalize what we are trying to say and to point out the most important common features of these and other important classes of operators.

**Definition 2.1 (cf. [29, 1.1.1]).** *An operator ideal $\mathcal{A}$ is a class of bounded linear operators acting between arbitrary Banach spaces such that*

1° *for all Banach spaces $X$ and $Y$ the set*

$$\mathcal{A}(X,Y) := \mathcal{A} \cap \mathcal{L}(X,Y)$$

*of operators from $\mathcal{A}$ acting from $X$ to $Y$, called a component of $\mathcal{A}$, is a linear subspace of $\mathcal{L}(X,Y)$;*
2° *$\mathcal{A}$ contains all finite-rank operators;*
3° *if $X,Y,Z,W$ are Banach spaces and $A \in \mathcal{L}(X,Y), T \in \mathcal{A}(Y,Z), B \in \mathcal{L}(Z,W)$, then $BTA \in \mathcal{A}(X,W)$, meaning that $\mathcal{A}$ is closed under the composition with the bounded linear operators.*

If $X = Y$, then we shall write $\mathcal{A}(X)$ for the component $\mathcal{A}(X,X)$ of $\mathcal{A}$.

For instance, $\mathcal{F}, \mathcal{K}, \mathcal{W}$, and $\mathcal{L}$ all are operator ideals.

Concerning the finite-rank operators, it would be useful to recall the following. By definition, $T \in \mathcal{F}(X,Y)$ if $T \in \mathcal{L}(X,Y)$ and the range space ran$T = T(X)$ is finite dimensional. Let $T : X \to Y$ be a mapping. It is well known and straightforward to verify (using the basis of ran$T$) that $T \in \mathcal{F}(X,Y)$ if and only if there exist $n \in \mathbb{N}$, functionals $x_k^* \in X^*$, and elements $y_k \in Y, k = 1, \ldots, n$, such that

$$Tx = \sum_{k=1}^n x_k^*(x) y_k, \quad x \in X.$$

For a functional $x^* \in X^*$ and an element $y \in Y$ the operator $x \mapsto x^*(x)y, x \in X$, is denoted by $x^* \otimes y$. Clearly, $x^* \otimes y$ has *rank one*, i.e. dim ran $x^* \otimes y = 1$, if and only if $x^* \neq 0$ and $y \neq 0$. Hence, a mapping $T : X \to Y$ belongs to $\mathcal{F}(X,Y)$ if and only if $T$ can be represented as a finite sum of rank one operators

$$T = \sum_{k=1}^n x_k^* \otimes y_k.$$

Therefore, condition 2° above may be replaced by "$\mathcal{A}$ contains all rank one operators" (see also condition 2° of Definition 2.2 below).

**Definition 2.2 (cf. [29, 6.2.2]).** *An operator ideal $\mathcal{A}$ is called a Banach operator ideal and denoted by $(\mathcal{A}, \|\cdot\|_{\mathcal{A}})$ if there exists a non-negative function $\|\cdot\|_{\mathcal{A}} : \mathcal{A} \to \mathbb{R}$, called a norm, such that*

1° *all components $\mathcal{A}(X,Y)$, equipped with this norm, are Banach spaces;*
2° *$\|x^* \otimes y\|_{\mathcal{A}} = \|x^*\| \|y\|$ for all rank one operators $x^* \otimes y$;*

$3^\circ$ if $X, Y, Z, W$ are Banach spaces and $A \in \mathcal{L}(X,Y), T \in \mathcal{A}(Y,Z), B \in \mathcal{L}(Z,W)$, then $\|BTA\|_\mathcal{A} \leq \|B\|\|T\|_\mathcal{A}\|A\|$.

For instance, $\mathcal{K} = (\mathcal{K}, \|\cdot\|), \mathcal{W} = (\mathcal{W}, \|\cdot\|)$, and $\mathcal{L} = (\mathcal{L}, \|\cdot\|)$ are Banach operator ideals, but $(\mathcal{F}, \|\cdot\|)$ is not, since $\mathcal{F}(X,Y)$ is not, in general, a closed subspace of $\mathcal{L}(X,Y)$. If the Banach ideal norm $\|\cdot\|_\mathcal{A}$ of $\mathcal{A}$ is just the usual operator norm $\|\cdot\|$, then $\mathcal{A}$ is called a *classical* Banach operator ideal or a *closed* Banach operator ideal. The Banach operator ideals $\mathcal{K}, \mathcal{W}$, and $\mathcal{L}$ are classical. Let us introduce now two important Banach operator ideals which are not classical.

*Nuclear operators* $\mathcal{N} = (\mathcal{N}, \|\cdot\|_\mathcal{N})$. A nuclear operator is the infinite analogue of finite-rank operators. An operator $T \in \mathcal{L}(X,Y)$ is said to be *nuclear* if there exist $x_n^* \in X^*$ and $y_n \in Y$ such that $\sum_{n=1}^\infty \|x_n^*\|\, \|y_n\| < \infty$ and

$$Tx = \sum_{n=1}^\infty x_n^*(x) y_n, \quad x \in X.$$

In this case, one writes

$$T = \sum_{n=1}^\infty x_n^* \otimes y_n$$

and calls the latter sum a *nuclear representation* of $T$. Let us denote by $\mathcal{N}(X,Y)$ the collection of all nuclear operators from $X$ to $Y$. The *nuclear norm* $\|T\|_\mathcal{N}$ of a nuclear operator $T \in \mathcal{N}(X,Y)$ is defined by the equality

$$\|T\|_\mathcal{N} = \inf\left\{\sum_{n=1}^\infty \|x_n^*\|\, \|y_n\| : T = \sum_{n=1}^\infty x_n^* \otimes y_n\right\}$$

where the infimum is taken over all possible nuclear representations of $T$. Thus, in fact, $T = \sum_{n=1}^\infty x_n^* \otimes y_n$ is the infinite sum of an absolutely convergent series of rank one operators, since

$$\|T - \sum_{n=1}^m x_n^* \otimes y_n\|_\mathcal{N} \leq \sum_{n=m+1}^\infty \|x_n^*\|\|y_n\| \underset{m}{\to} 0.$$

It is straightforward to verify that $\mathcal{N} = (\mathcal{N}, \|\cdot\|_\mathcal{N})$ is a Banach operator ideal (see, e.g., [5, Theorem 5.25]).

*Integral operators* $\mathcal{I} = (\mathcal{I}, \|\cdot\|_\mathcal{I})$. They are usually defined through factorization schemes. An operator $T \in \mathcal{L}(X,Y)$ is said to be *integral* if there exist a probability measure space (with the measure $\mu$) and bounded linear operators $A : X \to L_\infty(\mu)$ and $B : L_1(\mu) \to Y^{**}$ such that $j_Y T =$

$Bj_1A$, where $j_1 : L_\infty(\mu) \to L_1(\mu)$ denotes the identity embedding, meaning that the diagram

$$\begin{array}{ccccc} X & \xrightarrow{T \in \mathcal{I}} & Y & \xrightarrow{j_Y} & Y^{**} \\ {\scriptstyle A \in \mathcal{L}} \downarrow & & & & \uparrow {\scriptstyle B \in \mathcal{L}} \\ L_\infty(\mu) & & \xrightarrow{j_1} & & L_1(\mu) \end{array}$$

commutes. Let us denote by $\mathcal{I}(X,Y)$ the collection of all integral operators from $X$ to $Y$. The *integral norm* $\|T\|_\mathcal{I}$ of an integral operator $T \in \mathcal{I}(X,Y)$ is defined by the equality

$$\|T\|_\mathcal{I} = \inf \|A\| \|B\|$$

where the infimum is taken over all possible factorizations of $T$ as above. It is straightforward to verify that $\mathcal{I} = (\mathcal{I}, \|\cdot\|_\mathcal{I})$ is a Banach operator ideal (see, e.g., [5, Theorem 5.2]).

*Strictly integral operators* $\mathcal{SI} = (\mathcal{SI}, \|\cdot\|_{\mathcal{SI}})$ (or *Pietsch integral operators*) are defined similarly with the only difference that $T$ itself can be factorized as above, instead of $j_Y T$. Although the definition of $\mathcal{SI}$ is a bit simpler than the definition of $\mathcal{I}$, the Banach operator ideal properties of $\mathcal{I}$ are much better than those of $\mathcal{SI}$.

It can be shown (see, e.g., [31, pp. 64–65]) that

$$\mathcal{N} \subset \mathcal{SI} \subset \mathcal{I} \subset \mathcal{W}.$$

At this point we should recall that, for Banach operator ideals $\mathcal{A}$ and $\mathcal{B}$, the inclusion $\mathcal{A} \subset \mathcal{B}$ means that $\mathcal{A}(X,Y) \subset \mathcal{B}(X,Y)$ and $\|T\|_\mathcal{A} \geq \|T\|_\mathcal{B}$ for all Banach spaces $X$ and $Y$, and for all $T \in \mathcal{A}(X,Y)$. In particular, the corresponding inequalities for norms are

$$\|\cdot\|_\mathcal{N} \geq \|\cdot\|_{\mathcal{SI}} \geq \|\cdot\|_\mathcal{I} \geq \|\cdot\|.$$

In fact, $\mathcal{N}, \mathcal{SI}$, and $\mathcal{I}$ are only the "first" members of the big families of Banach operator ideals of $p$-nuclear, strictly $p$-integral, and $p$-integral operators $\mathcal{N}_p, \mathcal{SI}_p$, and $\mathcal{I}_p$, where $1 \leq p \leq \infty$, with $\mathcal{N} = \mathcal{N}_1, \mathcal{SI} = \mathcal{SI}_1$, and $\mathcal{I} = \mathcal{I}_1$. Their definitions are quite similar. For $\mathcal{SI}_p$ and $\mathcal{I}_p$, just replace

$L_1(\mu)$ by $L_p(\mu)$ (and $j_1$ by the identity embedding $j_p : L_\infty(\mu) \to L_p(\mu)$). For $\mathcal{N}_p$, it is a bit more complicated. Here one has to replace the requirement that $\sum_{n=1}^\infty \|x_n^*\| \|y_n\| < \infty$ by the conditions

$$\|(x_n^*)\|_p := (\sum_{n=1}^\infty \|x_n^*\|^p)^{1/p} < \infty$$

and

$$\|(y_n)\|_{p^*}^w := \sup\{(\sum_{n=1}^\infty |y^*(y_n)|^p)^{1/p} : \quad y^* \in B_{Y^*}\} < \infty,$$

taking inf $\|(x_n^*)\|_p \|(y_n)\|_{p^*}^w$ to be the norm, where $p^*$ is the conjugate index of $p$, that is, $1/p + 1/p^* = 1$. For more details on $\mathcal{N}_p$, $\mathcal{SI}_p$, and $\mathcal{I}_p$, and for other Banach operator ideals, we refer to the books by J. Diestel, H. Jarchow, and A. Tonge [5], R. A. Ryan [31], and A. Pietsch [29].

The concept of a Banach operator ideal was introduced by Pietsch around 1969 (see [29]) relying, in particular, on ideas from Grothendieck's theory of tensor products. In the framework of this theory, Grothendieck introduced nuclear and integral operators in early 1950s (however, nuclear operators in Hilbert spaces were studied much earlier under the name of trace class operators). Grothendieck [10] showed, among others, that $\mathcal{N}(X,Y)$ coincides with the projective tensor product of $X^*$ and $Y$ whenever $X^*$ or $Y$ has the approximation property (see, e.g., [31, Corollary 4.8]). And he introduced the integral operators in [10, Chapter I, Proposition 27, pp. 124–127] with the aim to describe the dual space of an injective tensor product (see, e.g., [31, Section 3.4]). Without entering into the theory of tensor products, we may formulate this very important Grothendieck's description in the following way.

**Theorem 2.1 (Grothendieck).** *Let $X$ and $Y$ be Banach spaces. Then the dual space $(\mathcal{F}(X,Y), \|\cdot\|)^*$ is linearly isometric with $\mathcal{I}(Y, X^{**})$ under the duality*

$$\langle T, \sum_{k=1}^n x_n^* \otimes y_n \rangle = \sum_{k=1}^n (Ty_n)(x_n^*),$$

*and also with $\mathcal{I}(X^*, Y^*)$ under the duality*

$$\langle T, \sum_{k=1}^n x_n^* \otimes y_n \rangle = \sum_{k=1}^n (Tx_n^*)(y_n).$$

One usually expresses Theorem 2.1 by writing $(\mathcal{F}(X,Y), \|\cdot\|)^* = \mathcal{I}(Y, X^{**})$ and $(\mathcal{F}(X,Y), \|\cdot\|)^* = \mathcal{I}(X^*, Y^*)$.

The link with the tensor products of Banach spaces is that, in fact, $\mathcal{F}(X,Y)$ is algebraically the same as the algebraic tensor product $X^* \otimes Y$ with the rank one operator $x^* \otimes y$ corresponding to the elementary tensor $x^* \otimes y$. And the link with the *injective tensor norm* $\varepsilon = \|\cdot\|_\varepsilon$ is that, in fact, $\|T\| = \|\sum_{k=1}^n x_k^* \otimes y_k\|_\varepsilon$ for all $T \in \mathcal{F}(X,Y)$, $T = \sum_{k=1}^n x_k^* \otimes y_k$. By the same way, each tensor norm $\alpha = \|\cdot\|_\alpha$ {which is a certain method of ascribing to all tensor products $X \otimes Y$ of all Banach spaces $X$ and $Y$ a norm $\|\cdot\|_\alpha$ on $X \otimes Y$, converting $X \otimes_\alpha Y := (X \otimes Y, \|\cdot\|_\alpha)$ into a normed space; the interested reader is refered to Ryan's book [31]} converts $\mathcal{F}(X,Y)$ into a normed space $(\mathcal{F}(X,Y), \|\cdot\|_\alpha)$. In fact, as we shall see below, for common tensor norms, one is not needed to recall about tensor products at all, as we saw in the case of the injective tensor norm.

Grothendieck's description (see Theorem 2.1), which is from 1955, had its historical counterpart from 1946, due to R. Schatten [34]. Schatten's description (see Theorem 2.2 below) uses the projective tensor norm instead of the injective tensor norm, and $\mathcal{L}$ instead of $\mathcal{I}$.

Let us recall the definition of the *projective tensor norm* $\pi = \|\cdot\|_\pi$ using finite-rank operators: if $T \in \mathcal{F}(X,Y)$, then $\|T\|_\pi$ is defined by the equality

$$\|T\|_\pi = \inf\{\sum_{k=1}^n \|x_k^*\|\|y_k\| : T = \sum_{k=1}^n x_k^* \otimes y_k\}$$

where the infimum is taken over all possible representations of $T$. Thus, $\|\cdot\|_\pi$ may be viewed as a finite version of the nuclear norm $\|\cdot\|_\mathcal{N}$. It is straightforward to verify that $(\mathcal{F}(X,Y), \|\cdot\|_\pi)$ is a normed space (in fact, it is precisely $X^* \otimes_\pi Y$) (see, e.g., [31, Proposition 2.1]).

**Theorem 2.2 (Schatten).** *Let $X$ and $Y$ be Banach spaces. Then the dual space $(\mathcal{F}(X,Y), \|\cdot\|_\pi)^*$ is linearly isometric with $\mathcal{L}(Y, X^{**})$ and also with $\mathcal{L}(X^*, Y^*)$ under the dualities described in Theorem 2.1.*

One usually expresses Theorem 2.2 by writing $(\mathcal{F}(X,Y), \|\cdot\|_\pi)^* = \mathcal{L}(Y, X^{**})$ and $(\mathcal{F}(X,Y), \|\cdot\|_\pi)^* = \mathcal{L}(X^*, Y^*)$.

There are several different procedures to make from a given Banach operator ideal new Banach operator ideals (see [29, 8.1–8.7, 9.1]). In the next Section, we shall need one of them, namely $\mathcal{A}^{\text{dual}}$. We shall also need to know about regular Banach operator ideals.

Let $\mathcal{A}$ be a Banach operator ideal. The components of the *dual operator ideal* $\mathcal{A}^{\text{dual}}$ are defined by $\mathcal{A}^{\text{dual}}(X,Y) = \{T \in \mathcal{L}(X,Y) : T^* \in \mathcal{A}(Y^*, X^*)\}$ with $\|T\|_{\mathcal{A}^{\text{dual}}} = \|T^*\|_\mathcal{A}$.

If $\mathcal{A} \subset \mathcal{B}$, then clearly $\mathcal{A}^{\text{dual}} \subset \mathcal{B}^{\text{dual}}$. The Banach operator ideals $\mathcal{A}$ and $\mathcal{A}^{\text{dual}}$ are, in general, incomparable. Even $\mathcal{A}$ and $(\mathcal{A}^{\text{dual}})^{\text{dual}}$ are (see [29, 4.4.11]). One has the equality $\mathcal{A} = \mathcal{A}^{\text{dual}}$, and hence also the equality $\mathcal{A} = (\mathcal{A}^{\text{dual}})^{\text{dual}}$, if $\mathcal{A}$ is, for instance, $\mathcal{K}$ (the Schauder theorem), $\mathcal{W}$ (the Gantmacher theorem), $\mathcal{L}$ (obvious), or $\mathcal{I}$ (see, e.g., [5, Theorem 5.15]). In general, $\mathcal{I}_p \neq \mathcal{I}_p^{\text{dual}}$ for $p > 1$ (recall that $\mathcal{I} = \mathcal{I}_1$), but $\mathcal{I}_p = (\mathcal{I}_p^{\text{dual}})^{\text{dual}}$ for all $p, 1 \leq p \leq \infty$ (see, e.g., [5, Theorem 5.14]).

A Banach operator ideal $\mathcal{A}$ is said to be *regular* if $T \in \mathcal{A}(X, Y)$ and $\|T\|_{\mathcal{A}} = \|j_Y T\|_{\mathcal{A}}$ whenever $T \in \mathcal{L}(X, Y)$ and $j_Y T \in \mathcal{A}(X, Y^{**})$. The regularity of $\mathcal{A}$ means that if $T \in \mathcal{L}(X, Y) \cap \mathcal{A}(X, Y^{**})$, then $T \in \mathcal{A}(X, Y)$ (and has the same $\mathcal{A}$-norm).

It can be easily verified that $\mathcal{A}^{\text{dual}}$ is regular for any $\mathcal{A}$ (see [29, 8.3.5]). In particular, $\mathcal{I}_p$ is regular for all $p, 1 \leq p \leq \infty$. Concerning the $p$-nuclear operators, $\mathcal{N}_p$ is regular if and only if $p = 2$ (see [8] for $p = 1$ and [30] for the general case).

The important result that the Banach operator ideal $\mathcal{N}$ of nuclear operators is not regular was obtained in 1973 by Figiel and Johnson [8]. But the possible regularity of nuclear operators was studied already by Grothendieck [10] in 1955. He proved a regularity condition in the presence of the approximation property: if $T \in \mathcal{L}(X, Y) \cap \mathcal{N}(X, Y^{**})$, then $T \in \mathcal{N}(X, Y)$, whenever $X^*$ has the AP (see [10, Chapter I, pp. 85–86]). He also claimed (see [ibidem, p. 86] and [11, p. 17]) that the same holds whenever the second dual space $Y^{**}$ has the AP. In [26] (see also [27]) a counterexample to this claim of Grothendieck was constructed: there exists a Banach space $Y$ such that $Y^{**}$ has a basis, $Y^{***}$ is separable and fails the AP, and there exists an operator $T \in \mathcal{L}(Y^{**}, Y) \cap \mathcal{N}(Y^{**}, Y^{**})$ such that $T \notin \mathcal{N}(Y^{**}, Y)$. In [27] (see [20] for a simpler proof), it is proven that Grothendieck's claim holds whenever the third dual $Y^{***}$ has the AP. Thus, the above example shows that the assumptions about the AP of $X^*$ and the AP of $Y^{***}$ are essential in this regularity condition.

## 3. Bounded approximation properties via inner and outer inequalities

The first result characterizing the BAP, in fact, the MAP via a Banach operator ideal is due to Grothendieck (see [10, Chapter I, Proposition 39, p. 179]). By a similar proof, Grothendieck's characterization of the MAP extends to the BAP. This is essentially Theorem 3.1 below. Grothendieck formulated his result in terms of projective tensor products and integral bilinear forms. Our formulation of the Grothendieck theorem will use the recent concepts of inner and outer inequalities [24].

For introducing inner and outer inequalities, let us recall that $\mathcal{F}(X,Y)$ is algebraically the same as the algebraic tensor product $X^* \otimes Y$ (see Section 2). Hence, for any tensor norm $\alpha = \|\cdot\|_\alpha$, we may consider the normed space $(\mathcal{F}(X,Y), \|\cdot\|_\alpha)$. If $\mathcal{A} = (\mathcal{A}, \|\cdot\|_\mathcal{A})$ is a Banach operator ideal, then $\mathcal{F}(X,Y) \subset \mathcal{A}(X,Y)$ and $(\mathcal{F}(X,Y), \|\cdot\|_\mathcal{A})$ is again a normed space. This observation is needed in the following definition and in the subsequent results.

**Definition 3.1 (see [24]).** *Let $X$ be a closed subspace of a Banach space $W$. If $\alpha = \|\cdot\|_\alpha$ is a tensor norm, $\mathcal{A}$ is a Banach operator ideal, and $\lambda > 0$, then we call the condition*

$$\|S\|_\alpha \leq \lambda \|S\|_{\mathcal{A}(X,W)} \text{ for all } S \in \mathcal{F}(X)$$

*an inner inequality, and we call the condition*

$$\|T\|_\alpha \leq \lambda \|T\|_{\mathcal{A}(Y,W)} \text{ for all Banach spaces } Y \text{ and for all } T \in \mathcal{F}(Y,X)$$

*an outer inequality.*

Our reformulation of the Grothendieck theorem is as follows.

**Theorem 3.1 (Grothendieck).** *Let $X$ be a Banach space and let $\lambda \geq 1$. The following assertions are equivalent.*

*(a) $X$ has the $\lambda$-BAP.*
*(b) $\|S\|_\pi \leq \lambda \|S\|_{\mathcal{I}(X)}$ for all $S \in \mathcal{F}(X)$.*
*(c) $\|T\|_\pi \leq \lambda \|T\|_{\mathcal{I}(Y,X)}$ for all Banach spaces $Y$ and for all $T \in \mathcal{F}(Y,X)$.*

**Proof.** The proof relies on Theorems 2.1 and 2.2. Keeping in mind what was said after Theorem 2.1, it is not difficult to adapt the proof of [31, Theorem 4.14] to our case. □

Theorem 3.1 tells us that the BAP can be characterized through the inner and outer inequalities involving the pair $(\|\cdot\|_\pi, \mathcal{I})$ of the projective tensor norm $\|\cdot\|_\pi$ and the Banach operator ideal $\mathcal{I}$ of integral operators. What about some other pair $(\|\cdot\|_\alpha, \mathcal{A})$? More precisely, we ask the following.

**Problem 3.1.** *Does Theorem 3.1 hold if one replaces, in conditions (b) and (c), the pair $(\|\cdot\|_\pi, \mathcal{I})$ with some other pair $(\|\cdot\|_\alpha, \mathcal{A})$ of a tensor norm $\|\cdot\|_\alpha$ and a Banach operator ideal $\mathcal{A}$?*

We can say at once that for some pairs $(\|\cdot\|_\alpha, \mathcal{A})$, it is not the case: they cannot be used to characterize the BAP. To explain this, we first need to recall the *Chevet–Saphar tensor norms* $g_p = \|\cdot\|_{g_p}$ (cf., e.g., [31, p. 135]), which are finite versions of $p$-nuclear norms $\|\cdot\|_{\mathcal{N}_p}$.

Let $1 \leq p \leq \infty$. If $T \in \mathcal{F}(X,Y)$, then $\|T\|_{g_p}$ is defined by the equality

$$\|T\|_{g_p} = \inf\{\|(x_k^*)_{k=1}^n\|_p \, \|(y_k)_{k=1}^n\|_{p^*}^w : T = \sum_{k=1}^n x_k^* \otimes y_k\}$$

where the infimum is taken as for $\|T\|_\pi$. In fact, obviously, $\|T\|_\pi = \|T\|_{g_1}$. Recalling also that $\mathcal{I} = \mathcal{I}_1$, one can recognize a general form of the outer inequality of Theorem 3.1 in the following definition.

**Definition 3.2 (Saphar).** *Let $1 \leq p \leq \infty$ and let $1 \leq \lambda < \infty$. A Banach space $X$ is said to have the $\lambda$-BAP of order $p$ ($\lambda$-$BAP_p$) if*

$\|T\|_{g_p} \leq \lambda \|T\|_{\mathcal{I}_p(Y,X)}$ for all Banach spaces $Y$ and for all $T \in \mathcal{F}(Y,X)$.

*The 1-$BAP_p$ is called the $MAP_p$.*

The $\lambda$-$BAP_p$ was essentially introduced in 1970 by P. Saphar in his seminal paper [32] (we gave an equivalent reformulation using finite-rank operators). (In [33], quite exceptionally, instead of $p$, the conjugate index $p^*$ was used for this notion; and in [4], the term "the bounded $g_p$-approximation property with constant $\lambda$" is used for it.) Since $\|\cdot\|_{g_1} = \|\cdot\|_\pi$ and $\mathcal{I}_1 = \mathcal{I}$, the $\lambda$-$BAP_1$ coincides with the $\lambda$-BAP and the $MAP_1$ with the MAP.

As shown by Saphar [33, Proposition 3], the $\lambda$-BAP implies the $\lambda$-$BAP_p$ for all $p$. As also shown by Saphar, every Banach space $X$ has the $MAP_2$ (see [32, Theorem 3.5] or [33, Corollary 1 of Theorem 5] or, e.g., [4, 21.7]). Since there exist Banach spaces without the BAP, the pair $(\|\cdot\|_{g_2}, \mathcal{I}_2)$ cannot characterize the BAP. More generally, no pair $(\|\cdot\|_{g_p}, \mathcal{I}_p)$ with $p \geq 2$ can characterize the BAP, because the BAP and $BAP_p$ are different properties whenever $p > 2$. The latter fact was observed by O. Reinov [30, p. 126]. Reinov [30] also proved that there exist Banach spaces with the AP which fail the $BAP_p$ for all $p, p \neq 2$.

As Definition 3.2 tells us, the $\lambda$-$BAP_p$ is defined by an outer inequality. It seems to be unknown whether this could be done using the corresponding inner inequality, as in the case $p = 1$. Thus, we are asking the following.

**Problem 3.2.** *Let $1 < p \leq \infty, p \neq 2$. Does the inner inequality $\|\cdot\|_{g_p} \leq \lambda \|\cdot\|_{\mathcal{I}_p}$ imply the corresponding outer inequality?*

Recently, in [24] we have been studying two rather general questions related, in particular, to Problem 3.2. (1) When does an inner inequality $\|\cdot\|_\alpha \leq \|\cdot\|_\mathcal{A}$ imply the corresponding outer inequality (or just its restriction to all reflexive Banach spaces)? (2) When is an outer inequality determined by a subclass of Banach spaces (by reflexive spaces, for instance)? Let us look at the following sample results on (1) and (2), which among others, point out why Problem 3.2 may be not an easy one.

**Proposition 3.1 (see [24, Proposition 2.1]).** *Let $X$ be a closed subspace of a Banach space $W$ and let $\mathcal{A}$ be a Banach operator ideal. If*

$$\|S\|_\pi \leq \lambda \|S\|_{\mathcal{A}(X,W)}$$

*for all $S \in \mathcal{F}(X)$, then*

$$\|T\|_\pi \leq \lambda \|T\|_{\mathcal{A}(Z,W)}$$

*for all reflexive Banach spaces $Z$ and for all $T \in \mathcal{F}(Z, X)$.*
*If, moreover, $\mathcal{A}$ is regular and satisfies $\mathcal{A} \subset (\mathcal{A}^{\text{dual}})^{\text{dual}}$, then*

$$\|T\|_\pi \leq \lambda \|T\|_{\mathcal{A}(Y,W)}$$

*for all Banach spaces $Y$ and for all $T \in \mathcal{F}(Y, X)$.*

**Theorem 3.2 (see [24, Theorem 2.3]).** *Let $X$ be a closed subspace of a Banach space $W$. Let $\alpha = \|\cdot\|_\alpha$ be a tensor norm and let $1 \leq p \leq \infty$. If*

$$\|S\|_\alpha \leq \lambda \|S\|_{\mathcal{N}_p(Z,W)}$$

*for all separable reflexive Banach spaces $Z$ and for all $S \in \mathcal{F}(Z, X)$, then*

$$\|T\|_\alpha \leq \lambda \|T\|_{\mathcal{N}_p(Y,W)}$$

*for all Banach spaces $Y$ and for all $T \in \mathcal{F}(Y, X)$.*

As we saw in Section 2, the assumptions of the "moreover" part of Proposition 3.1 are satisfied for $\mathcal{A} = \mathcal{I}_p, 1 \leq p \leq \infty$. In particular, taking $W = X$ and $\mathcal{A} = \mathcal{I}$ in Proposition 3.1 yields the implication (b)⇒(c) of Theorem 3.1. This might be considered as an alternative direct proof of it since the proofs of the equivalent implication in the literature pass through condition (a) of Theorem 3.1 (see, e.g., [4, 16.3], [6, pp. 243–244], or [31, pp. 80–81]) as the original proof in [10, Chapter I, pp. 179–180] does.

As we also saw in Section 2, $\mathcal{N}_p, 1 \leq p \leq \infty, p \neq 2$, is not regular. In this case, Proposition 3.1 enables us to pass from the inner inequality $\|\cdot\|_\pi \leq \lambda \|\cdot\|_{\mathcal{N}_p}$ to the outer inequality only with respect to the reflexive

Banach spaces. But now we are able to go further and obtain the outer inequality with respect to all Banach spaces by applying Theorem 3.2.

Basing on Theorem 3.2 and other related results, the following information concerning the $\mathrm{BAP}_p$ and Problem 3.2 was obtained in [24].

**Theorem 3.3 (see [24, Theorem 4.3]).** *Let $X$ be a Banach space, let $1 < p < \infty$, and let $\lambda \geq 1$. Then the following assertions are equivalent.*

*(a) $X$ has the $\lambda$-$BAP_p$.*

*(b) $\|T\|_{\mathcal{G}_p} \leq \lambda \|T\|_{\mathcal{I}_p(Z,X)}$ for all separable reflexive Banach spaces $Z$ and $T \in \mathcal{F}(Z,X)$.*

*(c) $\|T\|_{\mathcal{G}_p} \leq \lambda \|T\|_{\mathcal{N}_p(Y,X^{**})}$ for all Banach spaces $Y$ and for all $T \in \mathcal{F}(Y,X)$.*

In 1985, J. Bourgain and O. Reinov [2] (see, e.g., [4, 21.9]) proved that it is sufficient to check Definition 3.2 only for reflexive Banach spaces $Y$ whenever $1 < p < \infty$. The implication (b)$\Rightarrow$(a) of Theorem 3.3 tells that the Bourgain–Reinov characterization of the $\lambda$-$\mathrm{BAP}_p$ holds with separable reflexive spaces instead of reflexive spaces.

One of the most beautiful results concerning the $\mathrm{BAP}_p$ is the Bourgain–Reinov theorem from [2] asserting that the Hardy space $H_\infty$ (of bounded analytic functions on the open unit disc) has the $\mathrm{MAP}_p$ whenever $1 < p < \infty$ (see [24] for an alternative proof). The related long-standing problem whether $H_\infty$ has the AP (see, e.g., [17, Problem 1.e.10]) is still open. Since $H_\infty$ is a dual space, this also may concern the AP-implies-MAP problem (Problem 1.1).

Trying to approach the AP-implies-MAP problem, a new notion "the weak BAP" was introduced and studied by Å. Lima and E. Oja [16] in 2005.

**Definition 3.3 (see [16]).** *A Banach space $X$ has the weak $\lambda$-BAP if for every Banach space $Y$ and every operator $T \in \mathcal{W}(X,Y)$, there exists a net $(S_\alpha) \subset \mathcal{F}(X)$ such that $S_\alpha \to I_X$ uniformly on compact subsets of $X$ and $\limsup_\alpha \|TS_\alpha\| \leq \lambda \|T\|$. The weak MAP is the weak 1-BAP. The weak BAP is the weak $\lambda$-BAP for some $\lambda$.*

The $\lambda$-BAP clearly implies the weak $\lambda$-BAP, which in turn implies the AP (take $T = 0$ in Definition 3.3). In [21], it was established (for a simpler proof, see [24]) that the weak $\lambda$-BAP and the $\lambda$-BAP are equivalent for $X$ whenever $X^*$ or $X^{**}$ has the Radon–Nikodým property. It remains open whether the weak $\lambda$-BAP is strictly weaker than the $\lambda$-BAP. If they were equivalent, then the answer to the AP-implies-MAP problem would be "yes" because, by [16], the AP of a dual space is always weakly metric.

We conclude this Section by characterizing the weak BAP through inner and outer inequalities. The proof of this characterization in [24] uses both Proposition 3.1 and Theorem 3.2.

**Theorem 3.4 (see [24, Theorem 3.6]).** *Let $X$ be a Banach space and let $\lambda \geq 1$. Then the following assertions are equivalent.*

*(a) $X$ has the weak $\lambda$-BAP.*
*(b) $\|S\|_\pi \leq \lambda \|S\|_{\mathcal{N}(X, X^{**})}$ for all $S \in \mathcal{F}(X)$.*
*(c) $\|T\|_\pi \leq \lambda \|T\|_{\mathcal{N}(Y, X^{**})}$ for all Banach spaces $Y$ and for all $T \in \mathcal{F}(Y, X)$.*

For $p = 1$, condition (c) of Theorem 3.3 would be exactly the same as condition (c) of Theorem 3.4. This shows that, for $1 < p < \infty$, "the weak $\lambda$-BAP$_p$" would be exactly the $\lambda$-BAP$_p$.

## 4. Approximation properties which are bounded with respect to some Banach operator ideal

Definition 3.3 of the weak BAP involves the Banach operator ideal $\mathcal{W}$ of weakly compact operators. If one replaces $\mathcal{W}$ with an arbitrary Banach operator ideal $\mathcal{A}$, then one obtains the following notion which could be used to unify various ideas related to different variants of the bounded approximation property.

**Definition 4.1 (see [14]).** *Let $\mathcal{A} = (\mathcal{A}, \|\cdot\|_\mathcal{A})$ be a Banach operator ideal. A Banach space $X$ has the $\lambda$-bounded approximation property for $\mathcal{A}$ ($\lambda$-BAP for $\mathcal{A}$) if for every Banach space $Y$ and every operator $T \in \mathcal{A}(X, Y)$, there exists a net $(S_\alpha) \subset \mathcal{F}(X)$ such that $S_\alpha \to I_X$ uniformly on compact subsets of $X$ and*

$$\limsup_\alpha \|TS_\alpha\|_\mathcal{A} \leq \lambda \|T\|_\mathcal{A}.$$

*The BAP for $\mathcal{A}$ is the $\lambda$-BAP for $\mathcal{A}$ for some $\lambda$.*

The BAP of $\mathcal{A}$ implies the AP (take $T = 0$ in Definition 4.1). In fact, one may say that the BAP for $\mathcal{A}$ is the AP which is bounded with respect to the given Banach operator ideal $\mathcal{A}$. The Banach space $X_W$, constructed by G. Willis [35], has the metric compact approximation property (MCAP) (in fact, it has the commuting MCAP (see [28])), but fails the AP. Therefore $X_W$ fails the BAP for any Banach operator ideal $\mathcal{A}$.

On the other hand, the $\lambda$-BAP implies the $\lambda$-BAP for all Banach operator ideals $\mathcal{A}$. Indeed, if $X$ has the $\lambda$-BAP, then the net $(S_\alpha)$ is bounded in norm by $\lambda$, and, by $\|TS_\alpha\|_\mathcal{A} \leq \|T\|_\mathcal{A}\|S_\alpha\| \leq \lambda\|T\|_\mathcal{A}$, one has the inequality of Definition 4.1. If $X$ has the $\lambda$-BAP for $\mathcal{L}$, then taking $T = I_X$ in Definition 4.1, we see that $X$ has the $\lambda$-BAP. Thus, the $\lambda$-BAP is precisely the $\lambda$-BAP for $\mathcal{L}$. But it is also precisely the $\lambda$-BAP for the Banach operator ideals $\mathcal{I}$ and $\mathcal{SI}$ of integral and strictly integral operators.

**Theorem 4.1 ([14, Theorem 2.1]).** *A Banach space $X$ has the $\lambda$-BAP if and only if $X$ has the $\lambda$-BAP for $\mathcal{I}$ if and only if $X$ has the $\lambda$-BAP for $\mathcal{SI}$.*

The "only if" parts of Theorem 4.1 are clear from the above. The *proof* of the "if" parts of *Theorem* 4.1 in [14] uses condition (b) of Theorem 3.1 to establish the $\lambda$-BAP, it also uses the fact that $X$ is a quotient space of some $\ell_1(\Gamma)$-space combined with the following lemma and with some factorization of finite-rank operators (see [14, Lemma 2.4]).

**Lemma 4.1 (see [14, Lemma 2.3]).** *Let $X$ be a Banach space and let $\lambda \geq 1$. If a Banach space $Y$ has the property that for every $T \in \mathcal{I}(X, Y^{**})$ there exists a net $(S_\alpha) \subset \mathcal{F}(X)$ such that $S_\alpha \to I_X$ pointwise and $\limsup_\alpha \|TS_\alpha\|_\pi \leq \lambda\|T\|_\mathcal{I}$, then every quotient space of $Y$ has the same property.*

By definition, the weak $\lambda$-BAP is precisely the $\lambda$-BAP for $\mathcal{W}$. But it is also precisely the $\lambda$-BAP for the Banach operator ideal $\mathcal{N}$ of nuclear operators.

**Theorem 4.2 (see [14, Theorem 3.1]).** *A Banach space $X$ has the weak $\lambda$-BAP if and only if $X$ has the $\lambda$-BAP for $\mathcal{N}$.*

The *proof* of *Theorem* 4.2 in [14], concerning the "if" part, is quite similar: to show that $X$ has the weak $\lambda$-BAP, one relies on its characterization (b) in Theorem 3.4. For the "only if" part, one proceeds from another recent criterion of the weak BAP, which is formulated in terms of extension operators. Since this criterion would give a different insight into the concept of the weak BAP than Theorem 3.4, we include it for completeness.

Let $Y$ be a closed subspace of a Banach space $Z$. An operator $\Phi \in \mathcal{L}(Y^*, Z^*)$ is called an *extension operator* if $(\Phi y^*)(y) = y^*(y)$ for all $y^* \in Y^*$ and all $y \in Y$. If $Y$ admits an extension operator $\Phi \in \mathcal{L}(Y^*, Z^*)$, which is norm-preserving (i.e., $\|\Phi\| = 1$), then $Y$ is called an *ideal* in $Z$. This is equivalent to the annihilator $Y^\perp$ of $Y$ being the kernel of a norm one projection in $Z^*$. The concept of ideal in Banach spaces was introduced by G. Godefroy, N. J. Kalton, and P. D. Saphar [9] in 1993.

**Theorem 4.3 (see [13, Props. 2.1, 2.3, 2.5] and [24, Cor. 3.18]).**
*A Banach space $X$ has the weak $\lambda$-BAP if and only if there exists an extension operator $\Phi \in \mathcal{L}(X^*, X^{***})$ such that*
$$\Phi \in \overline{\{T^* : T \in \mathcal{F}(X)\}}^{w^*} \subset (\mathcal{F}(X^*), \|\cdot\|_\pi)^*$$
*and $\|\Phi\| \leq \lambda$.*

Remark that the formulation of Theorem 4.3 uses Schatten's description (Theorem 2.2) that $(\mathcal{F}(X^*), \|\cdot\|_\pi)^* = \mathcal{L}(X^*, X^{***})$.

There is an ample choice of Banach operator ideals (have a look at the Index of Operator Ideals in Pietsch's book [29]!). Fixing a Banach operator ideal $\mathcal{A}$ and returning to the concept of the BAP for $\mathcal{A}$, one may wonder how the BAP for $\mathcal{A}$ "actually" looks like. If $\mathcal{A} = \mathcal{L}, \mathcal{W}, \mathcal{I}, \mathcal{SI}, \mathcal{N}$, then, as we saw, the BAP for $\mathcal{A}$ is precisely either the BAP or the weak BAP. What else is known? By [16, Theorem 2.4], the BAP for $\mathcal{K}$ (compact operators) is precisely the weak BAP, and, by [14, Corollary 4.4], the BAP for $\mathcal{P}^{\text{dual}}$ is precisely the BAP. We denote by $\mathcal{P}$ the Banach operator ideal of absolutely summing operators (1-summing in [5]) (those operators take unconditionally summable sequences to absolutely summing sequences). This is all we know.

Let us try to make a Résumé of the all we know, recalling that the inclusion $\mathcal{A} \subset \mathcal{B}$ for Banach operator ideals $\mathcal{A}$ and $\mathcal{B}$ (see Section 2) provides a natural partial ordering on the family of all Banach operator ideals. In this partially ordered set, $\mathcal{N}$ is the smallest element and $\mathcal{L}$ is the largest element. Our Résumé is as follows (we write wBAP for the weak BAP):

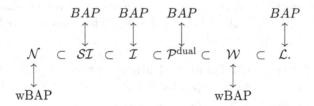

Here only $\mathcal{K}$ (together with wBAP) is omitted: $\mathcal{K}$ cannot be included in the above chain.

As one may notice, even such an important operator ideal as $\mathcal{P}$ is missing in our Résumé, not speaking about the counterparts $\mathcal{N}_p, \mathcal{SI}_p, \mathcal{I}_p$, and $\mathcal{P}_p$

of $\mathcal{N} = \mathcal{N}_1, \mathcal{SI} = \mathcal{SI}_1, \mathcal{I} = \mathcal{I}_1$, and $\mathcal{P} = \mathcal{P}_1$ for the general $p$ (to be precise, $\mathcal{P}_\infty = \mathcal{L}$ is present). The relevant problems, and several others, are pointed out in [14 and 25]. Here we would like to recall a problem which concerns the definition of the BAP for $\mathcal{A}$.

**Problem 4.1 (see [14, Problem 5.5]).** *Let $\mathcal{A}$ be an arbitrary Banach operator ideal. Could the $\lambda$-BAP for $\mathcal{A}$ of a Banach space $X$ be equivalently defined by the following (at least formally) weaker condition: for every Banach space $Y$ and for every operator $T \in \mathcal{A}(X,Y)$ there exists a net $(T_\alpha) \subset \mathcal{F}(X,Y)$ such that $T_\alpha \to T$ pointwise and*

$$\limsup_\alpha \|T_\alpha\|_\mathcal{A} \leq \lambda \|T\|_\mathcal{A}?$$

The answer is obviously "yes" if $\mathcal{A} = \mathcal{L}$, and also if $\mathcal{A} = \mathcal{W}$ (see [23, Theorem 3.6]).

Another related question could be: does one really need *all* Banach spaces in the definition of the BAP for $\mathcal{A}$? At least, for $\mathcal{A} = \mathcal{N}, \mathcal{I}$, and $\mathcal{SI}$, the answer is "no". For $\mathcal{A} = \mathcal{N}$, it sufficies to take $Y = c_0^*$ (see [14, Proposition 4.1]). This actually can be seen from the proof of Theorem 4.2 in [14]. By [14, Proposition 4.2], for $\mathcal{A} = \mathcal{I}$ and $\mathcal{SI}$, it suffices to take $Y = \ell_\infty^*$. To prove this result (see [14]), one develops further ideas of the proof of Theorem 4.1 in [14] and also relies on the fact that the BAP is separably determined with respect to ideals as expressed below.

**Proposition 4.1 (see [14, Proposition 4.3]).** *A Banach space $X$ has the $\lambda$-BAP if and only if every separable ideal $Y$ in $X$ has the $\lambda$-BAP.*

Everybody knows that the classical spaces $c_0$ and $\ell_\infty$ are very different from each other. A natural question would be: can the spaces $c_0$ and $\ell_\infty$ be replaced by *one* classical Banach space, preferably separable, which would characterize both the BAP and the weak BAP? Theorems 4.4 and 4.5 below show that the space $C[0,1]$ of continuous functions fits for the both BAPs.

**Theorem 4.4 (see [15, Theorem 1.3]).** *A Banach space $X$ has the $\lambda$-BAP if and only if for every operator $T \in \mathcal{I}(X, C[0,1]^*)$ there exists a net $(S_\alpha) \subset \mathcal{F}(X)$ such that $S_\alpha \to I_X$ pointwise and*

$$\limsup_\alpha \|TS_\alpha\|_\mathcal{I} \leq \lambda \|T\|_\mathcal{I}.$$

**Theorem 4.5 (see [15, Theorem 1.4]).** *A Banach space $X$ has the weak $\lambda$-BAP if and only if for every operator $T \in \mathcal{N}(X, C[0,1]^*)$ there* exists a net $(S_\alpha) \subset \mathcal{F}(X)$ such that $S_\alpha \to I_X$ pointwise and

$$\limsup_\alpha \|TS_\alpha\|_\mathcal{N} \leq \lambda \|T\|_\mathcal{N}.$$

The "only if" parts of Theorems 4.4 and 4.5 are clear from Theorems 4.1 and 4.2. The *proof* of the "if" parts of *Theorem* 4.5 and the *separable case* of *Theorem* 4.4 in [15] use, for instance, the fact that the Banach operator ideal $\mathcal{I}$ is injective with respect to norm-preserving extension operators (see [15, Proposition 2.1]) and the Grothendieck–Sakai characterization of $L_1$-preduals (or Lindenstrauss's spaces). The *proof* of the *non-separable case* in *Theorem* 4.4 (see [15]) is rather involved. It is quite natural that the $\lambda$-BAP of $X$ is obtained by showing that every separable ideal of $X$ has the $\lambda$-BAP (see Proposition 4.1) and using the separable case of Theorem 4.4. However to reach this point requires a round-about way involving the weak BAP in the proof (!).

Although it is not known whether the weak BAP is strictly weaker than the BAP, the weak BAP has already shown up to be a quite efficient mean in proofs of several results on approximation properties (whose statements do not mention the weak BAP at all). Such an example is the new criterion of the BAP in terms of $C[0,1]$ from Theorem 4.4. Other examples are, for instance, the (new) result that a Banach space $X$ has the $\text{MAP}_p$, $1 < p < \infty$, whenever $X^*$ has the AP (see [24, Corollary 4.3]), and the main theorem of [22]. The weak BAP has also served to enlighten the essence of the old AP-implies-MAP problem (Problem 1.1): in all known cases when the AP of a dual space $X^*$ is metric, it is so *because* $X$ has the MAP in all its equivalent norms (see [21]).

## References

1. S. Banach, *Théorie des opérations linéaires*, Monografje Matematyczne, Warszawa, 1932.
2. J. Bourgain and O. Reinov, *On the approximation properties for the space $H^\infty$*, Math. Nachr. **122** (1985) 19–27.
3. P.G. Casazza, *Approximation properties*, in: W.B. Johnson and J. Lindenstrauss (eds.), Handbook of the Geometry of Banach Spaces, Vol. 1, Elsevier (2001) 271–316.
4. A. Defant and K. Floret, *Tensor Norms and Operator Ideals*, North-Holland Mathematics Studies **176** (1993).
5. J. Diestel, H. Jarchow, and A. Tonge, *Absolutely Summing Operators*, Cambridge Univ. Press, Cambridge (1995).
6. J. Diestel and J.J. Uhl, Jr. *Vector Measures*, Mathematical Surveys **15**, Amer. Math. Soc., Providence, Rhode Island (1977).

7. P. Enflo, *A counterexample to the approximation problem in Banach spaces*, Acta Math. **130** (1973) 309–317.
8. T. Figiel and W. B. Johnson, *The approximation property does not imply the bounded approximation property*, Proc. Amer. Math. Soc. **41** (1973) 197–200.
9. G. Godefroy, N.J.Kalton, and P.D. Saphar, *Unconditional ideals in Banach spaces*, Studia Math. **104** (1993) 13–59.
10. A. Grothendieck, *Produits tensoriels topologiques et espaces nucléaires*, Mem. Amer. Math. Soc. **16** (1955).
11. A. Grothendieck, *Résumé de la théorie métrique des produits tensoriels topologiques*, Bol. Soc. Mat. São Paulo **8** (1953/56) 1–79.
12. R. Kałuża, *Through a Reporter's Eyes: the Life of Stefan Banach*, Birkhäuser, Boston (1996).
13. V. Lima, *The weak metric approximation property and ideals of operators*, J. Math. Anal. Appl. **334** (2007) 593–603.
14. Å. Lima, V. Lima, and E. Oja, *Bounded approximation properties via integral and nuclear operators*, Proc. Amer. Math. Soc. **138** (2010) 287–297.
15. Å. Lima, V. Lima, and E. Oja, *Bounded approximation properties in terms of $C[0,1]$*, Math. Scand. (to appear).
16. Å. Lima and E. Oja, *The weak metric approximation property*, Math. Ann. **333** (2005) 471–484.
17. J. Lindenstrauss and L. Tzafriri, *Classical Banach Spaces I*, Ergebnisse der Mathematik und ihrer Grenzgebiete **92**, Springer-Verlag, Berlin-Heidelberg-New York (1977).
18. A. Lissitsin, K. Mikkor, and E. Oja, *Approximation properties defined by spaces of operators and approximability in operator topologies*, Illinois J. Math. **52** (2008) 563–582.
19. R. D. Mauldin (ed.), *The Scottish Book: Mathematics from the Scottish Café*, Birkhäuser, Boston (1981).
20. E. Oja, *Operators that are nuclear whenever they are nuclear for a larger range space*, Proc. Edinburgh Math. Soc. **47** (2004) 679–694.
21. E. Oja, *The impact of the Radon-Nikodým property on the weak bounded approximation property*, Rev. R. Acad. Cien. Serie A. Mat. **100** (2006) 325–331.
22. E. Oja, *Lifting of the approximation property from Banach spaces to their dual spaces*, Proc. Amer. Math. Soc. **135** (2007) 3581–3587.
23. E. Oja, *The strong approximation property*. J. Math. Anal. Appl. **338** (2008) 407–415.
24. E. Oja, *Inner and outer inequalities with applications to approximation properties*, Trans. Amer. Math. Soc. (to appear).
25. E. Oja, *On bounded approximation properties of Banach spaces*. Banach Center Publ. (to appear).
26. E. Oja and O. Reinov, *Un contre-exemple à une affirmation de A. Grothendieck*, C. R. Acad. Sci. Paris, Sér. I, **305** (1987) 121–122.
27. E. Oja and O. Reinov, *A counterexample to A. Grothendieck*, Proc. Acad. Sci. Estonian SSR, Phys.-Math., **37** (1988) 14–17. (Russian; Estonian and English summaries.)

28. E. Oja and I. Zolk, *On commuting approximation properties of Banach spaces*, Proc. Royal Soc. Edinburgh **139A** (2009) 551–565.
29. A. Pietsch, *Operator Ideals*, North-Holland Publishing Company, Amsterdam - New York - Oxford (1980).
30. O. Reinov, *Approximation properties of order p and the existence of non-p-nuclear operators with p-nuclear second adjoints*, Math. Nachr. **109** (1982) 125–134.
31. R.A. Ryan, *Introduction to Tensor Products of Banach Spaces*, Springer Monographs in Mathematics, Springer-Verlag, London (2002).
32. P. Saphar, *Produits tensoriels d'espaces de Banach et classes d'applications linéaires*, Studia Math. **38** (1970) 71–100.
33. P. Saphar, *Hypothèse d'approximation à l'ordre p dans les espaces de Banach et approximation d'applications p absolument sommantes*, Israel J. Math. **13** (1972) 379–399.
34. R. Schatten, *The cross-space of linear transformations*, Ann. Math. **47** (1946) 73–84.
35. G. Willis, *The compact approximation property does not imply the approximation property*, Studia Math. **103** (1992) 99–108.

# Linear or bilinear mappings between spaces of continuous or Lipschitz functions

Fernando Rambla-Barreno

Department of Mathematics
University of Cádiz, Spain
e-mail: fernando.rambla@uca.es

## 1. Introduction

Let $\mathbb{K}$ be $\mathbb{R}$ or $\mathbb{C}$. The Banach-Stone theorem asserts that, given two locally compact Hausdorff spaces $L, M$, the surjective linear isometries from $\mathcal{C}_0(L)$ to $\mathcal{C}_0(M)$ are precisely those mapping of the form $Tf(t) = \rho(t)f(\sigma(t))$, where $\rho : M \to S_{\mathbb{K}}$ is continuous and $\sigma : M \to L$ is a homeomorphism. The main consequence of this is that $\mathcal{C}_0(M)$ and $\mathcal{C}_0(L)$ are linearly isometric if and only if $L$ and $M$ are homeomorphic.

Several generalizations have been given of the above result. In these work we present some unfinished results by Antonio Aizpuru, who sadly passed away during their realization, and the author. They deal with the four possible combinations of considering continuous or Lipschitz functions in the linear or in the bilinear setting. The proofs of the theorems will appear elsewhere. Besides, the concept of separating mapping plays a central role. Recall that a mapping $T$ is separating if $fg = 0$ implies $(Tf)(Tg) = 0$.

## 2. Continuous functions

### 2.1. *Preliminaries*

In this section $X, Y, Z$ will always be compact Hausdorff spaces and $E, F, G$ will be unital commutative Banach algebras.

Given $f \in \mathcal{C}(X), \mathrm{z}(f)$ is the zero set of $f$ and $\mathrm{cz}(f)$ its complement. We will say that $f$ is a unit if $\mathrm{z}(f) = \emptyset$. The function constantly equal to 1 will be denoted by $1_X$ or simply 1 if no confusion can arise.

Jarosz in [2] proves that if the mapping $T : \mathcal{C}(X) \to \mathcal{C}(Z)$ is linear and separating, if we consider

- $Z_1 = \{z \in Z : \delta_z T \neq 0 \text{ and } \delta_z T \text{ is continuous}\}$
- $Z_2 = \{z \in Z : \delta_z T \text{ is discontinuous}\}$
- $Z_3 = \{z \in Z : \delta_z T = 0\}$

then there exists a continuous mapping $t : Z_1 \cup Z_2 \to X$ such that $Tf(z) = l(z)f(t(z))$ if $(z,f) \in Z_1 \times C(X)$ and $Tf(z) = 0$ if $(z,f) \in Z_3 \times C(X)$, there exists a finite set of isolated points $F \subseteq X$ such that $t(Z_2) = F$ and moreover the sets $Z_1 \cup Z_3$ and $Z_3$ are closed subsets of $Z$. From this it can be deduced that if $T$ is bijective and separating then $T$ is continuous and induces a homeomorphism between $X$ and $Y$. There is a very enlightening proof of the latter result in [1].

In [4] Moreno and Rodríguez-Palacios obtain the following result: Let $T : C(X) \times C(Y) \to C(Z)$ be a bilinear mapping satisfying $\|T(f,g)\| = \|f\|\|g\|$ for every $(f,g) \in C(X) \times C(Y)$. Then there exists a closed subset $Z_0$ of $Z$, a continuous mapping $t : Z_0 \to X \times Y$ and a norm-one element $\alpha \in C(Z)$ satisfying $|\alpha(z)| = 1$ and $T(f,g)(z) = \alpha(z)f(t_1(z))g(t_2(z))$, for every $(z,f,g) \in Z_0 \times C(X) \times C(Y)$. Here $t_1$ and $t_2$ are the corresponding projections into $X$ and $Y$ respectively. Also from this result, the authors obtain some consequences on the theory of absolute-valued Banach spaces (recall that a Banach space $V$ is said to be absolute-valued (av) if there exists a bilinear mapping $T : V \times V \to V$ such that $\|T(a,b)\| = \|a\|\|b\|$ if $(a,b) \in V \times V$).

## 2.2. Linear mappings, scalar-valued functions

Let $g \in C(Y)$. Call $Y_0 = cz(g)$. Suppose that $t : Y_0 \to X$ is a continuous mapping and $T : C(X) \to C(Y)$ is defined by $Tf(y) = g(y)f(t(y))$ if $y \in Y_0$ and $Tf(y) = 0$ if $y \in Y \setminus Y_0$, it is clear that $T$ is linear and

(1) $T$ is continuous
(2) $z(T(f_1 f_2)) = z(Tf_1) \cup z(Tf_2)$
(3) $\{y \in Y : \delta_y T \neq 0\} = \{y \in Y : Tf(y) \neq 0 \text{ if } f \text{ is a unit}\}$.

Note that every linear mapping satisfying (2) must be separating but the converse is not true. In the next theorem we state that for every separating linear map $T : C(X) \to C(Y)$, conditions (1), (2) and (3) are equivalent.

**Theorem 2.1.** *Let $T : C(X) \to C(Y)$ be a linear, separating mapping. Then the following are equivalent:*

(1) $T$ is continuous
(2) $z(T(f_1 f_2)) = z(Tf_1) \cup z(Tf_2)$

(3) $\{y \in Y : \delta_y T \neq 0\} = \{y \in Y : Tf(y) \neq 0 \text{ if } f \text{ is a unit}\}$.

**Remark 2.1.** In the situation of the previous theorem we have

(1) $T$ is surjective if and only if $Y_0$ is dense in $Y$. Indeed, if $Y_0$ is not dense in $Y$ then there exists a nonempty open set $B$ such that $B \subseteq Y \setminus Y_0$. If $g \in C(Y)$ and $\emptyset \neq \text{cz}(g) \subseteq B$ then for every $f \in C(X)$ it is $Tf \neq g$.
(2) $T$ is injective if and only if $t(Y_0)$ is dense in $X$. Indeed, if $t(Y_0)$ is not dense in $X$ then there exists $f \in C(X)$ such that $f \neq 0$ and $\text{cz}(f) \subseteq X \setminus t(Y_0)$, thus $Tf = 0$.
(3) $T$ is a bijective mapping from $C(X)$ onto $C(Y)$ if and only $t$ is a homeomorphism from $Y$ onto $X$.

The next result follows immediately.

**Corollary 2.1.**

*(1) The following are equivalent:*

- *$X$ has a dense subset which is continuous image of a cozero set of $Y$.*
- *There exists a linear, injective mapping $T : C(X) \to C(Y)$ such that $\text{z}(T(f_1 f_2)) = \text{z}(Tf_1) \cup \text{z}(Tf_2)$ if $f_1, f_2 \in C(X)$.*
- *There exists a linear, injective mapping $T : C(X) \to C(Y)$ such that $\{y \in Y : \delta_y T \neq 0\} = \{y \in Y : Tf(y) \neq 0 \text{ if } f \text{ is a unit}\}$.*

*(2) The following are equivalent:*

- *$X$ is a continuous image of $Y$.*
- *There exists a linear mapping $T : C(X) \to C(Y)$ such that $Tf$ is a unit of $C(Y)$ for some $f \in C(X)$ and $\text{z}(T(f_1 f_2)) = \text{z}(Tf_1) \cup \text{z}(Tf_2)$ if $f_1, f_2 \in C(X)$.*
- *There exists a linear separating mapping $T : C(X) \to C(Y)$ such that $Tf$ is a unit of $C(Y)$ for some $f \in C(X)$.*

### 2.3. Bilinear mappings, scalar-valued functions

Let $T : C(X) \times C(Y) \to C(Z)$ a bilinear mapping, we will say that it is separating if it satisfies:

(1) If $f_1, f_2 \in C(X)$ and $\text{cz}(f_1) \cap \text{cz}(f_2) = \emptyset$ then $\text{cz}(T(f_1, g_1)) \cap \text{cz}(T(f_2, g_2)) = \emptyset$ for every $g_1, g_2 \in C(Y)$.
(2) If $g_1, g_2 \in C(Y)$ and $\text{cz}(g_1) \cap \text{cz}(g_2) = \emptyset$ then $\text{cz}(T(f_1, g_1)) \cap \text{cz}(T(f_2, g_2)) = \emptyset$ for every $f_1, f_2 \in C(X)$.

The following standard example was our motivation.

Let $l \in C(Z)$, call $Z_0 = cz(l)$ and assume $t : Z_0 \to X \times Y$ is a continuous mapping. Let $T : C(X) \times C(Y) \to C(Z)$ be defined by $T(f,g)(z) = l(z)f(t_1(z))g(t_2(z))$ if $z \in Z_0$ and $T(f,g)(z) = 0$ if $z \in Z \setminus Z_0$ for each $(f,g) \in C(X) \times C(Y)$. It is clear that $T$ is bilinear and also:

(1) $T$ is continuous
(2) $z(T(f_1 f_2, g_1 g_2)) = z(T(f_1 g_1)) \cup z(T(f_2 g_2))$ if $(f_1, g_1), (f_2, g_2) \in C(X) \times C(Y)$
(3) $\{z \in Z : \delta_z T \neq 0\} = \{z \in Z : T(f,g)(z) \neq 0 \text{ if } f \in C(X) \text{ and } g \in C(Y) \text{ are units}\}$.

Condition (2) implies that the bilinear mapping is separating, but the converse is not true. As in the linear case, we prove that if a bilinear mapping $T$ is separating then (1), (2) and (3) are equivalent. For this purpose we will need a bilinear version of Jarosz's result.

**Lemma 2.1.** Let $T : C(X) \times C(Y) \to C(Z)$ a bilinear separating mapping. Denote $Z_1 = \{z \in Z : \delta_z T \neq 0 \text{ and } \delta_z T \text{ is continuous}\}$, $Z_2 = \{z \in Z : \delta_z T \text{ is discontinuous}\}$ and $Z_3 = \{z \in Z : \delta_z T = 0\}$ then there exists a continuous mapping $t : Z_1 \cup Z_2 \to X \times Y$ such that $T(f,g)(z) = l(z)f(t_1(z))g(t_2(z))$ if $(z,f,g) \in Z_1 \times C(X) \times C(Y)$ and $T(f,g)(z) = 0$ if $(z,f,g) \in Z_3 \times C(X) \times C(Y)$ where $l \in C(Z)$ satisfies $Z_1 \subseteq cz(l)$. Moreover the sets $Z_1 \cup Z_3$ and $Z_3$ are closed.

**Remark 2.2.** Jarosz in [2] proves that there exists a compact Hausdorff space $X$ and a linear, separating map $\varphi : C(X) \to C(X)$ which is not continuous. Consider the mapping $T : C(X) \times C(X) \to C(X)$ defined by $T(f,g) = \varphi(f)\varphi(g)$, we have that $T$ is bilinear and separating but $T$ is not continuous.

**Theorem 2.2.** Let $T : C(X) \times C(Y) \to C(Z)$ be a bilinear, separating mapping. The following are equivalent:

(1) $T$ is continuous.
(2) $z(T(f_1 f_2, g_1 g_2)) = z(T(f_1, g_1)) \cup z(T(f_2, g_2))$ if $(f_1, g_1), (f_2, g_2) \in C(X) \times C(Y)$.
(3) $\{z \in Z : \delta_z T \neq 0\} = \{z \in Z : T(f,g)(z) \neq 0 \text{ if } f \in C(X) \text{ and } g \in C(Y) \text{ are units}\}$.

**Corollary 2.2.**
 The following are equivalent:

(1) $X \times Y$ is a continuous image of $Z$.
(2) There exists a bilinear mapping $T : C(X) \times C(Y) \to C(Z)$ such that
   (a) If $(f_1, g_1), (f_2, g_2) \in C(X) \times C(Y)$ then $z(T(f_1 f_2, g_1 g_2)) = z(T(f_1, g_1)) \cup z(T(f_2, g_2))$
   (b) If $(f, g) \in C(X) \times C(Y)$ and $f \neq 0$, $g \neq 0$ then $T(f, g) \neq 0$.
   (c) There exists $(f, g) \in C(X) \times C(Y)$ such that $T(f, g)$ is a unit in $C(Z)$.

**Corollary 2.3.**
If there exists a bilinear mapping $T : C(X) \times C(X) \to C(X)$ satisfying

(1) If $(f_1, g_1), (f_2, g_2) \in C(X) \times C(X)$ then $z(T(f_1 f_2, g_1 g_2)) = z(T(f_1, g_1)) \cup z(T(f_2, g_2))$
(2) If $(f, g) \in C(X) \times C(X)$ and $f \neq 0$, $g \neq 0$ then $T(f, g) \neq 0$.

Then $C(X)$ is absolute-valued.

### 2.4. Vector-valued functions

Assume there exists $x_0 \in X$ such that $\{x_0\}$ is a clopen set and define the mapping $T : C(X, E) \to C(X, E)$ by $Tf(x) = f(x)$ if $x \in X \setminus \{x_0\}$ and $Tf(x_0) = \varphi_{x_0}(f(x_0))$, where $\varphi_{x_0} : E \to E$ is a linear mapping. We have that $T$ is bijective, linear and $T, T^{-1}$ are separating, but if $\varphi_{x_0}$ is not continuous then neither is $T$.

Now we will give a vector-valued version of Jarosz's result.

**Lemma 2.2.** Let $T : C(X, E) \to C(Z, F)$ be a linear, separating mapping. Consider the sets $Z_1 = \{z \in Z : \delta_z T \text{ is continuous and nonzero}\}$, $Z_2 = \{z \in Z : \delta_z T \text{ is discontinuous}\}$ and $Z_3 = \{z \in Z : \delta_z T = 0\}$. Then there exists a continuous mapping $t : Z_1 \cup Z_2 \to X$ and a SOT-continuous mapping $k : Z_1 \to \mathcal{L}(E, F)$ such that for every $(z, f) \in Z_1 \times C(X, E)$ we have $Tf(z) = k(z)(f(t(z)))$, moreover $\varphi(Z_2)$ is a finite subset of $X$.

**Theorem 2.3.** Assume $X$ has no isolated points and let $T : C(X, E) \to C(Y, F)$ be a linear, bijective, separating mapping. The following are equivalent:

(1) If $f \in C(X, E)$ is such that $f(x) \neq 0$ for every $x \in X$ then $Tf(z) \neq 0$ for every $z \in Z$
(2) $T^{-1}$ is separating
(3) $T$ is continuous.

**Example 2.1.** Consider the linear mapping $T : C(X,c) \to C(X,c_0)$ defined by $Tf(x)(i) = \frac{1}{i}f(x)(i)$. We have that $Tf(x) = k(x)(f(x))$ where $k(x)$ is the linear mapping from $c$ into $c_0$ defined by $k(x)(a) = \frac{1}{i}a_i$, for each $x \in X$. Then $T$ is injective, linear and separating. Moreover,

(1) If $f \in C(X,c)$ and $f(x) \neq 0$ for each $x \in X$ then $Tf(x) \neq 0$ for every $x \in X$.
(2) If $x_1, x_2 \in X$ and $x_1 \neq x_2$ then there exists $f \in C(X,c)$ such that $Tf(x_1) = 0$ and $Tf(x_2) \neq 0$
(3) If $f_1, f_2 \in C(X,c)$ and $\operatorname{cz}(Tf_1) \cap \operatorname{cz}(Tf_2) = \emptyset$ then $\operatorname{cz}(f_1) \cap \operatorname{cz}(f_2) = \emptyset$.

Finally let us observe that $T$ is not surjective since for $a = \left(\frac{1}{\log i}\right)_i$, we have no preimage of $\xi_a$.

In the next theorem we will see that if a mapping is linear, separating and injective and satisfies (1) and (2) then it is continuous and satisfies also (3).

**Theorem 2.4.** *Let* $T : C(X,E) \to C(Y,F)$ *be a linear, injective, separating mapping satisfying*

*(1) If* $f \in C(X,E)$ *and* $f(x) \neq 0$ *for every* $x \in X$ *then* $Tf(z) \neq 0$ *for every* $z \in Z$.
*(2) If* $z_1, z_2 \in Z$ *and* $z_1 \neq z_2$ *then there exists* $f \in C(X,E)$ *such that* $Tf(z_1) = 0$ *and* $Tf(z_2) \neq 0$.

*Then $T$ is continuous and if* $\operatorname{cz}(Tf_1) \cap \operatorname{cz}(Tf_2) = \emptyset$ *then* $\operatorname{cz}(f_1) \cap \operatorname{cz}(f_2) = \emptyset$.

**Corollary 2.4.**

*(1) $Z$ is homeomorphic to $X$ if and only if there exist a commutative Banach algebra with unit $E$ and a mapping $T : C(X,E) \to C(Z,E)$ which is linear, injective and separating and satisfies (1) and (2) in the previous theorem.*

*(2) $Z$ is homeomorphic to $X$ if and only if there exist a commutative Banach algebra with unit $E$ and a mapping $T : C(X,E) \to C(Z,E)$ which is linear, bijective and separating and satisfies (1) in the previous theorem.*

Now we will give a bilinear, vector-valued version of Jarosz's result.

**Theorem 2.5.** *Let* $T : C(X,E) \times C(Y,F) \to C(Z,G)$ *be a linear, separating mapping and let* $Z_1 = \{z \in Z : \delta_z T \text{ is continuous and nonzero}\}$, $Z_2 = \{z \in Z : \delta_z T \text{ is discontinuous}\}$. *Then there exist a continuous mapping* $t : Z_1 \cup$

$Z_2 \to X \times Y$ and a SOT-continuous mapping $k : Z_1 \to \mathcal{L}(E \times F, G)$ such that if $(z, f, g) \in Z_1 \times C(X, E) \times C(Y, F)$ then $T(f,g)(z) = k_z(f(t_1(z)), g(t_2(z)))$ and $T(f,g)(z) = 0$ if $z \in Z_3 = Z \setminus (Z_1 \cup Z_2)$.

**Example 2.2.** Let $T : C(X, c) \times C(X, c) \to C(X, c_0)$ be the mapping defined by $T(f,g)(x)(i) = \frac{1}{i}f(x)(i)g(x)(i)$, we have that $T$ is bilinear and separating and we can write $T(f,g)(x) = k(x)(f(x), g(x))$ where $k(x)$ is the bilinear mapping $k(x) : c \times c \to c_0$ defined by $k(x)(a,b) = (\frac{1}{i}a(i)b(i))_i$ for every $x \in X$, $k(x)$ is bilinear and $T$ has the following properties:

(1) If $f \in C(X, c)$ and $f(x) \neq 0$ for every $x \in X$ then $T(f, \xi_e)(x) \neq 0$ for every $x \in X$.
(2) If $x_1 \neq x_2$, $x_1, x_2 \in X$ then there exists $f \in C(X, E)$ such that $T(f, \xi_e)(x_1) = 0$ and $T(f, \xi_e)(x_2) \neq 0$.
(3) For every $x \in X$ we have $\sup\{\|T(f,g)(x)\| : \|f\| = \|g\| = 1\} = \sup\{\|T(f, \xi_e)(x)\| : \|f\| = 1\}$.

**Theorem 2.6.** Assume that $X$ has no isolated points and let $T : C(X, E) \times C(Y, F) \to C(Z, G)$ be a bilinear, separating mapping such that (1): If $f \in C(X, E)$ and $f(x) \neq 0$ for every $x \in X$ then $T(f, \xi_{e'})(z) \neq 0$ for every $z \in Z$; (2): If $z_1, z_2 \in Z$ and $z_1 \neq z_2$ then there exists $f \in C(X, E)$ such that $T(f, \xi_{e'})(z_1) = 0$ and $T(f, \xi_{e'})(z_2) \neq 0$; and (3) For every $z \in Z$ it is $\sup\{\|T(f,g)(z)\| : \|f\| = \|g\| = 1, (f,g) \in C(X,E) \times C(Y,F)\} = \sup\{\|T(f, \xi_{e'})(z)\| : \|f\| = 1 \text{ and } f \in C(X, E)\}$; then $T$ is continuous and there exist a continuous, injective mapping $t : Z \to X \times Y$ and a SOT continuous mapping $k : Z \to \mathcal{L}_2(E, F, G)$ such that if $(z, f, g) \in Z \times C(X, E) \times C(Y, F)$ then $T(f,g)(z) = k(z)(f(t_1(z)), g(t_2(z)))$.

## 3. Lipschitz functions

### 3.1. Linear mappings

Assume $g \in Lip(Y)$ and $Y_0 = cz(g)$ and also $t : Y_0 \to X$ is a continuous mapping satisfying that there exists $M > 0$ such that if $y_1, y_2 \in Y$ then $|g(y_1)|d(t(y_1), t(y_2)) \leq Md(y_1, y_2)$. Let us consider the mapping $T : Lip(X) \to Lip(Y)$ given by $Tf(y) = g(y)f(t(y))$ if $(y, f) \in Y_0 \times C(X)$, it is easily seen to be well defined, linear and continuous. Moreover, $T$ satisfies:

(1) $z(T(f_1, f_2)) = z(Tf_1) \cup z(Tf_2)$ if $f_1, f_2 \in C(Y)$.
(2) There exists $H > 0$ such that for every $x \in X$ we have $L(Tf_x) \leq H$.

It is clear that condition (1) implies that $T$ is separating but the converse is not true in general. In the next theorem we prove that if a linear mapping $T: Lip(X) \to Lip(Y)$ satisfies (1) and (2) then $T$ is continuous.

**Theorem 3.1.** *Let $T: Lip(X) \to Lip(Y)$ be a linear mapping satisfying:*

*(1) $z(T(f_1, f_2)) = z(Tf_1) \cup z(Tf_2)$ if $f_1, f_2 \in Lip(X)$.*
*(2) There exists $H > 0$ such that for every $x \in X$ we have $L(Tf_x) \le H$.*

*Then there exist $g \in Lip(Y)$ and a continuous mapping $t$ from $Y_0 = cz(g)$ into $X$ such that $Tf(y) = g(y)f(t(y))$ if $(y, f) \in Y_0 \times Lip(X)$ and $Tf(y) = 0$ if $(y, f) \in (Y \setminus Y_0) \times Lip(X)$. Moreover, if $y_1, y_2 \in Y_0$ then $|g(y_1)|d(t(y_1), t(y_2))| \le H d(y_1, y_2)$.*

As a consequence, we deduce the next result from the previous remark and the result of Jiménez in [3].

**Theorem 3.2.** *Let $T: lip_\alpha(X) \to lip_\alpha(Y)$ be a linear separating mapping, then $T$ is continuous if and only if $T$ satisfies*

*(1) $z(T(f_1, f_2)) = z(Tf_1) \cup z(Tf_2)$ if $f_1, f_2 \in lip_\alpha(X)$*
*(2) There exists $H > 0$ such that $L_\alpha(Tf_x) \le H$ for each $x \in X$.*

**Example 3.1.** Consider $X = [0,1]$ with $d(x,y) = |x - y|$ and let $t: X \to X$ be the mapping defined by $t(x) = \sqrt{x}$, we have that $t \notin Lip(X)$, let $g(x)$ be the mapping defined in $X$ by $g(x) = x^2$ and consider the linear mapping $T: Lip(X) \to Lip(X)$ defined by $Tf(x) = g(x)f(t(x))$. Then $T$ is well defined and if $f \in Lip(X)$ then $Tf \in Lip(X)$. Thus it is possible to have a weighted composition operator between $Lip(X)$ spaces without the corresponding mapping $t: Y \to X$ being Lipschitz. Let us observe that, in this example, $cz(T\xi_1) = [0,1]$. However, in the situation of the main theorem of this section, if the linear mapping $T$ is such that $cz(T\xi_1) = Y$ then necessarily the corresponding mapping $t: Y \to X$ is Lipschitz.

## 3.2. Bilinear mappings

Assume $l \in Lip(Y)$ and $Y_0 = cz(l)$. Suppose there exists a continuous mapping $t: Z_0 \to X \times Y$, $t(z) = (t_1(z), t_2(z))$ and there exists $H > 0$ such that if $z_1, z_2 \in Z_0$ then $|l(z_1)|d(t_i(z_1), t_i(z_2)) \le H d(z_1, z_2)$ if $i \in \{1, 2\}$. Consider the mapping $T: Lip(X) \times Lip(Y) \to Lip(Z)$ defined by $T(f, g)(z) = l(z)f(t_1(z))g(t_2(z))$, it is easily seen to be well defined, linear and continuous. Moreover, $T$ satisfies:

(1) $z(T(f_1 f_2, g_1 g_2)) = z(T(f_1, g_1)) \cup z(T(f_2, g_2))$ if $(f_1, g_1), (f_2, g_2) \in Lip(X) \times Lip(Y)$.
(2) If $z \in Z_0$, there exists $H > 0$ such that $\max\{L(T(f_{t_1(z)}, \xi_1)), L(T(\xi_1, f_{t_2(z)}))\} < H$

It is clear that the condition (1) implies that $T$ is separating on each component. Now let us see an example of this situation we are considering: Let $X = [0, 1]$ with the usual metric. Let $l \in Lip(X)$ be given by $l(x) = x^2$ and let $t : X \to X \times X$ be the mapping defined by $t(x) = (\sqrt{x}, \sqrt{x})$, we have that $t_1(x) = t_2(x) = \sqrt{x}$ is not a Lipschitz function. Consider the mapping $T : Lip(X) \times Lip(X) \to Lip(X)$ given by $T(f, g)(x) = l(x) f(t_1(x)) g(t_2(x))$. If $x_1, x_2 \in X$ then $|l(x_1)||t_i(x_1) - t_i(x_2)| \leq |x_1 - x_2|$ if $i \in \{1, 2\}$, from this we deduce that $T$ is well defined.

In the next theorem we will see that if a bilinear mapping $T : Lip(X) \times Lip(Y) \to Lip(Z)$ satisfies (1) and (2) then it is continuous.

**Theorem 3.3.** *Let $T : Lip(X) \times Lip(Y) \to Lip(Z)$ be a bilinear mapping satisfying:*

*(1) $z(T(f_1 f_2, g_1 g_2)) = z(T(f_1, g_1)) \cup z(T(f_2, g_2))$ if $(f_1, g_1), (f_2, g_2) \in Lip(X) \times Lip(Y)$.*
*(2) If $z \in Z_0$, there exists $H > 0$ such that $\max\{L(T(f_x, \xi_1)), L(T(\xi_1, f_y))\} < H$*

*Then there exists $l \in Lip(Z)$ and a continuous mapping $t$ from $Z_0 = cz(g)$ into $X \times Y$ such that $T(f, g)(z) = l(z) f(t_1(z)) g(t_2(z))$ if $(z, f, g) \in Z_0 \times Lip(X) \times Lip(Y)$. Also, if $z_1, z_2 \in Z_0$ then $|l(z_1)| d(t_i(z_1), t_i(z_2)) < H d(z_1, z_2)$ for each $i \in \{1, 2\}$.*

**Remark 3.1.**

(1) In the situation of the previous theorem, if $cz(T(\xi_1, \xi_1)) = cz(l) = Z$ then for the mapping $t : Z \to X \times Y$ we have that $t_1$ and $t_2$ are Lipschitz.
(2) In the situation of the previous theorem it can be proved that the following are equivalent:
   (a) $T$ satisfies that if $(f, g) \in Lip(X) \times Lip(Y)$ and $f \neq 0, g \neq 0$ then $T(f, g) \neq 0$.
   (b) $t(Z_0)$ is dense in $X \times Y$.

We also have that $\|T(f, g)\|_\infty = \|l\|_\infty \|f\|_\infty \|g\|_\infty$ if $(f, g) \in Lip(X) \times Lip(Y)$ and if $T_0 = \frac{1}{\|l\|_\infty} T$ then $\|T_0(f, g)\|_\infty = \|f\|_\infty \|g\|_\infty$.

(3) We deduce that if there exists a bilinear mapping $T : Lip(X) \times Lip(X) \to Lip(X)$ satisfying (1) and (2) in theorem 1 and also (3): If $(f,g) \in Lip(X) \times Lip(X)$, $f \neq 0$, $g \neq 0$ then $T(f,g) \neq 0$ then $Lip(X)$ is av for the norm $\|f\|_\infty$.

## References

1. GAU, H.L., JEANG, J.S. AND WONG, H.C., An algebraic approach to the Banach-Stone theorem, Taiwanese Journal of Mathematics **6** (2002), no.3, 399–403.
2. JAROSZ, K., Automatic continuity of separating linear isomorphisms, Bull. Canadian Math. Soc. **33** (1990), 139–144.
3. JIMÉNEZ VARGAS, A., Disjointness preserving operators between little Lipschitz algebras, J. Math. Anal. Appl. **337** (2008), 984–993.
4. MORENO GALINDO, A. AND RODRÍGUEZ PALACIOS, A., A bilinear version of Holsztynski's theorem on isometries of $\mathcal{C}(X)$ spaces, Studia Math. **166** (2005), no.1, 83–91.

# Summability and lineability in the work of Antonio Aizpuru Tomás

Juan B. Seoane-Sepúlveda*

*Departamento de Análisis Matemático. Universidad Complutense de Madrid*
*Plaza de Ciencias 3. 28040 Madrid, Spain*
*e-mail: jseoane@mat.ucm.es*

This survey note is dedicated to the memory of Prof. Antonio Aizpuru Tomás (Valencia, 1954 - Cádiz, 2008). We present an overview of some of Prof. Aizpuru's results on the applications of *lineability* to Series and Summability Methods. We will go through his last papers and some of his results relating these two concepts.

*"Archimedes will be remembered when Aeschylus is forgotten, because languages die and mathematical ideas do not. Immortality may be a silly word, but probably a mathematician has the best chance of whatever it may mean."*
G. H. Hardy (1877–1947)

## 1. Preliminaries

Lately it has become a sort of a trend in Mathematical Analysis the search for what are often large algebraic structures (vector spaces, algebras, among others) of functions on $\mathbb{K}$ ($\mathbb{R}$ or $\mathbb{C}$) enjoying certain *pathological* properties. Besides all of Antonio Aizpuru's knowledge in most branches of Mathematical Analysis (specially Summability Theory and the Theory of Series), he also got interested in this direction of research. Let us first recall a basic definition in this topic:

**Definition 1.1 (Aron, Gurariy, Seoane-Sepúlveda, 2004).** *A subset $M$ of a topological vector space is said to be lineable (resp. spaceable) if $M \cup \{0\}$ contains an infinite dimensional vector space (resp. infinite dimensional closed subspace).*

---
*Supported by the Spanish Ministry of Science and Innovation, grant 2009-07848.

Fig. 1. Prof. Antonio Aizpuru Tomás

Antonio got interested in this particular topic in 2005, and due to his wide knowledge of *Summability Theory*, he successfully linked both areas and obtained many interesting results involving these topics. In this note we will give a brief overview of some of Antonio's results in these directions and comment, as well, the many different directions in which other authors have been working on lately.

## 2. Lineability and the theory of series in Antonio's work

This topic was the most fruitful in Antonio's research when trying to link the theory of Series with the theory of Lineability. Let's enumerate a list of some of his contributions to the area.

If $V$ denotes the set of conditionally convergent series then, clearly, $V \cup \{0\}$ is not a vector space in $CS(\mathbb{K})$, the set of convergent series, but ([1]), on the other hand:

**Theorem 2.1 (Aizpuru, Pérez, Seoane-Sepúlveda, 2006).** $CS(\mathbb{K})$ *contains a vector space $E$ verifying the following properties:*

(1) *Every $x \in E \setminus \{0\}$ is a conditionally convergent series.*
(2) $dim(E) = c$.
(3) $span\{E \cup c_{00}\}$ *is an algebra and its elements are either elements of $c_{00}$ or conditionally convergent series.*

**Theorem 2.2 (Aizpuru, Pérez, Seoane-Sepúlveda, 2006).** *There exists a vector space $E \subset BS(\mathbb{K})$ (the set of all series with bounded partial sums) such that:*

*(1) $E$ is non-separable.*
*(2) $dim(E) = c$.*
*(3) Every $x \in E \setminus \{0\}$ is a divergent series.*
*(4) $span\{E \cup c_{00}\}$ is an algebra and every element of it is either a divergent series or an element of $c_{00}$.*

**Theorem 2.3 (Aizpuru, Pérez, Seoane-Sepúlveda, 2006).** *There exists a vector space $E \subset l_\infty$ such that:*

*(1) $dim(E) = c$.*
*(2) Every $x \in E \setminus \{0\}$ is a divergent sequence.*
*(3) $E \oplus c_0$ is an algebra.*
*(4) Every element in $cl(E) + c_0$ is either a divergent sequence or a sequence in $c_0$.*

Before carrying on with more results, we need to recall a couple of definitions that will be necessary in order to state the next results of Antonio.

**Remark 2.1.**

(1) If $X$ is a Banach space and $\sum_i x_i$ is a series in $X$, we say that:
  - $\sum_i x_i$ is *unconditionally convergent* (UC) if, for every permutation $\pi$ of $\mathbb{N}$, we have that $\sum_{i=1}^\infty x_{\pi(i)}$ converges.
  - $\sum_i x_i$ is *weakly unconditionally Cauchy* (WUC) if $\sum_{i=1}^\infty |f(x_i)| < \infty$ for every $f \in X^*$, the dual space of $X$.

(2) It is also known that if $X$ is a Banach space, then there exists a WUC series in $X$ which is convergent but which is not UC if and only if $X$ has a copy of $c_0$.

(3) A well known fact states that every infinite dimensional Banach space has a series $\sum_i x_i$ which is unconditionally convergent and so that $\sum_i \|x_i\| = \infty$.

**Theorem 2.4 (Aizpuru, Pérez, Seoane-Sepúlveda, 2006).** *There exists a vector space $E \subset l_1^\omega(c_0)$ (the space of all weakly unconditionally Cauchy series in $c_0$) verifying:*

*(1) $dim(E) = c$.*
*(2) If $x \in E \setminus \{0\}$ then $\sum_i x_i$ is not weakly convergent.*

**Theorem 2.5 (Aizpuru, Pérez, Seoane-Sepúlveda, 2006).** *Let $X$ be an infinite dimensional Banach space. Then there exists a vector subspace $E$ of $UC(X)$ such that $dim(E) = c$, and if $x \in E \setminus \{0\}$ then $\sum_i \|x_i\| = \infty$.*

Of course, Antonio was interested in many different areas of Mathematical Analysis, specially Real Analysis, where he also contributed to the lineability problem as we see in what follows.

## 3. More results of Antonio on lineability and other contributors to the theory

Antonio had a wide knowledge of many areas of mathematics, and also got very interested in applying the notion of lineability to other topics besides Summability Theory. A new notion was coined in an article coauthored by him (see [2]), namely:

**Definition 3.1 (Aizpuru, García, Pérez, Seoane-Sepúlveda, 2008).** *A set of functions in $\mathcal{F}(\mathbb{R}, \mathbb{R})$ is said to be **coneable** if it possesses a positive (or negative) cone containing an infinite linearly independent set.*

If we now denote by $\lambda(M)$ the maximum dimension (if it exists) of a vector space inside $M \cup \{0\}$, we have:

**Theorem 3.1 (Aizpuru, García, Pérez, S., 2008).** *Given a closed set $F$, the set $H$ of all functions whose set of points of discontinuity is $F$ is lineable with $\lambda(H) \geq c$. Moreover, if $\text{int}(F) \neq \varnothing$ then $\lambda(H) = 2^c$.*

**Theorem 3.2 (Aizpuru, García, Pérez, S., 2008).** *Given any non-closed $\mathcal{F}_\sigma$ set $F$, the set of functions whose set of points of discontinuity is exactly $F$ is coneable.*

**Theorem 3.3 (Aizpuru, García, Pérez, S., 2008).** *Let $I$ be any nontrivial interval and let $a \in I$. Let $K$ denote the set of all functions from $I$ to $\mathbb{R}$ having a removable discontinuity at $a$. We have that $\lambda(K) = 1$.*

As we have already mentioned, the term lineability was coined by Gurariy and first introduced by Aron, Gurariy and Seoane-Sepúlveda in [Proc. Amer. Math. Soc. **133** (2004) 795–803, see 6.] Around that time and since then, many authors have shown their interest in this topic, just to cite some in different areas of mathematics (see, e.g. [3–17]):

- Zeros of polynomials:
  - Plichko, Zagorodnyuk (1998).

Fig. 2. Richard M. Aron, Vladimir I. Gurariy et al. introduced the concept of lineability in 2004, concept coined originally by Gurariy some years earlier.

- Aron, Rueda (1997).
- Aron, Gonzalo, Zagorodnyuk (2000).
- Aron, Boyd, Ryan, Zalduendo (2003).
- Aron, Hajék (2006).
- Aron, García, Maestre (2001).

• Hypercyclicity:

- Godefroy, Shapiro (1991).
- Montes-Rodríguez (1996).
- Aron, García, Maestre (2001).
- Aron, Bès, León, Peris (2005).
- Aron, Conejero, Peris, Seoane-Sepúlveda (2007).

• Spaces of nowhere differentiable functions

- Rodríguez-Piazza (1995).
- Fonf, Gurariy, Kadeč (1999).
- Aron, García, Maestre (2001).
- Bernal-González (2008).

• Norm-attaining mappings:

- Aron, García, Maestre (2001).
- Acosta, Aizpuru, Aron, García (2007).
- Pellegrino, Teixeira (2009).

• Sets of functions in general:

- Aron, Gurariy, Seoane-Sepúlveda (2004).
- Gurariy, Quarta (2004).
- Bayart, Quarta (2007).
- Aron, Seoane-Sepúlveda (2007).

- Aron, Gorkin (2007).
- García, Palmberg, Seoane-Sepúlveda (2007).
- Aizpuru, García, Pérez, Seoane-Sepúlveda (2008).
- Bernal-González (2008).
- Azagra, Muñoz-Fernández, Sánchez, Seoane-Sepúlveda (2009).
- Gámez-Merino, Muñoz-Fernández, Sánchez, Seoane-Sepúlveda (2010).

• Sets of convergence/divergence of series:
  - Aizpuru, Pérez, Seoane-Sepúlveda (2005).
  - Aron, Pérez-García, Seoane-Sepúlveda (2006).

• Sets of measures/measure spaces
  - García, Seoane-Sepúlveda (2006).
  - Muñoz-Fernández, Palmberg, Puglisi, Seoane-Sepúlveda (2008).
  - García, Pérez, Seoane-Sepúlveda (2010).

• Bounded linear non-absolutely summing operators:
  - Puglisi, Seoane-Sepúlveda (2008).
  - Botelho, Diniz, Pellegrino (2009).

As a personal note, I would like to emphasize the role of Prof. Antonio Aizpuru Tomás in my mathematical education. Antonio was a supportive, caring, and friendly advisor. The day he passed away the world lost an outstanding mathematician and I lost a great friend.

**References**

1. A. Aizpuru, C. Pérez-Eslava & J. B. Seoane-Sepúlveda, *Linear structure of sets of divergent sequences and series*, Linear Algebra Appl. **418** (2006), 2–3, 595–598.
2. A. Aizpuru, F. J. García-Pacheco, C. Pérez-Eslava & J. B. Seoane-Sepúlveda, *Lineability and coneability of discontinuous functions on* $\mathbb{R}$, Publ. Math. Debrecen **72** (2008), 1–2, 129–139.
3. R. M. Aron, D. García & M. Maestre, *Linearity in non-linear problems*, RACSAM Rev. R. Acad. Cienc. Exactas Fís. Nat. Ser. A Mat. **95**, (2001), 1, 7–12.
4. R. M. Aron, J. A. Conejero, A. Peris & J. B. Seoane-Sepúlveda, *Uncountably generated algebras of everywhere surjective functions*, Bull. Belg. Math. Soc. Simon Stevin, to appear.
5. R. M. Aron, F. J. García-Pacheco, D. Pérez-García & J. B. Seoane-Sepúlveda, *On dense-lineability of sets of functions on* $\mathbb{R}$, Topology, to appear.
6. R. M. Aron, V. I. Gurariy & J. B. Seoane-Sepúlveda, *Lineability and spaceability of sets of functions on* $\mathbb{R}$, Proc. Amer. Math. Soc. **133** (2005), 3, 795–803.
7. R. M. Aron, D. Pérez-García & J. B. Seoane-Sepúlveda, *Algebrability of the set of non-convergent Fourier series*, Studia Math. **175** (2006), 1, 83–90.
8. R. M. Aron & J. B. Seoane-Sepúlveda, *Algebrability of the set of everywhere surjective functions on* $\mathbb{C}$, Bull. Belg. Math. Soc. Simon Stevin **14** (2007), 1, 25–31.
9. F. Bayart & L. Quarta, *Algebras in sets of queer functions*, Israel J. Math. **158** (2007), 285–296.
10. L. Bernal-González, *Dense-lineability in spaces of continuous functions*, Proc. Amer. Math. Soc. **136** (2008), 9, 3163–3169.
11. G. Botelho, D. Diniz & M. Pellegrino, *Lineability of the set of bounded linear non-absolutely summing operators*, J. Math. Anal. Appl. **357** (2009), 1, 171–175.
12. G. Botelho, M. Matos & D. Pellegrino, *Lineability of summing sets of homogeneous polynomials*, Linear Multilinear Algebra, to appear.
13. D. García, B. C. Grecu, M. Maestre & J. B. Seoane-Sepúlveda, *Infinite dimensional Banach spaces of functions with nonlinear properties*, Math. Nachr., to appear.
14. V. I. Gurariy, *Subspaces and bases in spaces of continuous functions (Russian)*, Dokl. Akad. Nauk SSSR **167** (1966), 971–973.
15. V. I. Gurariy, *Linear spaces composed of nondifferentiable functions*, C.R. Acad. Bulgare Sci. **44** (1991), 13–16.

16. G. A. Muñoz-Fernández, N. Palmberg, D. Puglisi & J. B. Seoane-Sepúlveda, *Lineability in subsets of measure and function spaces*, Linear Algebra Appl. **428** (2008), 11-12, 2805–2812.
17. D. Puglisi & J. B. Seoane-Sepúlveda, *Bounded linear non-absolutely summing operators*, J. Math. Anal. Appl. **338** (2008), 1, 292–298.

# Optimal bounds for the Hardy operator minus the Identity

Javier Soria*

*Department of Applied Mathematics and Analysis*
*University of Barcelona*
*E-08007 Barcelona, Spain*
*e-mail: soria@ub.edu*

*En memoria del Profesor Antonio Aizpuru.*

We survey some recent works ([3,11,13]) concerning the study of the norm of the Hardy operator minus the Identity. In particular, optimal estimates are found for the cone of radially decreasing functions on the minimal Lorentz spaces $\Lambda(X)$ (restricted type estimates), and a new Banach function space, closely related to $\Lambda(X)$, is also naturally defined. We also mention a recent solution to a conjecture of Kruglyak and Setterqvist.

*Keywords*: Hardy operator, restricted type, Lorentz spaces, rearrangement invariant spaces, weighted inequalities.

## 1. Introduction

The study of the Hardy operator

$$Sf(t) = \frac{1}{t}\int_0^t f(r)\,dr,$$

on monotone functions has its origins in the works of Ariño-Muckenhoupt [1] and Sawyer [12], dealing with the characterization of the boundedness on the weighted Lorentz spaces $\Lambda^p(w)$,

$$\Lambda^p(w) = \left\{ f : \|f\|_{\Lambda^p(w)} = \left(\int_0^\infty (f^*(t))^p w(t)\,dt\right)^{1/p} < \infty \right\},$$

of the Hardy–Littlewood maximal operator, and the normability properties for these spaces, extending the well-known results of Lorentz [10].

---
*Research partially supported by grant MTM2007-60500.

We recall that the nonincreasing rearrangement of $f$ is defined as:

$$f^*(t) = \inf\{s > 0 : \lambda_f(s) \leq t\},$$

and the distribution function of $f$ is

$$\lambda_f(t) = |\{x \in \mathbb{R}^n : |f(x)| > t\}|.$$

**Definition 1.1.**

- $X$ will denote a rearrangement invariant (r.i.) Banach function space in $\mathbb{R}^n$ (see [2]). In particular this means that if $f$ and $g$ are equimeasurable ($\lambda_f = \lambda_g$), then $\|f\|_X = \|g\|_X$.
- The fundamental function of an r.i. space $X$:

$$\varphi_X(t) = \|\chi_E\|_X, \qquad \text{where } |E| = t.$$

- If $X$ is an r.i. space, the Lorentz space $\Lambda(X)$ is the minimal r.i. space having the same fundamental function than $X$:

$$\Lambda(X) = \left\{ f \in \mathcal{M}(\mathbb{R}^n) : \int_0^\infty f^*(t)\, d\varphi_X(t) < \infty \right\}.$$

- $X_{\mathrm{rd}} = \{f \in X : f \text{ is a positive radially decreasing function}\}$; that is, $f(x) = \bar{f}(|x|) \in X$ and $\bar{f}$ is a positive decreasing function on $\mathbb{R}^+$.
- The maximal function of $f$ is

$$f^{**}(t) = S(f^*)(t) = \frac{1}{t}\int_0^t f^*(s)\, ds.$$

It can be proved ([2, II.Theorem 4.10]) that given an r.i. space $X$, there exists a so called *representation space* $\overline{X}$ such that $\|f\|_X = \|f^*\|_{\overline{X}}$.

In recent years many authors have considered the study of the difference operator

$$f^{**} - f^* = S(f^*) - \mathrm{Id}\,(f^*),$$

which is equivalent to the Hardy operator minus the Identity acting on decreasing functions. This difference measures the oscillation of the function $f$, and finding good estimates for this operator has applications to, e.g., Sobolev-type embeddings and the Pólya–Szegö symmetrization principle (Bastero, Bennett, DeVore, Kolyada, Sharpley, Ul'yanov, etc.)

In particular, Kruglyak and Setterqvist [9] have studied the norm of the Hardy operator minus the Identity, restricted to $C_p(\mathbb{R}^+)$, the cone of decreasing functions on $L^p(\mathbb{R}^+)$, and were able to calculate the exact constant for the integer case $p \in \{2, 3, 4, \cdots\}$: If $f \in C_p(\mathbb{R}^+)$,

$$\|Sf - f\|_p \leq \frac{1}{(p-1)^{1/p}} \|f\|_p, \tag{1}$$

and conjectured the result should be true for every $p \geq 2$.

Motivated by this conjecture, we study the same kind of questions (see Section 2), but now we look at estimates of restricted type, and we obtain a complete answer in the more general setting of r.i. spaces $X$, and for the $n$−dimensional Hardy operator:

$$S_n f(x) = \frac{1}{|B(0,|x|)|} \int_{B(0,|x|)} f(y)\, dy.$$

This result introduces a new class of functions $R(X)$ which, in most cases, is the minimal Lorentz space $\Lambda(X)$ (see Section 3).

Just recently, in a joint work with Santiago Boza [3], we have been able to give a positive answer to the Kruglyak-Setterqvist's conjecture (1), even obtaining a further extension to more general weighted estimates (see Section 4).

This is a brief summary of the main topics that we will develop in this work (we refer to [3,11,13] for further information):

- Optimal estimates for $S_n - \mathrm{Id}$ on different spaces of restricted type $\Lambda(X)$ [13].
- Definition of the new space $R(X)$, functional properties and relationship with $\Lambda(X)$ (joint work with Salvador Rodríguez [11]).
- Solution to the Kruglyak-Setterqvist's conjecture for the norm of $S-\mathrm{Id}$ defined on $C_p(\mathbb{R}^+)$, the cone of decreasing functions on $L^p(\mathbb{R}^+)$, $p \geq 2$ (joint work with Santiago Boza [3]).

## 2. Restricted type estimates

**Lemma 2.1.** *If $f(x) = \bar{f}(|x|)$ is a positive radially decreasing function on $\mathbb{R}^n$, then:*

$$S_n f(x) - f(x) = \frac{1}{|x|^n} \int_{\bar{f}(|x|)}^{\infty} \lambda_{\bar{f}}^n(t)\, dt = \frac{1}{v_n |x|^n} \int_{f(x)}^{\infty} \lambda_f(t)\, dt,$$

*where $v_n$ is the measure of the unit ball of $\mathbb{R}^n$ and $x \in \mathbb{R}^n$.*

This result was proved by Carro, Gogatishvili, Martín, and Pick [5] for the case $n = 1$.

**Theorem 2.1.** *Let $X$ be an r.i. Banach function space in $\mathbb{R}^n$. Then*

$$\|S_n f - f\|_X \leq \int_0^\infty \lambda_{\bar{f}}^n(t) \left\| \frac{1}{\lambda_{\bar{f}}^n(t) + |\cdot|^n} \right\|_X dt, \qquad (2)$$

*for every $f \in X_{rd}$. Moreover, the inequality is sharp.*

**Proof.** Using Lemma 2.1 and Minkowski's integral inequality, we have

$$\|S_n f - f\|_X \leq \int_0^\infty \lambda_{\bar{f}}^n(t) \left\| \frac{\chi_{(\bar{f}(|\cdot|),\infty)}(t)}{|\cdot|^n} \right\|_X dt.$$

Set now

$$g_t(x) = \frac{1}{\lambda_{\bar{f}}^n(t) + |x|^n},$$

and

$$h_t(x) = \frac{\chi_{(\bar{f}(|x|),\infty)}(t)}{|x|^n}.$$

We need to prove that $g_t$ and $h_t$ are equimeasurable functions. But,

$$|\{x : g_t(x) > s\}| = \left| \left\{ x : \frac{1}{s} > \lambda_{\bar{f}}^n(t) + |x|^n \right\} \right|$$

$$= \begin{cases} 0, & \lambda_{\bar{f}}^n(t) \geq \frac{1}{s} \\ v_n\left(\frac{1}{s} - \lambda_{\bar{f}}^n(t)\right), & \lambda_{\bar{f}}^n(t) < \frac{1}{s}. \end{cases}$$

Similarly,

$$|\{x : h_t(x) > s\}| = \left| \left\{ x : \bar{f}(|x|) < t, \frac{1}{|x|^n} > s \right\} \right|$$

$$= \left| \left\{ x : \lambda_{\bar{f}}(t) \leq |x| < \frac{1}{s^{1/n}} \right\} \right|$$

$$= \begin{cases} 0, & \lambda_{\bar{f}}^n(t) \geq \frac{1}{s} \\ v_n\left(\frac{1}{s} - \lambda_{\bar{f}}^n(t)\right), & \lambda_{\bar{f}}^n(t) < \frac{1}{s}. \end{cases}$$

Therefore, $g_t$ and $h_t$ are equimeasurable functions and hence,

$$\left\| \frac{\chi_{(\bar{f}(|\cdot|),\infty)}(t)}{|\cdot|^n} \right\|_X = \left\| \frac{1}{\lambda_{\bar{f}}^n(t) + |\cdot|^n} \right\|_X.$$

The optimality of the inequality follows by choosing the function $f(x) = \chi_B(x) = \chi_{[0,1)}(|x|)$:

$$S_n f(x) - f(x) = \begin{cases} 0, & |x| < 1 \\ \dfrac{1}{|x|^n}, & |x| \geq 1, \end{cases}$$

and we get

$$\int_0^\infty \lambda_f^n(t) \left\| \frac{1}{\lambda_f^n(t) + |\cdot|^n} \right\|_X dt = \left\| \frac{1}{1 + |\cdot|^n} \right\|_X = \|S_n f - f\|_X. \qquad \square$$

The previous theorem motivates the study of the class of functions for which the right-hand side of (2) is finite:

**Definition 2.1.** Let $X$ be an r.i. space in $\mathbb{R}^n$. We define the class of functions:

$$R(X) = \left\{ f \in \mathcal{M}(\mathbb{R}^n) : \|f\|_{R(X)} < \infty \right\},$$

where

$$\|f\|_{R(X)} = v_n^{-1} \int_0^\infty \lambda_f(t) \left\| \frac{1}{v_n^{-1} \lambda_f(t) + |\cdot|^n} \right\|_X dt < \infty.$$

**Remark 2.1.**

(i) Observe that if $f \in X_{\mathrm{rd}}$, then

$$\|f\|_{R(X)} = \int_0^\infty \lambda_f^n(t) \left\| \frac{1}{\lambda_f^n(t) + |\cdot|^n} \right\|_X dt.$$

(ii) It is easy to see that $\|f\|_{R(X)} = 0$ if and only if $f \equiv 0$, and $\|af\|_{R(X)} = |a| \|f\|_{R(X)}$.

(iii) (See Proposition 3.1 and [11]). A necessary (and sufficient) condition to get that $R(X) \neq \{0\}$ is that there exists a $\delta > 0$ such that $1/(|x|^n + \delta) \in X$, which is equivalent to the fact that $(L^{1,\infty} \cap L^\infty) \subset X$. In particular $R(L^1) = \{0\}$.

(iv) Theorem 2.1 can now be rewritten as

$$\|S_n f - f\|_X \leq \|f\|_{R(X)},$$

for $f \in X_{\mathrm{rd}}$.

We are now going to identify $R(X)$ for some particular cases, like the Lorentz space $L^{p,q}(\mathbb{R}^n)$, with exact norms.

We will also prove a more general result for an r.i. space $X$, but only up to equivalence of norms.

We consider first the definition of $L^{p,q}(\mathbb{R}^n)$ in terms of the standard norm, which is only valid in the range $1 \leq p < \infty$, $1 \leq q \leq p$:

$$\|f\|_{p,q} = p^{1/q}\left(\int_0^\infty \left(t\lambda_f^{1/p}(t)\right)^q \frac{dt}{t}\right)^{1/q} = \left(\int_0^\infty \left(t^{1/p}f^*(t)\right)^q \frac{dt}{t}\right)^{1/q},$$

where $f^*$ is the nonincreasing rearrangement of $f$.

**Proposition 2.1.** *If $1 < p < \infty$, $1 \leq q \leq p$, and $L^{p,q}(\mathbb{R}^n)$ is endowed with the norm $\|\cdot\|_{p,q}$, then*

$$\|f\|_{R(L^{p,q})} = p^{-1/q'}\left(\frac{\Gamma\left(\frac{(p-1)q}{p}\right)\Gamma\left(\frac{p+q}{p}\right)}{\Gamma(q+1)}\right)^{1/q}\|f\|_{p,1}.$$

As a consequence of this result, we can find the best constant for the boundedness of $S_n - \text{Id}$, for radially decreasing functions on $L^{p,1}(\mathbb{R}^n)$.

**Proposition 2.2.** *If $1 < p < \infty$, $1 \leq q \leq p$, and $L^{p,q}(\mathbb{R}^n)$ is endowed with the norm $\|\cdot\|_{p,q}$, then for a radially decreasing function $f \in L^{p,1}(\mathbb{R}^n)$:*

$$\|S_n f - f\|_{p,q} \leq p^{-1/q'}\left(\frac{\Gamma\left(\frac{(p-1)q}{p}\right)\Gamma\left(\frac{p+q}{p}\right)}{\Gamma(q+1)}\right)^{1/q}\|f\|_{p,1},$$

*and the inequality is sharp. In particular:*

(i) *if $q = p$ and $f \in L^{p,1}_{rd}(\mathbb{R}^n)$, then*

$$\|S_n f - f\|_p \leq \frac{1}{p(p-1)^{1/p}}\|f\|_{p,1};$$

(ii) *if $q = 1$ and $f \in L^{p,1}_{rd}(\mathbb{R}^n)$, then*

$$\|S_n f - f\|_{p,1} \leq \frac{\pi}{p\sin\left(\frac{\pi}{p}\right)}\|f\|_{p,1},$$

*and the inequalities are sharp.*

**Remark 2.2.**

(i) It is important to observe that the previous estimate does not follow from the case $q=1$ and the embedding $L^{p,1} \subset L^{p,q}$. In fact, it is known that $\|f\|_{p,q} \leq \frac{1}{p}(\frac{p}{q})^{1/q}\|f\|_{p,1}$ (with optimal bound). But, if $1 < q \leq p$, then

$$p^{-1/q'}\left(\frac{\Gamma(\frac{(p-1)q}{p})\Gamma(\frac{p+q}{p})}{\Gamma(q+1)}\right)^{1/q} < \frac{1}{p}\left(\frac{p}{q}\right)^{1/q} \frac{\pi}{p\sin\left(\frac{\pi}{p}\right)}.$$

(ii) If $f \in C_p(\mathbb{R}^+)$ and $p > 1$, we have proved that

$$\|Sf - f\|_{L^p(\mathbb{R}^+)} \leq \frac{1}{p(p-1)^{1/p}}\|f\|_{L^{p,1}(\mathbb{R}^+)}.$$

This result should be compared with the strong type inequality proved by Kruglyak and Setterqvist [9] under the hypothesis that $f \in C_p(\mathbb{R}^+)$ and $p \in \{2, 3, \cdots\}$,

$$\|Sf - f\|_{L^p(\mathbb{R}^+)} \leq \frac{1}{(p-1)^{1/p}}\|f\|_{L^p(\mathbb{R}^+)}.$$

Similarly to what we have done for the norm $\|\cdot\|_{p,q}$, $1 < p < \infty$, $1 \leq q \leq p$, we now consider the usual renorming of the $L^{p,q}(\mathbb{R}^n)$ spaces, for the whole range of indices $1 < p < \infty$, $1 \leq q \leq \infty$, in terms of the maximal norm:

$$\|f\|_{p,q}^* = \left(\int_0^\infty (t^{1/p}f^{**}(t))^q \frac{dt}{t}\right)^{1/q},$$

where $f^{**}(t)$ is the maximal function of $f$. We will denote by $L_*^{p,q}(\mathbb{R}^n)$ the space $L^{p,q}(\mathbb{R}^n)$ endowed with the norm $\|\cdot\|_{p,q}^*$.

**Proposition 2.3.** If $1 < p < \infty$, $1 \leq q \leq \infty$, and $f \in L_*^{p,q}(\mathbb{R}^n)$, then

$$\|f\|_{R(L_*^{p,q})} = \frac{1}{pp'}\left(\int_0^\infty \log^q(1+s)s^{-q/p'-1}ds\right)^{1/q}\|f\|_{p,1}^*,$$

if $1 \leq q < \infty$, and

$$\|f\|_{R(L_*^{p,\infty})} = \frac{(p-1)^{1/p'}}{pp'}\log p' \|f\|_{p,1}^*.$$

**Proposition 2.4.** Let $1 < p < \infty$, $1 \le q \le \infty$, and $f \in L^{p,q}_*(\mathbb{R}^n)$ be a radially decreasing function. Then,

$$\|S_n f - f\|^*_{p,q} \le \frac{1}{pp'} \left( \int_0^\infty \log^q(1+s) s^{-q/p'-1} ds \right)^{1/q} \|f\|^*_{p,1},$$

if $1 \le q < \infty$, and

$$\|S_n f - f\|^*_{p,\infty} \le \frac{(p-1)^{1/p'}}{pp'} \log p' \|f\|^*_{p,1},$$

and both inequalities are sharp. In particular:

(i) if $q = p'$ and $f \in L^{p,p'}_{*,\mathrm{rd}}(\mathbb{R}^n)$, then

$$\|S_n f - f\|^*_{p,p'} \le \frac{(\Gamma(p'+1)\zeta(p'))^{1/p'}}{pp'} \|f\|^*_{p,1};$$

(ii) if $q = 1$ and $f \in L^{p,1}_{*,\mathrm{rd}}(\mathbb{R}^n)$, then

$$\|S_n f - f\|^*_{p,1} \le \frac{\pi}{p \sin\left(\frac{\pi}{p}\right)} \|f\|^*_{p,1}.$$

## 3. The space $R(X)$

Even though we have shown that $R(L^1) = \{0\}$, we see that, in many other cases, $R(X)$ is isometric, up to a multiplicative constant, with the minimal Lorentz space $\Lambda(X)$. This is the case for $X = L^{p,q}$ and $X = L^{p,q}_*$ ($1 < p < \infty$, $1 \le q \le \infty$), for which the minimal space corresponds to the index $q = 1$. We will now study this observation for a general $X$, and show that, under some weak assumptions, $R(X)$ and $\Lambda(X)$ agree. Recall that

$$\Lambda(X) = \left\{ f \in \mathcal{M}(\mathbb{R}^n) : \int_0^\infty f^*(t) \, d\varphi_X(t) < \infty \right\},$$

where $\varphi_X(t) = \|\chi_E\|_X$ ($|E| = t$), is the fundamental function of $X$. To simplify the calculations we are going to assume that $\lim_{t \to 0^+} \varphi_X(t) = 0$, which is always true for the $L^{p,q}$ spaces (observe that we exclude the nonseparable case of $L^\infty$). Then, we can write

$$\|f\|_{\Lambda(X)} = \int_0^\infty f^*(t) \varphi'_X(t) \, dt = \int_0^\infty \varphi_X(\lambda_f(t)) \, dt.$$

We can assume, without loss of generality, that $\varphi_X$ is a (nonnegative and nondecreasing) concave function, so that $\|\cdot\|_{\Lambda(X)}$ is a norm. We also set

$$W_X(t) = \left\| \frac{1}{1+\frac{\cdot}{t}} \right\|_X.$$

With this notation it is easy to see that

$$\|f\|_{R(X)} = \int_0^\infty W_X(\lambda_f(t))\,dt.$$

**Proposition 3.1.** *Let $X$ be an r.i. space. The following are equivalent:*

(1) $R(X) \neq \{0\}$.
(2) There exists $r > 0$ such that $W_X(r) < +\infty$.
(3) $W_X(r) < +\infty$, for every $r > 0$.
(4) $g^*(s) = 1/(1+s) \in \overline{X}$.
(5) $(L^{1,\infty} \cap L^\infty) \subset X$.

We will also make use of the fact that

$$D_X \equiv \sup_{t>0} \frac{\varphi_X(2t)}{\varphi_X(t)} \in [1,2]$$

(this is known as the $\Delta_2$–condition): fix $t > 0$ and choose two disjoint sets with $|A| = |B| = t$. Then,

$$\varphi_X(2t) = \|\chi_{A\cup B}\|_X = \|\chi_A + \chi_B\|_X \leq \|\chi_A\|_X + \|\chi_B\|_X = 2\varphi_X(t).$$

E.g., if $X = L^{p,q}(\mathbb{R}^n)$ ($1 \leq q \leq p < \infty$) or $X = L^{p,q}_*(\mathbb{R}^n)$ ($1 < p < \infty$, $1 \leq q \leq \infty$ or $p = q = 1$), then $D_X = 2^{1/p}$.

**Proposition 3.2.** *Let $X$ be an r.i. space.*

(i) *For every $f \in R(X)$:*

$$\|f\|_{\Lambda(X)} \leq 2\|f\|_{R(X)}.$$

(ii) *If $D_X \in [1,2)$, then for every $f \in \Lambda(X)$:*

$$\|f\|_{R(X)} \leq \frac{2}{2 - D_X}\|f\|_{\Lambda(X)}.$$

*In particular, if $D_X \in [1,2)$ then $R(X) = \Lambda(X)$.*

**Proof.** To prove (i), we consider the ball $B_t = B(0, (v_n^{-1}t)^{1/n})$, so that $|B_t| = t$. Then, $\chi_{B_t}(x) \leq 2t/(t + v_n|x|^n)$ and hence

$$\varphi_X(t) \leq 2v_n^{-1}t \left\| \frac{1}{v_n^{-1}t + |\cdot|^n} \right\|_X.$$

Therefore,

$$\|f\|_{\Lambda(X)} = \int_0^\infty \varphi_X(\lambda_f(t))\,dt$$

$$\leq 2v_n^{-1} \int_0^\infty \lambda_f(t) \left\| \frac{1}{v_n^{-1}\lambda_f(t) + |\cdot|^n} \right\|_X dt = 2\|f\|_{R(X)}.$$

(ii) If we now assume that $D_X \in [1,2)$, set $S_r = B_{2r} \setminus B_r$. Then,

$$\frac{t}{t+v_n|x|^n} = \sum_{k=1}^{\infty} \frac{t}{t+v_n|x|^n}\chi_{S_{2^k t}}(x) + \frac{t}{t+v_n|x|^n}\chi_{B_t}(x)$$

$$\leq \sum_{k=1}^{\infty} \frac{1}{1+2^k}\chi_{S_{2^k t}}(x) + \chi_{B_t}(x).$$

Thus,

$$\left\|\frac{t}{t+v_n|\cdot|^n}\right\|_X \leq \sum_{k=1}^{\infty} \frac{1}{1+2^k}\|\chi_{S_{2^k t}}\|_X + \|\chi_{B_t}\|_X$$

$$= \sum_{k=1}^{\infty} \frac{1}{1+2^k}\varphi_X(2^k t) + \varphi_X(t)$$

$$\leq \sum_{k=0}^{\infty} \frac{D_X^k}{2^k}\varphi_X(t) = \frac{2}{2-D_X}\varphi_X(t).$$

Hence,

$$\|f\|_{R(X)} \leq \frac{2}{2-D_X}\|f\|_{\Lambda(X)}. \qquad \Box$$

This result has been extended to a more general case. Recall that if $X$ is an r.i. space, we define [2]

$$\overline{\varphi}_X(s) := \sup_{t>0} \frac{\varphi_X(st)}{\varphi_X(t)},$$

and the upper fundamental index

$$\overline{\beta}_X = \inf_{s>1} \frac{\log \overline{\varphi}_X(s)}{\log s}.$$

**Theorem 3.1 ([11]).** *Let $X$ be an r.i. space. The following are equivalent:*

(1) $\Lambda(X) = R(X)$.
(2) $\overline{\beta}_X < 1$.

**Remark 3.1.**

(i) Since $2^{\overline{\beta}_X} \leq D_X$, this implies that if $D_X < 2$, then $\overline{\beta}_X < 1$.
(ii) As we have already observed, $R(L^1) = \{0\} \neq \Lambda(L^1) = L^1$, and hence, the embedding $\Lambda(X) \subset R(X)$ is not true, in general, for the case $D_X = 2$.

(iii) We have seen that, for the $L^{p,q}$ spaces, $\|\cdot\|_{R(X)}$ is a multiple of $\|\cdot\|_{\Lambda(X)}$. However, this is not true in general: Consider $X = \Lambda^2(w)$, $w = \chi_{(0,1)}$, and $f_r^*(t) = \chi_{(0,r)}(t)$. Then,

$$\|f_r\|_{\Lambda(X)} = \begin{cases} \sqrt{r} & \text{if } r < 1, \\ 1 & \text{if } r > 1, \end{cases}$$

but

$$\|f_r\|_{R(X)} = \sqrt{\frac{r}{1+r}}.$$

In fact, it can be proved that ([11]) $\|f\|_{R(X)} = c\|f\|_{\Lambda(X)}$, if and only if $W_X(t) = c\varphi_X(t)$.

(iv) If $w$ is a decreasing weight (so that $\|\cdot\|_{\Lambda^p(w)}$ is a norm), then $D_{\Lambda^p(w)} \leq 2^{1/p}$, and hence $D_{\Lambda^p(w)} < 2$, whenever $p > 1$. Therefore, the previous theorem applies.

## 4. Kruglyak-Setterqvist's conjecture

For $p \geq 1$, we recall that a weight $w$ is in the Ariño-Muckenhoupt $B_p$-class if there exists a positive constant $C > 0$ such that, for every $r > 0$,

$$r^p \int_r^\infty \frac{w(x)}{x^p} \, dx \leq C \int_0^r w(x) \, dx.$$

If $w \in B_p$ we denote by $\|w\|_{B_p}$ the best constant in the above inequality.

$B_p$ weights give a simple characterization of the boundedness of the Hardy–Littlewood maximal operator on $\Lambda^p(w)$ (Ariño-Muckenhoupt [1]), and give also necessary and sufficient conditions for the Lorentz spaces $\Lambda^p(w)$ to be normable when $p > 1$, see [12] (the case $p = 1$ was characterized in [4]).

In what follows, we will use the standard notation

$$\|f\|_{L^p(w)} = \left( \int_0^\infty |f(x)|^p w(x) \, dx \right)^{1/p}.$$

**Theorem 4.1.** *Let $p \geq 2$ and $w$ be a weight in the $B_p$-class satisfying that the function*

$$r \longrightarrow r^{p-1} \int_r^\infty \frac{w(x)}{x^p} \, dx, \tag{3}$$

*is decreasing for $r > 0$. Then, for any decreasing function $f$, we have*

$$\|Sf - f\|_{L^p(w)}^p \leq \|w\|_{B_p} \|f\|_{L^p(w)}^p,$$

and $\|w\|_{B_p}$ is the best constant.

It is easy to prove that Theorem 4.1 is false if $1 < p < 2$. There are also counterexamples in the case $p \geq 2$, if no extra condition like (3) is assumed on $w$. However, the case $p = 1$ follows without any other condition than $B_1$:

**Proposition 4.1.** *Let $w \in B_1$. Then, for any decreasing function $f$, we have the following sharp inequality*

$$\|Sf - f\|_{L^1(w)} \leq \|w\|_{B_1} \|f\|_{L^1(w)}.$$

**Proof.** First we use Fubini's theorem to write

$$\|Sf - f\|_{1,w} = \int_0^\infty \left( \frac{1}{t} \int_0^t f(s)\,ds - f(t) \right) w(t)\,dt$$

$$= \int_0^\infty \left( \int_s^\infty \frac{w(t)}{t}\,dt \right) f(s)\,ds - \int_0^\infty f(t)\,w(t)\,dt. \quad (4)$$

We observe that the first integral in the last equality is the norm of $f$ in the Lorentz space $\Lambda^1(v)$ where $v(s) = \int_s^\infty \frac{w(t)}{t}\,dt$. It is known ([6]) that the best constant for the embedding $\Lambda^1(w) \hookrightarrow \Lambda^1(v)$ is given by

$$\sup_{t>0} \frac{\int_0^t v(s)\,ds}{\int_0^t w(s)\,ds}.$$

Hence, taking into account the expression in (4), the best constant $C_w$ in the inequality $\|Sf - f\|_{1,w} \leq C_w \|f\|_{1,w}$ is given by

$$C_w = \sup_{t>0} \frac{\int_0^t v(s)\,ds}{\int_0^t w(s)\,ds} - 1 = \sup_{t>0} \frac{\int_0^t \int_s^\infty \frac{w(x)}{x}\,dx\,ds}{\int_0^t w(s)\,ds} - 1$$

$$= \sup_{t>0} \frac{\int_0^t w(s)\,ds + t \int_t^\infty \frac{w(s)}{s}\,ds}{\int_0^t w(s)\,ds} - 1 = \|w\|_{B_1}. \qquad \square$$

**Corollary 4.1.** *(Kruglyak-Setterqvist's conjecture). For all $p \geq 2$, and any decreasing function $f$,*

$$\|Sf - f\|_p \leq \frac{1}{(p-1)^{1/p}} \|f\|_p,$$

and the inequality is sharp.

**Proof.** Just take $w = 1$ in Theorem 4.1. □

### References

1. M.A. Ariño and B. Muckenhoupt, *Maximal functions on classical Lorentz spaces and Hardy's inequality with weights for nonincreasing functions*, Trans. Amer. Math. Soc. **320** (1990), 727–735.
2. C. Bennett and R. Sharpley, *Interpolation of Operators*, Academic Press, Inc. 1988.
3. S. Boza and J. Soria, *Solution to a conjecture on the norm of the Hardy operator minus the Identity*, to appear in J. Funct. Anal.
4. M.J. Carro, A. García del Amo, and J. Soria, *Weak-type weights and normable Lorentz spaces*, Proc. Amer. Math. Soc. **124** (1996), 849–857.
5. M.J. Carro, A. Gogatishvili, J. Martín, and L. Pick, *Functional properties of rearrangement invariant spaces defined in terms of oscillations*, J. Funct. Anal. **229** (2005), no. 2, 375–404.
6. M.J. Carro, L. Pick, J. Soria, and V. D. Stepanov, *On embeddings between classical Lorentz spaces*, Math. Inequal. Appl. **4** (2001), 397–428.
7. M.J. Carro, J.A. Raposo, and J. Soria, *Recent Developments in the Theory of Lorentz Spaces and Weighted Inequalities*, Mem. Amer. Math. Soc. **187**, Providence, RI, 2007.
8. M.J. Carro and J. Soria, *Weighted Lorentz spaces and the Hardy operator*, J. Funct. Anal. **112** (1993), no. 2, 480–494.
9. N. Kruglyak and E. Setterqvist, *Sharp estimates for the identity minus Hardy operator on the cone of decreasing functions*, Proc. Amer. Math. Soc. **136** (2008), no. 7, 2505–2513.
10. G.G. Lorentz, *Some new functional spaces*, Ann. of Math. **51** (1950), 37–55.
11. S. Rodríguez-López and J. Soria, *A new class of restricted type spaces*, to appear in Proc. Edinb. Math. Soc. (2)
12. E. Sawyer, *Boundedness of classical operators on classical Lorentz spaces*, Studia Math. **96** (1990), 145–158.
13. J. Soria, *Optimal bounds of restricted type for the Hardy operator minus the Identity on the cone of radially decreasing functions*, Studia Math. **197** (2010), 69–79.
14. E.M. Stein and G.Weiss, *Introduction to Fourier Analysis on Euclidean Spaces*, Princeton Univ. Press, 1971.

# AUTHOR INDEX

Aiena, P., 13
Araujo, J., 145
Aron, R. M., 158

Diestel, J., 58

Galindo, P., 158
García-Pacheco, F. J., 165

Mbekhta, M., 174

Oja, E., 196

Pérez-Fernández, F. J., 1

Rambla-Barreno, F., 216

Schlumprecht, Th., 116
Seoane-Sepúlveda, J. B., 226
Soria, J., 234
Spalsbury, A., 58